Energy, Environment, and Sustainability

Series editors

This books series publishes cutting edge monographs and professional books focused on all aspects of energy and environmental sustainability, especially as it relates to energy concerns. The Series is published in partnership with the International Society for Energy, Environment, and Sustainability. The books in these series are editor or authored by top researchers and professional across the globe. The series aims at publishing state-of-the-art research and development in areas including, but not limited to:

- Renewable Energy
- Alternative Fuels
- Engines and Locomotives
- Combustion and Propulsion
- Fossil Fuels
- Carbon Capture
- Control and Automation for Energy
- Environmental Pollution
- Waste Management
- Transportation Sustainability

More information about this series at http://www.springer.com/series/15901

Shantanu Bhattacharya · Avinash Kumar Agarwal
T. Rajagopalan · Vinay K. Patel
Editors

Nano-Energetic Materials

Springer

Editors
Shantanu Bhattacharya
Department of Mechanical Engineering
Indian Institute of Technology Kanpur
Kanpur, Uttar Pradesh, India

T. Rajagopalan
Department of Mechanical Engineering
Amrita School of Engineering
Coimbatore, Tamil Nadu, India

Avinash Kumar Agarwal
Department of Mechanical Engineering
Indian Institute of Technology Kanpur
Kanpur, Uttar Pradesh, India

Vinay K. Patel
Department of Mechanical Engineering
Govind Ballabh Pant Institute
of Engineering and Technology
Pauri, Uttarakhand, India

ISSN 2522-8366 ISSN 2522-8374 (electronic)
Energy, Environment, and Sustainability
ISBN 978-981-13-3268-5 ISBN 978-981-13-3269-2 (eBook)
https://doi.org/10.1007/978-981-13-3269-2

Library of Congress Control Number: 2018961693

This Springer imprint is published by the registered company Springer Nature Singapore Pte Ltd.
The registered company address is: 152 Beach Road, #21-01/04 Gateway East, Singapore 189721, Singapore

Preface

Energy demand has been rising remarkably due to increasing population and urbanization. Global economy and society are significantly dependent on the energy availability because it touches every facet of human life and its activities. Transportation and power generation are two major examples. Without the transportation by millions of personalized and mass transport vehicles and availability of 24 × 7 power, human civilization would not have reached contemporary living standards.

The International Society for Energy, Environment and Sustainability (ISEES) was founded at Indian Institute of Technology Kanpur (IIT Kanpur), India, in January 2014 with the aim of spreading knowledge/awareness and catalyzing research activities in the fields of energy, environment, sustainability, and combustion. The society's goal is to contribute to the development of clean, affordable, and secure energy resources and a sustainable environment for the society and to spread knowledge in the above-mentioned areas and create awareness about the environmental challenges, which the world is facing today. The unique way adopted by the society was to break the conventional silos of specializations (engineering, science, environment, agriculture, biotechnology, materials, fuels, etc.) to tackle the problems related to energy, environment, and sustainability in a holistic manner. This is quite evident by the participation of experts from all fields to resolve these issues. ISEES is involved in various activities such as conducting workshops, seminars, and conferences in the domains of its interest. The society also recognizes the outstanding works done by the young scientists and engineers for their contributions in these fields by conferring them awards under various categories.

The second international conference on "Sustainable Energy and Environmental Challenges" (SEEC-2018) was organized under the auspices of ISEES from December 31, 2017, to January 3, 2018, at J N Tata Auditorium, Indian Institute of Science Bangalore. This conference provided a platform for discussions between eminent scientists and engineers from various countries including India, USA, South Korea, Norway, Finland, Malaysia, Austria, Saudi Arabia, and Australia. In this conference, eminent speakers from all over the world presented their views related to different aspects of energy, combustion, emissions, and alternative energy resources for sustainable development and a cleaner environment. The conference presented five high-voltage plenary talks from globally renowned experts on topical themes, namely "Is It Really the End of Combustion Engines and Petroleum?" by Prof. Gautam Kalghatgi, Saudi Aramco; "Energy Sustainability in India: Challenges and Opportunities" by Prof. Baldev Raj, NIAS Bangalore; "Methanol Economy: An Option for Sustainable Energy and Environmental Challenges" by Dr. Vijay Kumar Saraswat, Hon. Member (S&T), NITI Aayog, Government of India; "Supercritical Carbon Dioxide Brayton Cycle for Power Generation" by Prof. Pradip Dutta, IISc Bangalore; and "Role of Nuclear Fusion for Environmental Sustainability of Energy in Future" by Prof. J. S. Rao, Altair Engineering.

The conference included 27 technical sessions on topics related to energy and environmental sustainability including 5 plenary talks, 40 keynote talks, and 18 invited talks from prominent scientists, in addition to 142 contributed talks, and 74 poster presentations by students and researchers. The technical sessions in the conference included Advances in IC Engines: SI Engines, Solar Energy: Storage, Fundamentals of Combustion, Environmental Protection and Sustainability, Environmental Biotechnology, Coal and Biomass Combustion/Gasification, Air Pollution and Control, Biomass to Fuels/Chemicals: Clean Fuels, Advances in IC Engines: CI Engines, Solar Energy: Performance, Biomass to Fuels/Chemicals: Production, Advances in IC Engines: Fuels, Energy Sustainability, Environmental Biotechnology, Atomization and Sprays, Combustion/Gas Turbines/Fluid Flow/Sprays, Biomass to Fuels/Chemicals, Advances in IC Engines: New Concepts, Energy Sustainability, Waste to Wealth, Conventional and Alternate Fuels, Solar Energy, Wastewater Remediation and Air Pollution. One of the highlights of the conference was the rapid-fire poster sessions in (i) Energy Engineering, (ii) Environment and Sustainability, and (iii) Biotechnology, where more than 75 students participated with great enthusiasm and won many prizes in a fiercely competitive environment. More than 200 participants and speakers attended this four-day conference, which also hosted Dr. Vijay Kumar Saraswat, Hon. Member (S&T), NITI Aayog, Government of India, as the chief guest for the book release ceremony, where 16 ISEES books published by Springer, under a special dedicated series "Energy, Environment, and Sustainability" were released. This is the first time that such significant and high-quality outcome has been achieved by any society in India. The conference concluded with a panel discussion on "Challenges, Opportunities & Directions for Future Transportation Systems", where the panelists were Prof. Gautam Kalghatgi, Saudi Aramco; Dr. Ravi Prashanth, Caterpillar Inc.; Dr. Shankar Venugopal, Mahindra and Mahindra; Dr. Bharat Bhargava, DG, ONGC Energy Centre; and Dr. Umamaheshwar, GE Transportation, Bangalore. The panel discussion was moderated by Prof. Ashok Pandey, Chairman, ISEES. This conference laid out the road map for technology development, opportunities, and challenges in energy, environment, and sustainability domain. All these topics are very relevant for the country and the world in the present context. We acknowledge the support received from various funding agencies and organizations for the successful conduct of the second ISEES conference SEEC-2018, where these books germinated. We would therefore like to acknowledge SERB, Government of India (special thanks to Dr. Rajeev Sharma, Secretary); ONGC Energy Centre (special thanks to Dr. Bharat Bhargava), TAFE (special thanks to Sh. Anadrao Patil); Caterpillar (special thanks to Dr Ravi Prashanth); Progress Rail, TSI, India (special thanks to Dr. Deepak Sharma); Tesscorn, India (special thanks to Sh. Satyanarayana); GAIL, Volvo; and our publishing partner Springer (special thanks to Swati Meherishi).

The editors would like to express their sincere gratitude to a large number of authors from all over the world for submitting their high-quality work in a timely manner and revising it appropriately at short notice. We would like to express our special thanks to Dr. Saibal Banerjee, Dr. Rishi Kant, Dr. Vinay K. Patel, Mr. Aviru Kumar Basu, Ms. Anubhuti Saha, Ms.Geeta Bhatt, Mr. Pankaj Singh Chauhan, Ms. Punam Sundriyal, Mr. Mohit Panday, Mr. Kapil Manoharan who reviewed various chapters of this book and provided very valuable suggestions to the authors to improve their manuscript.

The book covers different aspects of nano-energetic materials and is broadly divided into four different domains, viz. current paradigm, fabrication of nano-energetic materials, tuning and characterization of these materials, and the emerging paradigm. The reason for outlaying in this format is to provide a comprehensive review in the domain of nano-energetic materials. This area of research is in fledgling stages, and there are very few comprehensive reviews in the domain. Major topics covered include aluminum-based nano-composite materials, nano-structured energetic composite, materials for defense application, nano-aluminum material for thermal decomposition, nano- and micro-electrode fabrications, tuning reactivity of nano-energetic gas generators. Various trends of existing technologies for fabrication and characterization related to this class of materials are presented in a series of the included chapters. The book also contains a separate section on recent advances in nano-energetic materials which includes applications within spacecraft, micro-thrusters. In yet another section called the emerging paradigm related to the field of nano-energetic material, there are chapters discussing charge storage domain and nanomaterials involved in such electrical energy containment. In yet another chapter, a discussion is initiated in some other form of energy storage such as energy storage through hydrogen generation wherein the process of photocatalysis is explored to make non-traditional energy storage. The focus of the emerging paradigm section is mostly on synthesis and characterization of materials associated with storage and generation of energy in forms other than primarily thermal energy which has been the trend of the nano-energetic materials so far.

Kanpur, India — Shantanu Bhattacharya
Kanpur, India — Avinash Kumar Agarwal
Coimbatore, India — T. Rajagopalan
Pauri, India — Vinay K. Patel

Contents

Editors and Contributors

About the Editors

Shantanu Bhattacharya (Ph.D.) is Professor of mechanical engineering and Head of Design Program at Indian Institute of Technology Kanpur. Prior to this, he completed his MS in mechanical engineering from the Texas Tech University, Lubbock, Texas, and a Ph.D. in bioengineering from the University of Missouri, Columbia, USA. He also completed a postdoctoral training at the Birck Nanotechnology Center at the Purdue University. His main research interests are design and development of micro- and nano-sensors and actuation platforms, nano-energetic materials, micro- and nano-fabrication technologies, water remediation using visible light photocatalysis, and product design and development. He has many awards and accolades to his credit which includes the Institution of Engineers Young Engineer Award, the Institute for Smart Structures and Systems Young Scientist Award, the Best Mechanical engineering design award (National Design Research Forum, IEI), fellowship from the high energetic materials institute at Australia, fellowship of the Institution of Engineers (India). He has guided many Ph.D. and master's students and has many international journal publications, patents, books, and conference proceedings.

Avinash Kumar Agarwal is Professor in the Department of Mechanical Engineering at Indian Institute of Technology Kanpur. His areas of interest are IC engines, combustion, alternative fuels, conventional fuels, optical diagnostics, laser ignition, HCCI, emission and particulate control, and large bore engines. He has published 24 books and more than 230 international journals and conference papers. He is Fellow of SAE (2012), ASME (2013), ISEES (2015), and INAE (2015). He received several awards such as prestigious Shanti Swarup Bhatnagar Award—2016 in engineering sciences, Rajib Goyal Prize—2015, NASI-Reliance Industries Platinum Jubilee Award—2012; INAE Silver Jubilee Young Engineer Award—2012; SAE International's Ralph R. Teetor Educational Award—2008; INSA Young Scientist Award—2007; UICT Young Scientist Award—2007; INAE Young Engineer Award—2005.

T. Rajagopalan (Ph.D.) is Associate Professor in the Department of Electronics and Communication Engineering, Amrita School of Engineering, Coimbatore, India, and Adjunct Professor in the Department of Electrical and Computer Engineering, University of Missouri, MO, USA. He has previously worked as Consultant and Senior Project Manager to NEMS/MEMS Works LLC, USA; Visiting Scientist in the Ben-Gurion University of the Negev, Israel; and others. His research interests include synthesis, processing, characterization, and application aspects of carbon nanotubes and graphene, self-assembly of nanomaterials, mesostructured materials, nanoenergetic materials, and 2D materials. He is Member of American Chemical Society, USA; Materials Research Society, USA; and IEEE, has published 36 peer-reviewed journal articles and 2 chapters, and holds 8 patents.

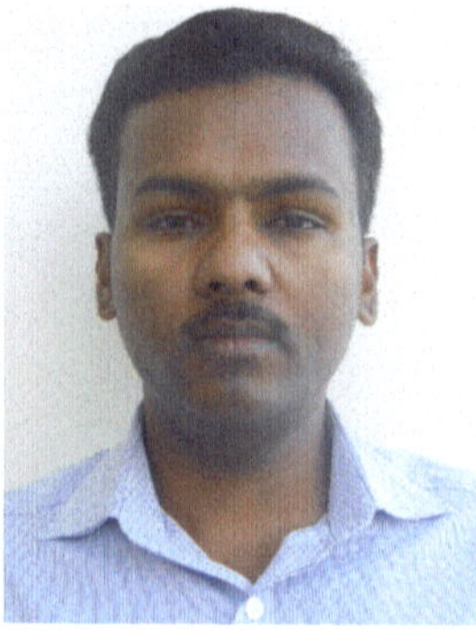

Vinay K. Patel is Assistant Professor in the Department of Mechanical Engineering, Govind Ballabh Pant Institute of Engineering & Technology, Pauri, Uttarakhand, India. He completed his Ph.D. from IIT Kanpur in 2015, specializing in nano-fabrication and characterization of high-energy composites. His research interests include nano-energetic materials, MEMS, high energy combustion, welding, and tribology. He has published 15 journal articles in high-impact journals, 7 conference papers, and 2 chapters.

Contributors

Avinash Kumar Agarwal Department of Mechanical Engineering, Indian Institute of Technology Kanpur, Kanpur, Uttar Pradesh, India

Aakash Ahuja Department of Energy Science and Engineering, Indian Institute of Technology Bombay, Powai, Mumbai, India

Shaibal Banerjee Organic Synthesis Laboratory, Department of Applied Chemistry, Defence Institute of Advanced Technology (DU), Pune, India

Aviru Kumar Basu Microsystems Fabrication Lab, Department of Mechanical Engineering, IIT Kanpur, Kanpur, India

Shantanu Bhattacharya Department of Mechanical Engineering, Indian Institute of Technology Kanpur, Kanpur, Uttar Pradesh, India

Mohamed S. A. Darwish Refining Department, Egyptian Petroleum Research Institute, Cairo, Egypt

Prasit Kumar Dutta Department of Energy Science and Engineering, Indian Institute of Technology Bombay, Powai, Mumbai, India

Ahmed M. A. El Naggar Refining Department, Egyptian Petroleum Research Institute, Cairo, Egypt

I. Emre Gunduz School of Mechanical Engineering, Purdue University, West Lafayette, IN, USA

Vishwas Goel Department of Energy Science and Engineering, Indian Institute of Technology Bombay, Powai, Mumbai, India

Ankur Gupta School of Mechanical Sciences, Indian Institute of Technology Bhubaneswar, Bhubaneswar, Odisha, India

Mkhitar A. Hobosyan Department of Physics and Astronomy, University of Texas at Rio Grande Valley, Brownsville, TX, USA

Sudarsana Jena School of Mechanical Sciences, Indian Institute of Technology Bhubaneswar, Bhubaneswar, Odisha, India; Defence Research, and Development Organization (DRDO), Bhubaneswar, Odisha, India

Amit Joshi Department of Mechanical Engineering, Govind Ballabh Pant Institute of Engineering and Technology, Pauri Garhwal, Uttarakhand, India

Jitendra Kumar Katiyar Department of Mechanical Engineering, SRM Institute of Science and Technology, Kattankulathur, Chennai, Tamil Nadu, India

Karen S. Martirosyan Department of Physics and Astronomy, University of Texas at Rio Grande Valley, Brownsville, TX, USA

K. K. S. Mer Department of Mechanical Engineering, Govind Ballabh Pant Institute of Engineering and Technology, Pauri Garhwal, Uttarakhand, India

Sagar Mitra Department of Energy Science and Engineering, Indian Institute of Technology Bombay, Powai, Mumbai, India

Asmaa S. Morshedy Refining Department, Egyptian Petroleum Research Institute, Cairo, Egypt

Ayotomi Olokun School of Aeronautics and Astronautics, Purdue University, West Lafayette, IN, USA

Vinay K. Patel Department of Mechanical Engineering, Govind Ballabh Pant Institute of Engineering and Technology, Pauri, Uttarakhand, India

Chandra Prakash School of Aeronautics and Astronautics, Purdue University, West Lafayette, IN, USA

Om Prakash Boeing India Corporation, New Delhi, India

P. Preetham Department of Energy Science and Engineering, Indian Institute of Technology Bombay, Powai, Mumbai, India

T. Raja Gopalan Department of Mechanical Engineering, Amrita School of Engineering, Coimbatore, India

Anubhuti Saha Microsystems Fabrication Lab, Department of Mechanical Engineering, IIT Kanpur, Kanpur, India

Megha Sahu Boeing India Corporation, New Delhi, India

Abhinanada Sengupta Department of Energy Science and Engineering, Indian Institute of Technology Bombay, Powai, Mumbai, India

Hema Singh Organic Synthesis Laboratory, Department of Applied Chemistry, Defence Institute of Advanced Technology (DU), Pune, India

Poonam Sundriyal Microsystems Fabrication Laboratory, Department of Mechanical Engineering, Indian Institute of Technology Kanpur, Kanpur, Uttar Pradesh, India

Rajagopalan Thiruvengadathan SIERS Research Laboratory, Department of Electronics and Communication Engineering, Amrita School of Engineering, Amrita Vishwa Vidyapeetham, Coimbatore, India

Vikas Tomar School of Aeronautics and Astronautics, Purdue University, West Lafayette, IN, USA

Part I
Nano-energetic Materials: The Current Paradigm

Chapter 1
Introduction to Nano-energetic Materials

Shantanu Bhattacharya, Avinash Kumar Agarwal, Vinay K. Patel, T. Raja Gopalan, Aviru Kumar Basu and Anubhuti Saha

Abstract With the advent of micro- and nano-scale devices, the energy management at molecular level is vital for performance enhancement. The field of nano-energetics focuses on the study of synthesis and fabrication of energetic materials or composites at nano-level. The nano-energetic materials may include almost all materials associated with the generation and storage of energy in all forms, viz., thermal, electrical, chemical, etc. The advantages of nano-scale are many which include characteristics like overall small particle size, large specific surface area, high surface energy and strong surface activity, and all these properties associated with the nano-scale provide a key to obtain an overall high energy turnover from such materials and composites and provide solutions to some very pressing current technology needs. The primary requirement of nano-energetic materials is to obtain an efficient energy release through combustion and other processes at the nano-scale. This is regulated by tuning the proportion of the oxidizer and fuel in combusting materials during the synthesis stage so that the thermite reaction can be stoichiometrically starved or over bred for different energy releases. These materials after synthesis are then interfaced with micro-/nano-scale electromechanical devices so that they can be put to use for concentrated blast release, pulse power generation, thrust generation, energy conversion and various other applications. These nano-structured energetic materials can be utilized as propellants, explosives and pyrotechnics on the basis of their specific spatial arrangements, enactments, and presentation spaces, etc. The various methods

S. Bhattacharya (✉) · A. K. Agarwal
Department of Mechanical Engineering, Indian Institute of Technology, Kanpur 208016, India

V. K. Patel
Department of Mechanical Engineering, Govind Ballabh Pant Institute of Engineering and Technology, Ghurdauri, Pauri Garhwal 246001, Uttarakhand, India

T. Raja Gopalan
Department of Mechanical Engineering, Amrita School of Engineering, Coimbatore 641105, India

A. K. Basu · A. Saha
Microsystems Fabrication Lab, Department of Mechanical Engineering, IIT Kanpur, Kanpur 208016, India

S. Bhattacharya et al. (eds.), *Nano-Energetic Materials*, Energy, Environment and Sustainability, https://doi.org/10.1007/978-981-13-3269-2_1

that are deployed to fabricate these energetic materials include wet chemical synthesis, DC reactive magnetron sputtering, electrocatalysis, molecular self-assembly. Since these nano-energetic materials and composites have wide scope in micro-/nano-energetic arena of applications, the corresponding book discusses some of the detailed and novel synthesis, fabrication, characterization, tunability, storage and application aspects of these materials.

Keywords Nano-energetics · Nano-eletromechanical devices · Nanocomposites Tenability · Characterization · High density storage · Electrodes, etc.

Nano-energetic materials have emerged as an important class of materials which find wide usage in high-density energy management at the molecular scale. These materials find wide-ranging usage in meeting high energy demand in ignition, propulsion and power generation platforms. The materials also host a big prospect in the future of charge storage devices mainly due to an overall high surface area. The benefits of possessing an overall larger surface area are many and may lead to faster reaction and shorter amount of energy release time. This shorter timescales for energy release is critical to make the usage of such materials vital for space applications, especially in case of micro- and nano-satellites, etc. The suspended objects in space under miniscule gravity conditions can be turned and propelled through digitally operated thrust forces. It is important to tune the energetic materials through stoichiometric changes, geometrical structuring and various other means so that they can be tuned to deliver a range of energy densities and also provide prolonged sustained energy at different rates.

Nano-energetic materials can be interfaced very well with micro-/nano-scale devices which can generate power, pulse or thurst. The art of fabrication of nano-energetic materials is the key to meet the energy demand created in micro-scale energy actuators and power drivers. Nano-energetic materials in principle provide a transformation of chemical energy stored at molecular level using external stimulations like heat, shock, or electrical current into heat energy, pressure energy, light energy. Similarly, another class of the nano-materials is also able to store high energy densities in form of electrical charge which can be successfully deployed in battery and capacitor applications. The main idea behind the nano-level energetic material synthesis is to engineer the chemistry/structure in a manner so that the energy management (such as release, storage and retention) can happen at very high energy densities. The ones in which the fuel and oxidizer are physically linked are called heterogenous (composite) nano-energetics, whereas the ones in which the fuel and oxidizers are chemically linked are called homogenous nano-energetics. Synthesis and chemical manipulation of the ingredients at micro-/nano-scale is the first step to obtain optimized/efficient nano-energetic materials. A variety of nanocomposites with structural and morphological differences like core–shell nano-structures, nano-foils, reactive nano-wires and directly assembled nano-particles have been manufactured time and again to meet the required level of efficiency of the materials. Efficacy in molecular level engineering, the ability to nano-manipulate materials using self-assembly and other thermodynamic processes and the packing methodologies of these nano-energetic materials (mostly available as powders and thins films) provide ways and means to eliminate drawbacks of ignition delay or ignition sensitivity faced very often by these nano-energetic materials. The combination of fuel and oxidizers at certain optimized proportion results in melt delays which is usually associated with diffusion of oxidizer and/or fuel through the protective layer of metal oxides. The chapters in this book discuss the various synthetic procedures that are deployed to assemble and fabricate nano-energetic materials using various techniques like self-assembly, cold spraying, ball milling, sol-gel, gas-phase processes. Further, it has been extensively discussed how the atomic-level engineering for organizing the nanocomposites is carried out and what are the outcomes of such engineering in terms of varied functionalities associated with the broad class of nano-energetic material. In other words, nano-level scaling of the otherwise available micro-scale reactants constitutes a key to obtain higher ignition sensitivity, high energy density and super reactivity creating an immediate response to stimuli. An indispensible requirement of all efficient nano-energetic systems is the tenability of combustion characteristics. This requires a proper choice of the proportion of fuel (aluminium, boron, magnesium, etc.) and the oxidizer (transition metal oxides like ferric oxide, cupric oxide, etc.). Among the set of various metallic fuels, aluminium is the most preferred one for theoretical and experimental formulation of nano-energetic fuel as aluminium at nano-scale exhibits an excellent catalytic activity, parallely maintaining a reduced level of decomposition temperature. Thus, this book has emphasized more aluminium-based nano-energetic materials and their applications in different areas. It also enunciates on how aluminium nano-powder can be useful in improving the ballistic performance and combustion behaviour of the corresponding nano-energetic materials. The emphasis on fashioning of the nano-energtic materials is an essential requirement for the parallel developments in many fields of research such as pulse power, micro-initiation, micro-ignition, micro-propulsion, micro-power generation, and pressure-mediated gene delivery/transfection. Thus, formulations and patterning of nano-energetic materials on silicon substrate are also discussed in this book with a future potential in microelectromechanical systems (MEMS) and energetic nano-devices.

Nano-energetic materials have proved to be one of the most novel materials in the last decade and have found various applications in energy delivery and energy storage devices. The extremely high demand of energy management within very less spatial domains coupled with the current, different micro-scale energy storage solutions provides energy management needs in terms of high storage/release densities, and thus researchers explore solutions away from the conventional nano-energetic (fuel-oxidizer composite) material domains for providing out of the box innovation for higher energy storage/release or rates related to these. Some examples have been the fast development in the supercapacitors/batteries and other forms of energy generation like hydrogen or fuels cells where the nano-structuring exercise is very commonplace. This book is also intended to provide a holistic view of those other domains which can be associated with the overall concept of nano-energy management and harnessing. When we talk about the integration of device technology to nano-energetic materials as the tunability of these materials plays a very important role, a variety of methods to characterize the energy density issues as well as the sensitivity, etc., need to be generated. To gauge a nano-energetic composite from a point of view of knowing the structures, various kinds of mechanical techniques are used to know about the defects generated in the materials which would cause flaws related to material performance. To study the failures in the interfaces of the structures, nano-mechanical Raman spectroscopic techniques are described. Other than above-mentioned characterization techniques, the nano-energetic materials can be tuned for the reactivity of energy generators, and such tuning can be deployed for various pulse power applications with release at different energy densities and rates.

The emerging paradigm of nano-energetic materials comes from processing of nano-materials for photocatalysis applications wherein energy generated can be stored as hydrogen. Such storages can be then be used combustively or in fuel cells to harness electrical energy. The same principles to obtain storage and retrieval of energy density are utilized for developing energy devices. Fabrication processes like Deskjet printing are utilized in the most modern technology for charge storage purposes. This book retains some of such emerging paradigms which may also be classified under the nano-energetic material domain. The thought that has been put in is related to materials that generate, harness, store, retrieve and manage energy in different forms utilizing the nano-scale concepts may be classified in the domain of nano-energetic materials. In this emerging paradigm for nano-scale energy materials, there are various end-of-the-road applications which may eventually be integrated into reduction in payload highly relevant to the space/aerospace applications. Of significant mention in this domain are the digital micro-thruster devices which are a nice integration of device technology to nano-scale energy generators. The digital micro-thrusters in the emerging paradigm are nozzleless and possess the ability, by virtue of self-assembling the material at certain shape and pattern to generate a high level of thrust in controllable directions.

The nano-energetic canvass started to get painted from the last decade and there have been significant initiatives, through which one can express nano-energetic materials from just being generators to being storage sources, to harness and manage different forms of energy, to apply to very cutting edge device technology so on so forth. In the current realm of technologies, the word nano is integrated with energy and materials to include processes which are surface dependant and which may eventually lead to perform management of high energy density and controlled energy releasing rate.

This research monograph thus provides both the current and the emerging paradigms of the field of nano-energetic materials providing the readership with a scope of delving deep into the various dimensions of this area. It summarizes some of the novel synthesis and fabrication techniques into four different sections, viz.,

(i) Nano-energetic materials: the current paradigm
(ii) Fabrication of nano-energetic materials
(iii) Tuning and characterization of nano-energetic materials
(iv) Nano-energetic materials: the emerging paradigm

This monograph is intended for the researchers, professionals and students whose area of interest is identical management of high energy density. The editors sincerely hope that the book brings out the necessary relevance.

Chapter 2
Aluminum-Based Nano-energetic Materials: State of the Art and Future Perspectives

Rajagopalan Thiruvengadathan

Abstract Technological innovations are indeed driven by enhanced abilities to understand and manipulate matter at molecular and atomic scale. Engineering energetic nanocomposites with tailored and tunable combustion characteristics is indispensable for their deployment in both civilian and defense applications. Specifically, a heterogeneous mixture of fuel [aluminum (Al), boron, magnesium, silicon, etc.] and oxidizer [cupric oxide, bismuth trioxide (Bi_2O_3), ferric oxide, etc.] with both the constituents having nanoscale dimensions constitutes a class of energetic material known as nanothermites. Among various fuels employed in nano-energetic formulations, the number of theoretical and experimental investigations on the utilization of Al outweighs that of any other metallic fuel. Knowledge on the physical and chemical characteristics of the constituents and their impact on combustion performance are fundamental to accelerate the pace of research and development in nano-energetic composites. Efforts to develop comprehensive understanding of the oxidation behavior are discussed in this article. Furthermore, the organization, intimacy, and dimensions of discrete fuels and oxidizers apart from their chemistry largely dictate the combustion kinetics exhibited by nanothermites. For a given nanocomposite, increasing the interfacial contact area between fuel and oxidizer improves its reaction rate by 3–5 orders of magnitude as a result of drastic reduction in mass and heat transport lengths. The bottom-up self-assembly process offers the most realistic solution to enhance the interfacial contacts between nanoscale constituents employing different approaches. This review summarizes the key findings in this area of research and lists the key challenges and opportunities for furthering the application aspects. Enhancement of combustion characteristics of energetic liquids through the utilization of Al and metal oxide nanoparticles as additives is another area of related research that continues to receive increasing attention (Sundaram et al. 2017). Energetic liquids possess unique characteristics such as lower activation temperature, higher pressure, and better volume expansion. Experimental research efforts have demon-

R. Thiruvengadathan (✉)
SIERS Research Laboratory, Department of Electronics and Communication Engineering, Amrita School of Engineering, Amrita Vishwa Vidyapeetham, Coimbatore 641112, India
e-mail: t_rajagopalan@cb.amrita.edu

S. Bhattacharya et al. (eds.), *Nano-Energetic Materials*, Energy, Environment and Sustainability, https://doi.org/10.1007/978-981-13-3269-2_2

strated ample promise for overcoming the inherent problems such as lower energy density and slow burn kinetics associated with energetic liquids. In gist, the central theme of this chapter is devoted to highlight and analyze the recent advancements on aluminum-based nano-energetic materials besides presenting the challenges and opportunities in the domain of nano-energetic materials development.

Keywords Nano-energetic materials · Aluminum · Oxidation · Surface passivation Energetic fuels · Combustion

2.1 Introduction

Energetic materials produce enormous amounts of heat energy during combustion as a result of chemical reaction between fuel and oxidizer (Son and Mason 2010; Martirosyan 2011; Rossi 2014; Zhou et al. 2014; Mukasyan and Rogachev 2016; Zarko 2016; Sundaram et al. 2017). Energetic materials can be broadly classified into two major types, namely monomolecular energetic materials and composite energetic materials. When the fuel and oxidizer groups are present within a single molecule, it is known as monomolecular energetic material. Explosives come under this class. On the other hand, when the fuel and the oxidizer are discrete components but physically mixed together, then the formed energetic material is known as composite energetic material. Examples of this class of energetic composites include nanothermites (defined as a mixture of metal as fuel and metal oxide as oxidizer with the dimensions of fuel and the oxidizer at nanoscale) and intermetallic compounds. Owing to the necessity to enhance the performance attributes such as energy density and of monomolecular energetic materials as well to develop novel energetic materials, metal and metal oxide nanoparticles are being used as additives in composite formulations (Son and Mason 2010; Martirosyan 2011; Rossi 2014; Zhou et al. 2014; Mukasyan and Rogachev 2016; Zarko 2016; Sundaram et al. 2017). For example, the energy density of conventional energetic materials can be substantially increased by the addition of metal particles (Sundaram et al. 2017).

Among the various metal fuels, aluminum (Al) is the best candidate in terms of energy release of about 31 kJ/g in reaction with oxygen as well due to other reasons including its abundance in Earth and being relatively cheaper compared to other fuels such as boron (B), silicon (Si), and magnesium (Mg). Aluminum particles are employed in a multitude of applications including propulsion (Ru et al. 2016b; Zarko 2016), explosion (Zhang et al. 2016; Kim et al. 2017), and pyrotechnics (Patel et al. 2015; Sundaram et al. 2017). The scope of this article is restricted to the discussion on the current state of the art of aluminum-based energetic materials. More specifically, this review is devoted to the fundamental aspects related to the utility of Al nanoparticles in energetic mixtures. Considering the various parameters responsible for the rate of energy release during the exothermic reaction, understanding the oxidation mechanism of Al nanoparticles is the key to enable tailoring and tuning the combustion characteristics. The first part of this paper is focused toward presenting the key findings and balanced views on the oxidation mechanism.

Apart from oxidation behavior, the combustion performance of nano-energetic materials is massively impacted by the physical and the chemical characteristics of Al nanoparticles such as the average size, nature of surface passivation (passivating material and thickness), and metal content of Al. The stability of the passivation layer and the stability of nanoparticles against oxidation are extremely important. Besides, knowing the degree of aggregation of Al nanoparticles pre - and post-combustion is vital. This chapter summarizes the key findings and dwells upon these aspects on Al nanoparticles. Though Al nanoparticles are highly reactive owing to its high surface area (Sundaram et al. 2017), there are problems related to its high sensitivity to electrostatic discharge (ESD) (Kelly et al. 2017a, b), impact (Wuillaume et al. 2014; Gordeev et al. 2017) and friction (Gibot et al. 2011; Kelly et al. 2017a, b), sintering, and aggregation due to solid-state diffusion (Jian et al. 2013; Chakraborty and Zachariah 2014; Wang et al. 2014a, b, 2015), and also to inherent processing challenges in liquid dispersions resulting from increased viscosity, causing erratic and unreliable combustion characteristics (Muthiah et al. 1992). Al nanoparticles are relatively expensive in comparison to their micron-sized Al. Therefore, substantial research efforts are being carried out by various groups to overcome the ignition problems associated with micron-sized Al particles. This review will also discuss about the recent developments associated with micron-sized Al particles.

Engineering energetic nanocomposites with tailored and tunable combustion characteristics is indispensable for their deployment in both civilian and defense applications. Specifically, a heterogeneous mixture of fuel (Al, B, Mg, Si, etc.) and oxidizer [cupric oxide (CuO), bismuth trioxide (Bi_2O_3), ferric oxide (Fe_2O_3), molybdenum oxide (MoO_3), etc.] with both the constituents having nanoscale dimensions constitutes a class of energetic material known as nanothermite. In general, the organization, intimacy, and dimensions of discrete fuels and oxidizers apart from their chemistry largely dictate the combustion kinetics exhibited by nanothermites. For a given nanocomposite, increasing the interfacial contact area between fuel and oxidizer improves its reaction rate by several orders of magnitude as a result of drastic reduction in mass and heat transfer lengths. The conventional method of preparing composite energetic materials consists of physical mixing of fuel and oxidizer particles by hand or by sonication. The combustion performance exhibited by such a random energetic mixture is not reliable and reproducible. Recent advances in molecular self-assembly, supramolecular chemistry, and synthesis techniques that connect nano- and macro-scale structures have paved the way for advanced next-generation nanoengineered energetic materials (Séverac et al. 2012; Yang et al. 2013; Rossi 2014; Thiruvengadathan et al. 2014; Zarko 2016; Geeson et al. 2018; Zakiyyan et al. 2018). Indeed, these self-assembled nano-energetic materials have been shown to possess higher energy density, improved chemical responsiveness, better controlled energy and mass generation rates, enhanced reliability and reproducibility, and reduced sensitivity with suitable additives. Another area of research that is receiving increased attention from nano-energetic community is the possibility to exploit the high-energy content of Al nanoparticles to enhance the inherently low energy densities of liquid fuels. A summary of recent findings is presented and discussed in this review. Finally, the challenges that still persist and the opportunities for conducting experimental and theoretical research for advancing the implementation of Al-based nano-energetic materials from fundamental and application points of view are listed and discussed.

2.2 Oxidation Mechanism

The motivation to realize reduced ignition delay time and improved burn rate of energetic mixtures by at least 2–3 orders of magnitude has undoubtedly contributed to the production and characterization of Al nanoparticles with average particle size (APS) ranging from 20 to 120 nm. Neat metallic Al nanoparticles without surface passivation are highly reactive and undergo spontaneous combustion in air. Therefore, Al nanoparticles are more often passivated with a native oxide (Al_2O_3) layer of 2–4 nm thickness, thereby forming a metallic core–oxide shell structure. Over the past decade, experimental works carried out by several research groups reveal that the APS, the size distribution of Al nanoparticles, the thickness of native oxide shell, and the metallic content of the Al core besides the dimension, surface area, morphology, and the chemical composition of metal oxides are the most critical parameters that determine the rate of exothermic reactions between fuel (Al) and oxidizer (e.g., metal oxides, nitrates, chlorates, perchlorates, fluoropolymers) (Sundaram et al. 2017). In fact, the total energy released and the rate of energy release are determined by the extent and the kinetics of oxidation of Al particles during the reaction with oxidizer. Therefore, knowledge of the oxidation reaction is of paramount importance. The two major mechanisms of oxidation that continue to dominate the scientific debate are diffusion (Jeurgens et al. 2002; Park et al. 2005; Rai et al. 2006) and melt-dispersion (Levitas et al. 2006, 2007, 2008; Watson et al. 2008). The two mechanisms are depicted in Fig. 2.1. In case of diffusion oxidation mechanism (DOM), both Al and oxygen atoms move toward each other through the growing oxide shell.

Trunov et al. (2005a, b, 2006a, b) suggested that upon heating, Al nanoparticles (passivated with Al_2O_3 shell) undergo oxidization in several steps. These steps (in order) are (i) growth of amorphous oxide shell, (ii) amorphous-to-QQγ phase transformation, (iii) growth of γ alumina, (iv) γ-to-α phase transformation, and (v) growth of alumina (Trunov et al. 2005a, b, 2006a, b).

Among these steps, amorphous-to-γ phase transformation is being regarded the crucial stage for the significant enhancement in reactivity around the onset of Al melting. The size-dependent oxidation kinetics of a single Al nanoparticle during isothermal heating was studied by Park et al. (2005), providing information about diffusion of Al and/or oxygen through the oxide shell. From these observations, it may be concluded that the mechanism of oxidation is indeed by diffusion when the material is subjected to a heating rate of 10^3 K/s. With such heating rates, the

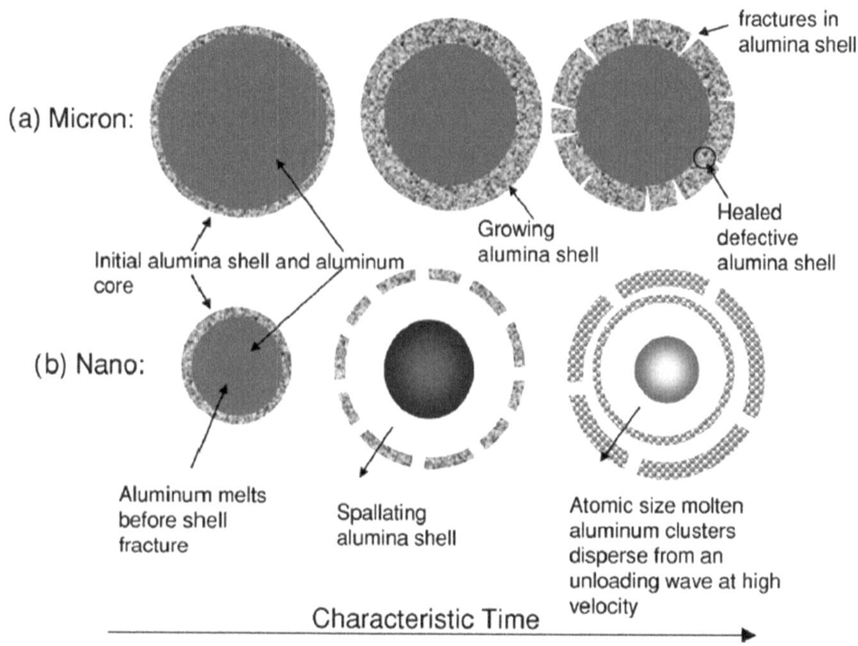

Fig. 2.1 Schematic representation of the two major mechanisms that describe the oxidation mechanism of Al particles. Reprinted from Levitas et al. (2007) with the permission of AIP Publishing

duration for oxidation is reported to be 1 s, which is reasonable to accept diffusion to be the governing mechanism.

However, the observed reaction rates of energetic mixtures formed with nano-Al and nano-oxidizers were at least three orders higher than that formed with their micron-sized counterparts. The immediate question then comes up is, can diffusion phenomenon explain the observed high reaction rates exhibited by nano-energetic mixtures? (Levitas et al. 2006, 2008). While the diffusion time required for the species to penetrate alumina shell with a typical thickness would be of the order of several tens of seconds (given the values of diffusion coefficient (Bergsmark et al. 1989) of the species in the order of 10^{-18}–10^{-19} $cm^2\ s^{-1}$), the typical reaction time in nano-energetic mixtures is of the order of 10–100 μs (Levitas 2009). Also, in case of high heating rates (say 10^8 K/s accomplished with hotwire or flash ignition), the duration for oxidation is significantly short. Furthermore, it has been experimentally found that various compositions of nanothermites formed by mixing nano-Al (typically with APS of 80 and 2–2.5 nm shell thickness) with a variety of metal oxide nanostructured powders exhibit high burn rates (1000–2400 m/s) (Apperson et al. 2007; Shende et al. 2008; Thiruvengadathan et al. 2011). These observations necessitated and spurred researchers to investigate and decipher the oxidation mechanism of Al nanoparticles beyond diffusion.

It was then that Pantoya et al. reported a mechanochemical mechanism, known as melt-dispersion mechanism (MDM) for high heating rates through theoretical and experimental efforts (Levitas et al. 2006, 2008; Watson et al. 2008; Levitas 2009). Here, it is reported that the melting of Al is concomitant with a 6% volumetric expansion strain, producing very high pressures (1–3 GPa in the molten Al core) and tensile hoop stresses in the Al oxide shell that exceeds 10 GPa and the ultimate strength of alumina (Al_2O_3). During fast heating, such stresses are suggested to cause dynamic fracture and spallation of the Al_2O_3 shell very rapidly. The resulting huge pressure variations between the bare Al surface and the molten Al core and the resulting pressure gradients lead to a spherical wave propagating to the center of the core. This, in turn, produces a reflected wave with a tensile pressure up to 8 GPa at the center of the molten core. Such a high-pressure wave causes dispersion of the Al core into smaller fragments that fly with a high velocity in the range of (100–250 m/s) (Levitas et al. 2011, 2015). Under these circumstances, the oxidation of these smaller metallic fragments is no longer limited by diffusion through the initial oxide shell (Levitas et al. 2011, 2015). At the same time, direct experimental evidences for explosive rupture of the shell and subsequent ejection of tiny Al clusters are not substantial and yet to be convincing.

Post the formulation of MDM theory, the fast heating experiments conducted on CuO (APS of 100 nm)/Al nanothermite (Al particles with APS of nearly 46 nm with a 2-nm-thick Al_2O_3 shell) at a heating rate of nearly 10^6 K/s using a hot wire enabled to determine the effective diffusion coefficient from ignition delay measurements to be around 10^{-10} cm^2/s (Chowdhury et al. 2010), which is 8–9 orders larger than the species diffusion coefficient values of 10^{-18}–10^{-19} $cm^2\ s^{-1}$ (Bergsmark et al. 1989). The increased value of effective diffusion coefficient was attributed to the built-in electric fields that promote ionic movement in the shell (Henz et al. 2010). Firmansyah et al. (2012) gave experimental evidence for DOM via high-resolution transmission electron microscopic imaging of the Al_2O_3 shell before and after the Al core melting (Fig. 2.2). Most remarkable aspect of this work is the observation of shell thickening even in the absence of oxygen flow. Besides, the fracturing of the shell suggests for the presence of dislocations near the amorphous-to-crystalline transition and finally leading phase transition completion (Firmansyah et al. 2012). Serious computational efforts have been undertaken by various research groups on different aspects of oxidation of Al nanoparticles. These computational methods are primarily based on molecular dynamics (MD) simulations. Clark et al. (2011) performed multimillion-atom reactive molecular dynamics simulations based on embedded atom method (EAM) potential for the aluminum bulk metal and a bond-order coupling scheme for aluminum/oxygen interactions. Campbell et al. (2005) employed electrostatic plus potential and also applied MD simulations of the oxidation of a neat aluminum cluster with a diameter of 20 nm. Henz et al. (2010) and Hong and Van Duin (2015) separately utilized MD simulation based on ReaxFF Reactive Force Field method. All of these methods mentioned here talk about diffusion of Al and O atoms to be dominating mechanism. The computational simulations also reveal that slower oxidation kinetics at lower temperature (typically well below Al melting) via diffusion and faster oxidation kinetics at higher temperatures (around the melting point of Al). Besides, the influencing factors for oxidation kinetics are reported to be (i) the temperature of the system, (ii) oxygen gas pressure, (iii) pressure differences within the oxide shell, and (iv) the induced electric field through the oxide shell (Campbell et al. 2005; Henz et al. 2010; Clark et al. 2011; Hong and Van Duin 2015). These findings strongly support the DOM theory.

However, there are certain questions that still remain unanswered. Researchers continue to endeavor to pursue thoughtful research to unearth the oxidation behavior of Al nanoparticles. Some of the additional important findings are summarized here. (Farley et al. 2014) reported an interesting study on the impact of atmospheric oxygen concentration on flame speeds of nano-energetic mixtures prepared with a variety of metal oxides mixed with Al nanoparticles (APS of 80 nm). It is reported here that the flame speeds improve only very slightly for high-speed reactions which activate the MDM. On the other hand, the flame speeds went up by 200% for slow-speed reactions operating under DOM in the presence of atmospheric oxygen. It is increasingly clear that experimental conditions such as

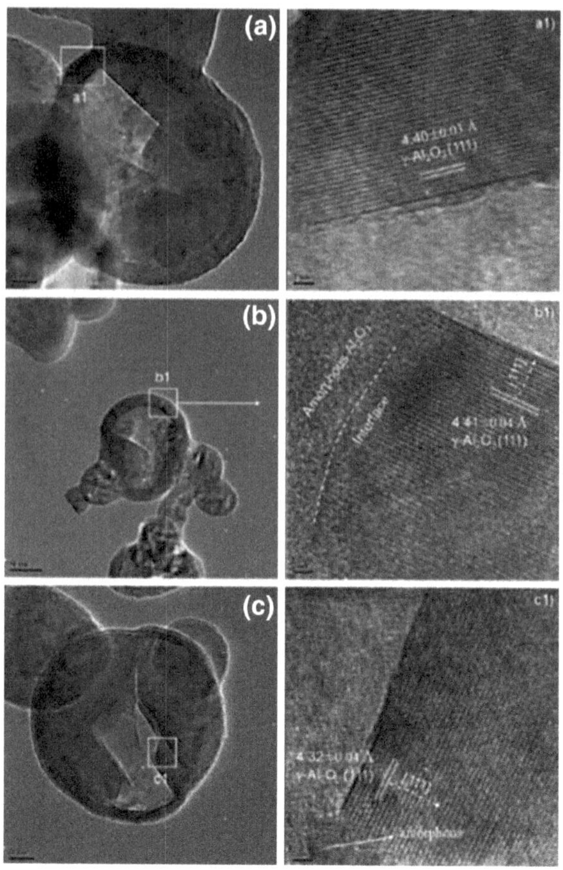

Fig. 2.2 High-resolution transmission electron microscopic imaging of the Al_2O_3 shell before and after the Al core melting presents evidence for inhomogeneous (local) crystallization of amorphous Al_2O_3 to γ-phase indicating the diffusion role. Reprinted from Firmansyah et al. (2012) with permission of American Chemical Society

heating rate, oxide shell thickness, ambient, and metallic content of Al core dictate whether MDM or DOM is operational. It is also evident that one single theory/ mechanism is not able to explain the oxidation behavior comprehensively under all experimental conditions that are yet to be well-defined and demarcated. Therefore, the debate on oxidation behavior under different experimental conditions still continues as also evident from the number of papers getting published every year till date, and the final word is yet to be said on oxidation behavior. In situ heating study of Al nanoparticles with advanced spectroscopic techniques and high-resolution transmission electron microscopy (HRTEM) may help to understand the microstructural behavior of the Al_2O_3 shell before and after melting of Al core in depth.

2.3 Micron- Versus Nano-Aluminum

There are merits and demerits associated with both micron- and nano-Al particles. Aluminum nanoparticles are quite attractive due to their unique and favorable properties, which are attributed to their high specific surface area and excess energy of surface atoms. As a result, there is a growing interest in employing metal nanoparticles in propulsion and energy conversion systems. There are issues connected with processing of Al nanoparticles in propellants as a result of increased viscosity. Also, Al nanoparticles are very expensive than micron-scale particles and pose safety and environmental issues. However, micron-sized particles present several drawbacks, such as high ignition temperatures and particle agglomeration, resulting in low energy release rates. In contrast, there are recent reports that suggest that reactivity of micron Al can be improved by suitable pre-treatments. A summary of key findings in this area of research is presented and discussed here.

Pre-stressed core–shell structures Al micron-scale particles were found to exhibit increased reactivity (Levitas et al. 2015). The alumina passivation shell surrounding Al core was subjected to compressive stressed when Al particles were annealed and quenched. The pretreatment of Al particles was tailored to meet the prerequisites of MDM theory. The experimentally measured flame propagation rates for CuO/Al (micron) were in quantitative agreement with the theoretical predictions based on the MDM. For the best thermal treatment determined through optimization processes, the flame rate went up by 36% in comparison to that of CuO/Al (micron) formed without pre-stressing. Most importantly, this value with micron Al is reported to be 68% of that for the best Al nanoparticles (Levitas et al. 2015)

2.4 Surface Passivation Efforts

It is now well reported in literature that incorporation of aluminum nanoparticles in explosive formulations enhances the reaction rates (Sundaram et al. 2017). However, nascent Al nanoparticles are highly pyrophoric and therefore, their surface is immediately passivated in situ during the synthesis under controlled oxygen environment to form an amorphous Al_2O_3 passivation layer around Al particles. The Al_2O_3 passivation layer with a typical thickness in the range of 2–4 nm can act as a heat sink as well as a diffusion barrier, which eventually slows down oxidation of Al when initiated. Ignition delay is therefore caused by the presence of this native oxide layer, and this is a major problem in case of aluminized explosives.

More often, Al nanoparticles are stored in glove box purged with high purity argon or nitrogen gas, which obviously provides controlled humidity conditions. Long-term storage of Al nanoparticles despite passivation is therefore a major issue, and the nature of the passivation layer becomes crucial. The Al_2O_3 shell is a dead weight in the sense that the dead layer does not contribute to the reaction. Rather, this layer is a hindrance to the reactivity of the energetic composite. The weight proportion of the Al_2O_3 shell is significant. For example, the active Al content in Al/Al_2O_3 core–shell nanoparticles with APS of 80 nm and nearly 2–2.5 nm shell is about 80 ± 2 wt% depending on shell thickness. To ensure safety during the processing of polymer-bonded explosives, water-based slurry is employed (He et al. 2016). For this purpose, fluorine polymer is being used as a polymeric binder in explosives whenever Al nanoparticles are used as an energetic additive (Yetter et al. 2009). So, the choice of polymer is generally restricted whenever Al nanoparticles are employed. Besides, high explosives are transported with some minimal water content for safe handling and processing at the receiver end. In this case, it is problematic to the use of Al nanoparticles in energetic formulations. The tendency of Al nanoparticles to agglomerate is another major issue. Though Al-based propellants show reduced ignition delay and enhanced burn rate, the reliability in the burning behavior is a question mark (Yetter et al. 2009).

In recent years, several works have been reported to form intermetallic structure with Al core being coated with another metal as a shell (Andrzejak et al. 2007). A few research groups have used fluorine-based coating as a passivation layer (Jouet et al. 2005, 2006; Dikici et al. 2009). The stability of these coatings and also that of the Al core needs to be investigated through systematic experimentation. Various groups have developed organic coating for passivation with a host of materials such as silanes (Zhou and Yu 2013) and phosphonic acids (Crouse et al. 2010). Some of these materials are inert, and therefore, the energy density of energetic materials is significantly lowered. Besides, the rate of energy release is poor upon coating with inert substances as the mass transport length scale is now increased.

Instead of passivating Al surface directly after etching the Al_2O_3 shell, some research groups have undertaken efforts to coat the nanothermites with superhydrophobic coatings in an attempt to enhance the long-term stability against aging.

Nixon et al. (2011) reported achieving superhydrophobic coating on Al/Fe_2O_3 nanothermites via sequential deposition of silsesquioxane layer as adhesion promoter and fluorocarbon self-assembled monolayer using a combination of chemical vapor deposition and atomic layer deposition system. These coatings were done on energetic pellets, and therefore, the surface coverage on individual Al nanoparticles may not be possible, for some of the surfaces would be inaccessible even for the vapors. Long-term stability against oxidation in such a case remains an issue. On laboratory scale, this coating methodology may be suitable for enhancing the stability of Al nanoparticles as evidenced by static contact angle measurement of 169°. The challenges associated with transition to large-scale technology need to be overcome for successful implementation in underwater application. Agglomeration of Al nanoparticles still remains an issue.

Sintering of nanoparticles during combustion causing aggregation is another major issue that prevents us from exploiting the full potential of Al nanoparticles' reactivity. Jacob et al. (2016) recently reported a novel approach of packing commercial nanoparticles into a meso-structure using nitrocellulose (NC). Low-temperature decomposition of the energetic binder would enhance the dispersion of the nanoparticles, thereby reducing sintering at the onset of combustion. Commercial nanoparticles were found to produce some very large spheres, which are a result of a large agglomerates sintering into a larger droplet, and subsequently burning in the oxidizing environment. In contrast, the nitrocellulose-coated meso-particles were found to burn better without suffering from aggregation owing to dispersion of nanoparticles as a result of thermal decomposition of NC and subsequent gas generation at lower temperature (Fig. 2.3).

Yu et al. (2018) investigated the ignition and aging characteristics of the superhydrophobic nanothermite film of the Co_3O_4 nanowires/Al coated with perfluorodecyltriethoxysilane (FAS) (Fig. 2.4) in aqueous environments.

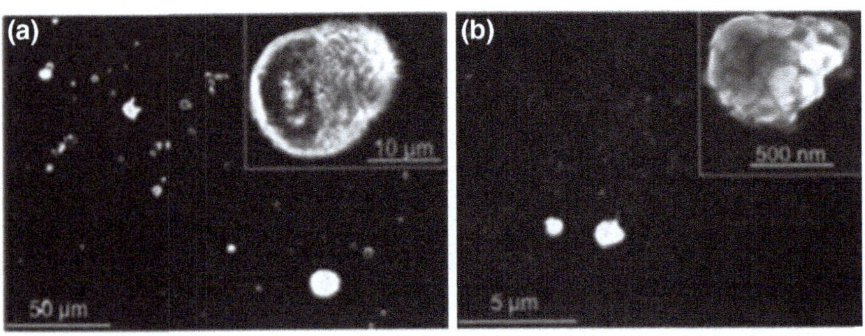

Fig. 2.3 SEM images of the products collected post-combustion: **a** Commercial aluminum nanoparticles and the inset reveals the larger size (nearly 20–25 µ) of the individual particle at high magnification. **b** Aluminum meso-particles formed with nitrocellulose and the inset shows smaller individual particle (<1 µ) at high magnification. A clear evidence for substantial reduction in the sintering effects is observed. Reprinted from Jacob et al. (2016) with permission of Elsevier

Fig. 2.4 SEM images of **a** neat Co_3O_4 nanowires and **b** an energetic film of Co_3O_4/Al. The thickness of Al film deposited over oxide nanowires was 600 nm. Reprinted from Yu et al. (2018) with permission of Elsevier

This nanothermite film exhibited a contact angle of 157 ± 2°. Furthermore, the authors here performed aging tests as a function of submersion times through comparisons of the heat release and ignition characteristics. While the neat Co_3O_4/Al film exhibits a total energy release of 2860 J/g, the Co_3O_4/Al coated with FAS releases about 2537 J/g without subjecting any underwater storage tests. However, the stability of the neat Co_3O_4/Al nanothermite film is extremely poor as it shows a heat release of only 72 J/g after an underwater storage for 6 h. In contrast, 'FAS-coated Co_3O_4/Al nanothermite' retained 50% of energy release (i.e., 1268 J/g) even after remaining submerged for 2 days (reference). These results are encouraging for underwater applications of nanothermites. As opposed to employing inert coatings for passivation, coating of Al nanoparticles with energetic material is an attractive strategy. Several groups have used different energetic materials for this purpose. Glycidyl azide polymer (GAP) is a well-known energetic material used as a polymeric binder widely. Coating of Al nanoparticles with GAP is useful in multiple ways. First of all, the hydrophobic shell provides protection to Al core from reaction with water. Secondly, the nitrogen functional groups in GAP participate in the reaction, thereby contributing to the overall energy content. Thirdly, the mechanical strength of GAP is so good that it can be used as a binder in Al-based explosive formulations.

Recently, Zeng et al. (2018) reported the synthesis of Al–GAP (core–shell) nanostructures via in situ grafting method. The contact angle measurements reveal the hydrophobic nature of these nanostructures with the contact angle changing from 20.2° to 142.4°, thereby protecting the Al core from water. More importantly, the flame propagation in Al–GAP core–shell nanoparticles is reported to be intense and violent. Though the findings of this work are noteworthy, the transmission electron microscopic image of GAP-coated Al nanoparticles still shows the presence of Al_2O_3 layer as indicated therein by the authors. Besides, it is not evident from images as well the analysis if all of the nanoparticles are separately coated with GAP. In other words, the issue of agglomeration of Al nanoparticles during coating needs to be clearly addressed. Smith et al. (2017a, b) demonstrated an interesting approach to remove the native oxide layer and replace with aluminum iodate hexahydrate (AIH) as a result of addition of Al particles to concentrated solutions of iodic acid. Most importantly, the Al (core)/AIH (shell) particles were highly reactive and exhibited flame speeds of 3200 m/s. Following this result, an interesting study was conducted by Gottfried et al. (2018) to understand AIH contribution to detonation and/or deflagration reactions using the laser-induced air shock from energetic materials (LASEM) technique.

For these tests, trinitrotoluene (TNT) was used in combination with Al/AIH core–shell particles. Laboratory-scale tests show enormous promise in enhancing the reaction rates of TNT/Al–AIH explosive both in the fast (detonation) and slow (deflagration) combustion modes. The enhancement in the reaction rate is attributed

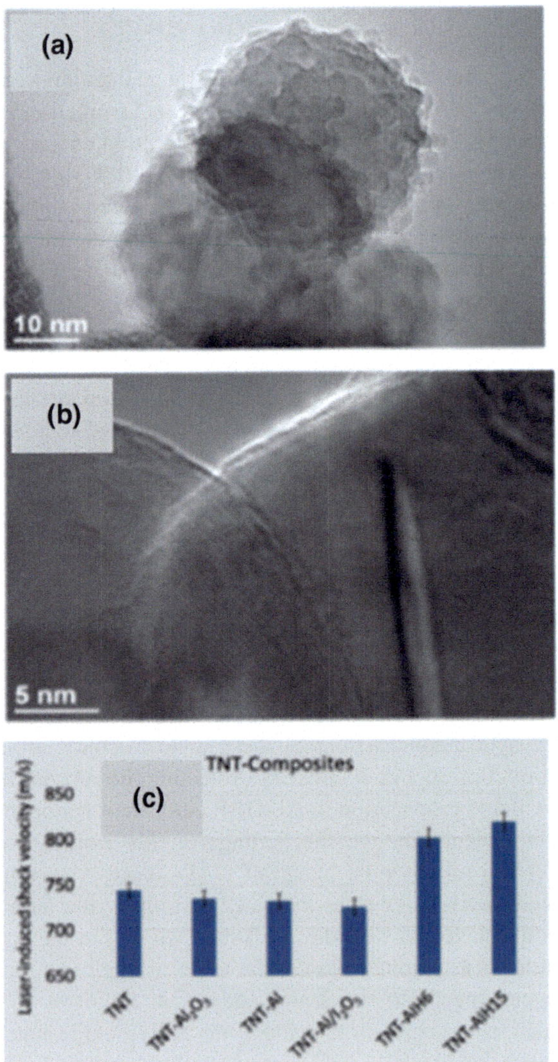

Fig. 2.5 **a** and **b** HRTEM images of Al(core)/AIH(shell) nanoparticle. The lower resolution image **a** clearly shows some roughness. The HRTEM image indicates crystalline nature of AIH shell. **c** Measurements of laser-induced shock velocity of TNT-based composites clearly reveal that the velocity is highest for the two samples prepared with Al(core)/AIH (shell) with different wt%. Reprinted from Gottfried et al. (2018) with permission of Nature Publishing group

to overcoming the oxygen balance problem in TNT (−74%) through participation in the reaction by AIH (Fig. 2.5). Besides, the hydrate layer in AIH dehydrates at low temperature, thereby providing the iodate oxidizer leading to the onset of fast reaction (Gottfried et al. 2018). Though these experiments were performed in laboratory scale, this work is perhaps a unique demonstration of combustion processes of Al-based explosive synthesized without the diffusion limited Al_2O_3 layer.

2.5 Self-assembled Nano-energetic Composites

Self-assembly is a process by which discrete components are driven to organize spontaneously into well-defined geometries by specific interactions (Grzelczak et al. 2010; Thiruvengadathan et al. 2013). These interactions may arise due to the intrinsic properties of the individual elements composing the system or under the influence of applied external fields. It is understood that the organization, intimacy, and dimensions of the discrete fuels and oxidizers in the nano-energetic composites largely impact the rate of energy release. Increasing fuel and oxidizer interfacial contact area enhances the reaction rate of a nanocomposite. Consequently, various ways to self-assemble the fuels and oxidizers into dense and arranged structures are being investigated in recent years. These approaches include: (i) electrostatic assembly (Kim and Zachariah 2004; Malchi et al. 2009; Zakiyyan et al. 2018), (ii) DNA-directed self-assembly (Séverac et al. 2012), (iii) polymer-mediated self-assembly (Shende et al. 2008; Thiruvengadathan et al. 2011), and (iv) functionalized graphene-directed self-assembly (Thiruvengadathan et al. 2014). Such self-assembled nano-energetic materials exhibit enhanced combustion performance by several orders over randomly mixed nano-energetic composites. Here we highlight and discuss some classic examples for the above-mentioned self-assembly schemes.

Electrostatic forces of interactions between constituents are exploited to form the self-assembled composite. Malchi et al. (2009) reported the synthesis of nanocomposite reactive microspheres with diameters of ~ 1–5 μm via electrostatic self-assembly of aluminum and cupric oxide nanoparticles. As soon as the dimethyl sulfoxide (DMSO) suspension of Al nanoparticles coated with an ω-functionalized alkanoic acid is added to the suspension of CuO nanoparticles coated with ω-functionalized alkanethiol in DMSO, the charge is neutralized and the two constituents self-assemble and precipitate out of solution over several hours. Image recorded using scanning electron microscope (SEM) of CuO/Al microsphere formed as a result of electrostatic assembly is shown in Fig. 2.6a. However, when they are in separate solutions of dimethyl sulfoxide (DMSO) (10 mM), the suspensions of functionalized particles are very stable without any precipitation. Upon ignition, the self-assembled material deposited in rectangular microchannels exhibits self-propagating combustion behavior, while the unassembled nanothermites does not get ignited.

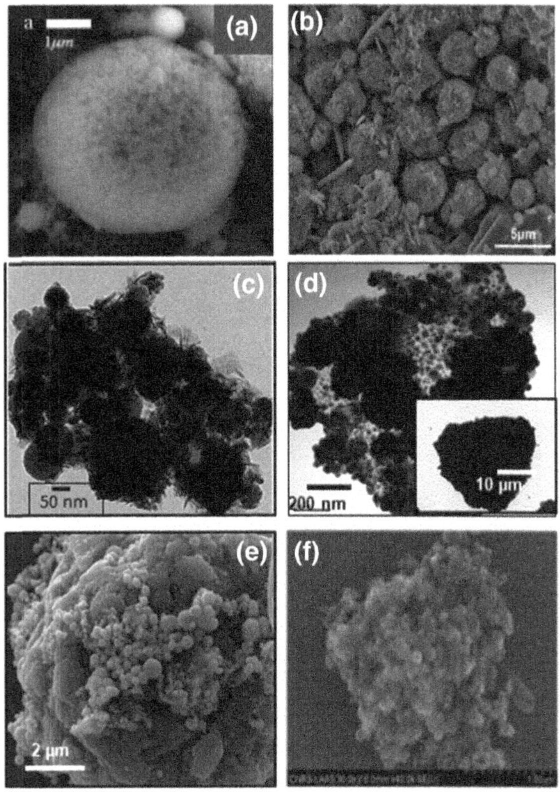

Fig. 2.6 **a** and **b** SEM images of self-assembled **a** CuO/Al microsphere and **b** MoO_3/Al clusters via electrostatic interactions. **c** TEM image of CuO nanorods/Al (80 nm) self-assembled macrostructure. **d** TEM image of graphene oxide (GO) densely decorated with Al first and then Bi_2O_3. The inset in **d** shows an ultra-dense assembly of Al and Bi_2O_3 on GO. **e** SEMs of ultra-dense macrostructures GO(5%)/Al/Bi_2O_3. **f** TEM image of DNA-directed self-assembled CuO/Al structure. Figure 6a from Malchi et al. (2009) and 6d, e from Thiruvengadathan et al. (2014) are reprinted with the permission of American Chemical Society. Figure 6b from Zakiyyan et al. (2018) and 6c from Thiruvengadathan et al. (2011) are reprinted with the permission of Elsevier. Figure 6f from Séverac et al. (2012) is reprinted with the permission of WILEY-VCH Verlag GmbH & Co. KGaA, Weinheim

More recently, Gangopadhyay et al. reported formation of self-assembled clusters of tightly packed Al nanoparticles and 2D sheets of MoO_3 with long-range order to many microns, harnessing the electrostatic attraction between Al NPs and MoO_3 nanosheets due to opposing surface charges (Zakiyyan et al. 2018). SEM image of self-assembled MoO_3/Al cluster shown in Fig. 2.6b reveals the long-range ordering of MoO_3 to Al and vice versa. Combustion measurements demonstrated peak pressures as high as 42.05 ± 1.86 MPa, pressurization rates up to 3.49 ± 0.31 MPa/μs, and linear combustion rates up to 1730 ± 98.1 m/s, the highest values reported to date for Al/MoO_3 composites (Zakiyyan et al. 2018).

Polymer-mediated self-assembly has been reported for CuO nanowires/Al via employing a self-assembled monolayer coating of poly(4-vinylpyridine) (PVP) (Shende et al. 2008). The linear combustion wave speed upon self-assembly increased from 1900 to 2400 m/s for CuO nanowires/Al system. In this work, CuO nanowires were first coated with PVP prior to mixing with Al nanoparticles. From thermogravimetric analysis, the total weight loss of the polymer is 2.13 wt%. Though this amount of PVP appears high, the total amount of moles of PVP and CuO are 1.25×10^{-6} and 6×10^{-3}, corresponding to >99.9% of CuO mol% in PVP-coated CuO. Thus, the presence of PVP layer did not increase the diffusional path length in a self-assembled composite. Following this work, a few research groups have used this methodology to self-assemble other nanothermites compositions and have realized improved combustion performance (Cheng et al. 2010a, b).

Another interesting method that enabled assembly of CuO nanorods around Al nanoparticles and vice versa based on a thin coating of poly(ethylene glycol) (PEG-400) has been reported recently (Thiruvengadathan et al. 2011). Here, the hydroxyl (OH) and the methyl and methylene CH_n (n = 2, 3) functional groups due to the presence of a thin coating of PEG on the surface of CuO nanorods is found to force self-assembly (Thiruvengadathan et al. 2011). Typical image recorded using transmission electron microscope (TEM) depicting the self-assembled CuO/Al nano-energetic composite is shown in Fig. 2.6c. The combustion performance of CuO nanorods/Al post-assembly was superior attributed to both the increased interfacial contacts between oxidizer and fuel that aids the heat transfer via conductive mechanism and enhanced gas generation as a result of functional groups that support heat transfer via convective mechanism. Séverac reported a DNA-directed assembly process to assemble CuO and Al nanoparticles, enabling micrometer-sized spheres of CuO/Al composite (Séverac et al. 2012). DNA-directed assembly is based on coating two types of nanoparticles with single-stranded DNA molecules of complementary sequences. SEM image of one individual Al/CuO aggregate of ≈2 μm through DNA-directed self-assembly is shown in Fig. 2.6f. This CuO/Al composite reveals significantly higher heat of reaction (1800 J/g with 80 nm APS for Al) compared with physically and randomly mixed l nanocomposites (1200 J/g with 80 nm APS for Al). Indeed, this value is the best value reported in literature for CuO/Al system till date. Besides, the onset temperature was reduced from 470 to 410 °C upon self-assembly. Most recently, graphene-based nano-energetic materials have shown a great promise for the development of advanced energetic material systems with tunable combustion performance (Thiruvengadathan et al. 2014, 2015). This is attributed to the dual use of graphene as an energetic additive as well its exploitation as a self-assembly-directing agent. Exploiting the functional groups present in functionalized graphene, multifunctional energetic materials through hierarchical self-assembly of nanoscale oxidizer and fuel were produced via long-range electrostatic attraction followed by short-range covalent [between graphene oxide (GO) and Al] and noncovalent interactions (between GO/Al and Bi_2O_3) (Thiruvengadathan et al. 2014, 2015). A methodology for directing the self-assembly of Al and Bi_2O_3 nanoparticles on functionalized graphene sheets

(FGS) leads to the formation of nanocomposite structures in a colloidal suspension phase that ultimately condense into ultra-dense macrostructures (Fig. 2.6d, e).

Significant enhancements in heat of reaction values up to 92% were observed when comparing randomly mixed (739 ± 18 J/g) to self-assembled nanocomposites (1421 ± 12 J/g) due to the benefits of self-assembly and role of GO as an energetic reactant. Furthermore, the self-assembled nano-energetic materials demonstrate significant combustion performance improvements in comparison to randomly mixed aluminum and bismuth trioxide nanoparticles with enhanced pressure generation from 60 to 200 MPa, reactivity from 3 to 16 MPa/μs, burn rate from 1.15 to 1.55 km/s, and specific impulse from 41 to 71 s (Thiruvengadathan et al. 2015). Figure 2.7a–d shows linear burn rate, sensitivity to ESD, force–time characteristics, and specific impulse measurements as a function of graphene oxide

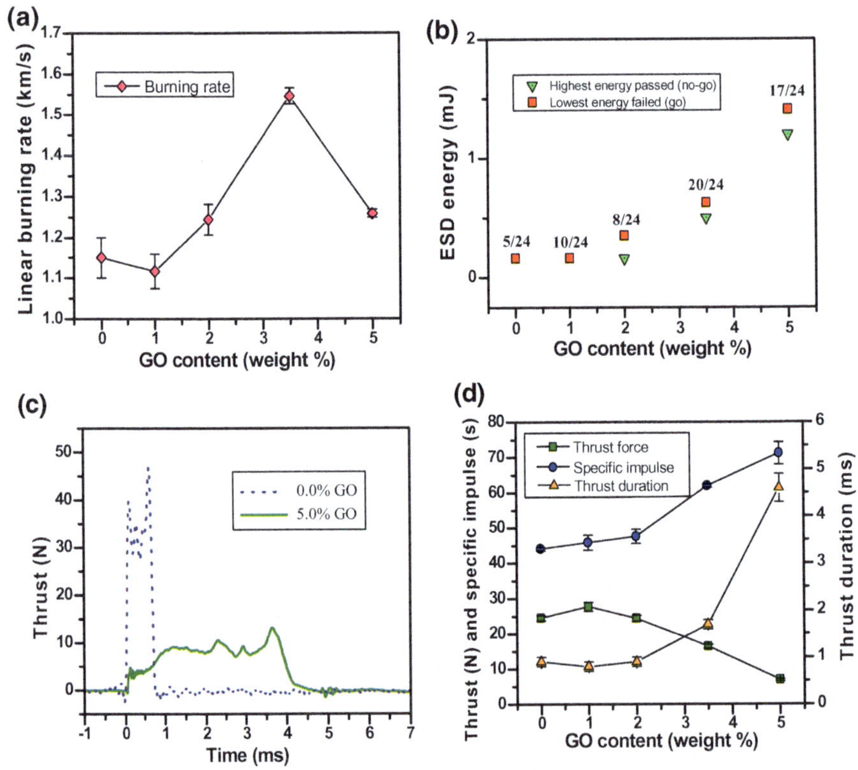

Fig. 2.7 Experimentally measured values of **a** linear burn rate, **b** sensitivity to ESD, **c** force–time characteristics, and **d** specific impulse, thrust force, and thrust duration as a function of concentration of functionalized graphene sheets. The obtained data reflects the usefulness of self-assembly and the resulting interfacial contacts between Al and Bi_2O_3 nanoparticles. Reprinted from Thiruvengadathan et al. (2015) with the permission of WILEY-VCH Verlag GmbH & Co. KGaA, Weinheim

content, respectively. Unquestionably, 2D materials and their functionalized forms with high surface area offer a profound opportunity to self-assemble oxidizer and fuel nanoparticles at multiple length scales through tailoring the various interactions.

2.6 Energetic Liquids

Employing energetic liquids over energetic solids for myriad applications is an attractive proposition owing to a number of reasons. Alkanes (Huber et al. 2005), nitrogen-rich ionic liquids (Singh et al. 2006), and combustible solvents (Sabourin et al. 2009; Liu et al. 2012; Basu and Miglani 2016) belong to the family of energetic liquids. These include lower activation temperature, higher pressure, and better volume expansion. Energetic liquids are generally employed at fuels as they possess low energy densities (typically 1 kg/g) and slow burn kinetics due to their innate chemical composition containing rich amounts of carbon, nitrogen, and hydrogen. In recent years, efforts have been undertaken toward improving both energy density and burn kinetics (Sabourin et al. 2009). Addition of metal nanoparticles such as boron and aluminum to liquid fuels is shown to improve the burn kinetics as well increase energy densities significantly. For example, synthesis of stable dispersions of Al and boron nanoparticles at low weight percent in nitrogen-rich ionic liquids and organic fuels has paved the way for enhanced energy output and improved burn characteristics. Very recently, Yetter et al. reported the enormous potential of functionalized graphene as an energetic additive in enhancing the linear burn rate of nitromethane. Reduction in the ignition temperature and improvement in the linear burning rates (by 175% in comparison to neat nitromethane) were the highlights of this interesting work. The biggest challenge involved in incorporating Al nanoparticles in energetic liquids is the observation of increased viscosity at a relative higher solid loading that imposes severe processing constraints. Another challenge is the observation of agglomeration of metal nanoparticles that eventually results in phase separation between solid and liquids. In other words, stability of the dispersions is very poor, leading to unreliable combustion performance.

In the past one decade, several works reported in literature have amply demonstrated the usefulness of nanothermites to a number of applications including chemical propulsion, enhanced blast with gas generation, biocidal, and pyrotechnics. It is understood that amounts of fuel and oxidizer in an energetic mixture should be optimum to realize theoretically predicted values of energy release during the exothermic reaction. Synthesis of stable dispersions of self-assembled nanothermites in liquid fuels at relatively high solid loading will have unparallel combustion performance both in terms of energy release and reaction rate.

In this respect, Slocik et al. (2017) demonstrated very recently a novel synthesis of producing energetic biothermite inks composed of high concentrations of Al nanoparticles (up to 20 wt%) using ferritin liquid protein without the use of stabilizers. In this energetic composition, the biologically derived iron oxide ferrihydrite nanoparticles (FeO(OH)) loaded within the protein cavity of ferritin is reported to serve as an excellent oxidizer, enabling increased energy output and combustion performance. For example, the combustion of neat ferritin protein ionic liquid monitored by performing weight loss and heat flow measurements shows an energy release of 1 kJ/g. In comparison, a sample of protein ionic liquid loaded with Al nanoparticles only (no FeO(OH)) shows a total heat release of 2.4 kJ/g. On the other hand, the energetic sample containing both FeO(OH) and Al nanoparticles mixed at an equivalence ratio of 8.2 in protein ionic liquid exhibits an energy release of 11.3 kJ/g while a mixture of nano-Al and ferritin biothermite produced a lower energy release of 8.9 kJ/g (Slocik et al. 2017). Furthermore, the authors reported lower ignition temperature and complete combustion of ferritin ionic liquid loaded with both Al nanoparticles and FeO(OH), which has been attributed to the stability of dispersed Al nanoparticles and homogeneity of dispersions that, in turn, is found to result in lower mass transport length scales. The remarkable feature of this work (Slocik et al. 2017) is the ability to form cylindrical crayon-like shape of energetic materials through freeze-casting of ferritin protein liquid ink containing Al nanoparticles in corresponding PDMS mold. The key results from this work are

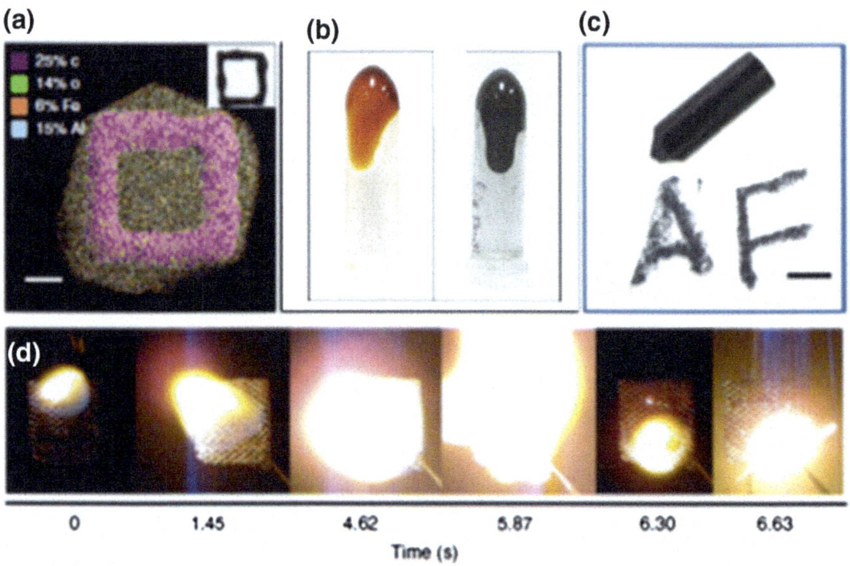

Fig. 2.8 Demonstration of the utility of energetic biothermite inks **a** elemental composition of the ink on glass. Scale bar = 1 mm. Inset shows optical image of a square drawn on a glass slide using ink. **b** Digital images of ferritin protein ionic liquid after synthesis (brown color) and (ii) ferritin protein ionic liquid loaded with nano-Al (20 wt%) (gray color). **c** Image of an energetic crayon created by filling a PDMS mold with the ink and freeze-casting and'AF' handwritten on paper with the crayon. Scale bar = 7 mm. **d** High-speed video frames at selected time intervals of a square pattern on stainless-steel fine mesh coated with ink. Reprinted from Slocik et al. (2017) with permission of Nature Publishing Group

presented in Fig. 2.8. The significance of this aspect is exemplified further by the ability to transfer and write energetic materials onto any contaminated surface and neutralize any biological/chemical agents by combustion of the ink. As bioenergetic liquids, they offer many advantages over traditional nanopowders in terms of increased energy output, stability against oxidation, increased dispersion stability at high wt%, lower activation temperatures, enhanced combustion kinetics, and greater functionality.

2.7 Perspectives

Several significant achievements have been realized in the past two decades of research on Al-based nano-energetic materials, thanks to the technological novelties in the fields of nanoscience and nanotechnology. In particular, the advancements in the synthesis approaches and the advent of tools for characterization of the material at multiple length scales (from molecular- to macro-scales) have contributed immensely to the growth of nano-energetic materials. The foremost achievement is the ability to synthesize designer-made nanostructures of both the Al fuel and a rich variety of oxidizers with superior control in their size, shape, and morphology. The procedures for the synthesis of nano-energetic composites employing a host of methods including physical mixing, arrested ball milling (Umbrajkar et al. 2008; Stamatis et al. 2009), and self-assembly approaches (Shende et al. 2008; Malchi et al. 2009; Séverac et al. 2012; Thiruvengadathan et al. 2014) are well laid and reported in literature. The merits of adopting bottom-up self-assembly approaches for enabling higher interfacial contacts are reasonably well understood by now. Combustion performance of self-assembled nano-energetic mixtures has been found to be always superior in comparison to that of random mixtures prepared by sonication or hand mixing. This observation is true and independent of the nature of self-assembly approach. Theoretical, experimental, and computational efforts have indeed enabled to develop a thorough qualitative and quantitative understanding on the dependence of combustion performance attributes of a variety of energetic formulations including solid and liquid propellants, metal–metal oxide systems, and intermetallic systems on the size of the constituents, their organization and stability. The combustion attributes include the reaction rates, total energy release, ignition delays, and sensitivity to external stimuli such as ESD, impact, and friction. Particularly, there is a deluge of original research articles published in the field of nano-energetic materials especially since 2005. Nano-energetic materials have been synthesized with a variety of forms such as powders (Shende et al. 2008; Thiruvengadathan et al. 2011, 2014; Piekiel et al. 2014), pellets (Pantoya et al. 2009; Puszynski et al. 2012), films (Fu et al. 2013; Zhang et al. 2013; Patel et al. 2015; Geeson et al. 2018; Yu et al. 2018; Zheng et al. 2018), foams (Comet et al. 2017), membrane (Yang et al. 2013; Zheng et al. 2016), and aerogel (Wang et al. 2018). Given the high sensitivity of nano-energetic powders, the safety issues come into play, and therefore, it is complex to implement powder form in practical applications. In this context, all of these forms have unique advantages. Advanced diagnostic characterization tools such as ultrafast laser-based spectrometer (Kim et al. 2002; Wang et al. 2017), time-of-flight mass spectrometer (Mahadevan et al. 2002; Zhou et al. 2009, 2010), and video capture with high-speed camera (Plummer et al. 2011; Jacob et al. 2018) have helped to develop better understanding of the combustion behavior including the initiation and the propagation mechanisms to a great extent. A variety of ignition initiation mechanisms including hot wire ignition (Tappan 2007), flash ignition (Ohkura et al. 2011), microchip-based ignition (Staley et al. 2011), and optical ignition with laser (Granier and Pantoya 2004) have been investigated and well-established. Presently, plasmonic gratings are being investigated as ignition platforms for nano-energetic materials (Chen et al. 2018). This ignition mechanism could be very attractive for patterned nano-energetic materials on microchip- or MEMS-based platforms as demonstrated independently by two research groups (Chen et al. 2018; Mutlu et al. 2018).

Many researchers have focused their research toward tailoring the combustion characteristics toward application developments. Nanothermites have shown enormous promise toward certain applications such as microthrusters (Apperson et al. 2009; Staley et al. 2013, 2014), propellant initiator (Bezmelnitsyn et al. 2010), MEMs-based igniters (Ru et al. 2014, 2016a; Xue et al. 2014; Chaalane et al. 2015; Oh et al. 2016), and biofilm removal (Lee et al. 2013) among others. Some progress has been made toward reducing the sensitivity of nanothermites to ESD, impact, heat, and friction through coating with materials like nitrocellulose and Viton. Also, use of electrically conductive additives like reduced graphene oxide, carbon black, and colloidal graphite in nano-energetic formulations have enabled to reduce ESD by three orders of magnitude. Undoubtedly, extensive fundamental and applied research has taken place in various laboratories across the world. However, the implementation of nano-energetic materials in practical applications demands establishing clear pathways for smooth transitioning of the fundamental research to technology development.

Some fundamental challenges continue to persist and, hence, demand careful attention and sustained research efforts. These challenges include: (i) prevention of particle agglomeration during synthesis as well during combustion due to sintering, (ii) achievement of long-term stability of Al nanoparticles, (iii) in situ surface passivation of Al nanoparticles with energetic material post-removal of Al_2O_3 layer to overcome the limitations imposed by diffusion, (iv) decreasing the sensitivity of nano-energetic formulations, (v) reduction in the production cost of Al nanoparticles, and (vi) establishment of standard operating procedures for scaled-up synthesis of nano-energetic materials. It is needless to emphasize the obvious that these issues need to be addressed and resolved comprehensively.

The key aspect that governs the reaction rate of nano-energetic mixture is dependent on the oxidation kinetics of Al particles. So far, both DOM and MDM have not been able to explain oxidation behavior of Al nanoparticles under all experimental conditions. In fact, the very definition of these experimental conditions needs to be clearly described. Currently, the prevailing thought is oxidation happens through diffusion mechanism at low heating rates. At high heating rates,

MDM describes the oxidation behavior. A final word is yet to be said about oxidation behavior under all experimental conditions. Computational research in the field of energetics seems to be mostly confined to elucidating oxidation behavior of Al particles and shock propagation in energetic materials. Nanostructured energetic materials present altogether different challenges. Computational model may be developed to determine an optimal approach to realize tunable combustion characteristics.

Quantification of interfacial contacts between fuel and oxidizer vis-à-vis combustion performance could be interesting. Determination of the effectiveness of self-assembly and the resulting interfacial contacts between fuel and oxidizer on combustion characteristics could be very useful. Though there are research works that talk about formation of ordered self-assembled energetic composite at multiple length scale from molecular (nanometric) to macro-scale (millimeter), these synthesis' processes are lengthy and requires meticulous attention vis-à-vis in controlling and optimizing experimental parameters. More importantly, scalability of self-assembly approach is yet to be demonstrated. Some research groups have renewed their research interests on the use of micron Al particles as fuel. Micron-sized Al particles offer realistic solution to reduce sensitivity of energetic materials by orders of magnitude besides bringing the cost down significantly. Some nano-energetic formulations exhibit detonation and deflagration behavior depending upon their chemical composition and initiation mechanism in laboratory-scale tests.

Exploitation of green chemistry to promote energetics domain is another interesting aspect that merits attention, and some works along this direction are already taking place. In situ passivation of Al nanoparticles with energetic monolayers is one of the hot areas of energetics research. Among various experimental efforts, the works on removal of Al_2O_3 shell and in situ passivation with AIH layer are highly commendable (Smith 2017a, b; Gottfried et al. 2018). However, the potential in enhancement of the reaction rates and the energy release through in situ passivation method is yet to be comprehensively demonstrated. The natural tendency of nanoparticles to agglomerate presents a great challenge in the processing of energetic powders at large scale. Phase separation of fuel and oxidizer nanoparticles is another serious issue that could hinder reliable and reproducible performance. Other concerns include handling of energetic powders at large scale, especially in the wake of their high sensitivity to electrostatic discharge (ESD), impact, and friction. Stability- and aging-related issues need to be addressed comprehensively through systematic experimental investigation.

Employing energetic liquids over energetic solids in any application is an attractive proposition owing to the high sensitivity of to external stimuli such as electrostatic discharge, friction, and impact. Recent works have shown beyond any doubt that incorporation of nano-Al with and without metal oxide nanoparticles in energetic liquids has enabled to overcome the inherent disadvantages such as lower energy density and slower combustion kinetics. Furthermore, the ability to use energetic liquids as inks for writing, printing, or stamping is an invaluable incentive apart from creation of 3D shapes and structures. Among the various candidates for energetic liquids reported in literature, ferritin ionic liquids with nano-Al inclusions are very promising for furthering the application aspects. The scope for the optimization of combustion performance of energetic liquids is massive. For example, one may choose a different metal oxide based on their potential to release higher energy upon reaction with fuel. Besides, the morphology and the dimensions of metal oxide nanostructures can be controlled to realize higher interfacial contacts with nano-Al. Controlled self-assembly of metal fuel with metal oxide at nanoscale without disrupting the stability of dispersions in energetic liquids is another potential route to enhance the reaction rates by orders of magnitude as well realizing higher energy content upon combustion. Implementation of energetic liquids in practical applications would require thorough testing of dispersions' stability and stability of Al nanoparticles against oxidation, beyond any doubt. Production of stable dispersions of metal and metal oxide nanoparticles over a period of long time at industrial scale is another direction of research that needs examination. Further investigation over longer self-life of such energetic liquids is therefore warranted.

References

Andrzejak TA, Shafirovich E, Varma A (2007) Ignition mechanism of nickel-coated aluminum particles. Combust Flame 150(1–2):60–70

Apperson S, Shende RV, Subramanian S, Tappmeyer D, Gangopadhyay S, Chen Z, Gangopadhyay K, Redner P, Nicholich S, Kapoor D (2007) Generation of fast propagating combustion and shock waves with copper oxide/aluminum nanothermite composites. Appl Phys Lett 91(24)

Apperson SJ, Bezmelnitsyn AV, Thiruvengadathan R, Gangopadhyay K, Gangopadhyay S, Balas WA, Anderson PE, Nicolich SM (2009) Characterization of nanothermite material for solid-fuel microthruster applications. J Propul Power 25(5):1086–1091

Basu S, Miglani A (2016) Combustion and heat transfer characteristics of nanofluid fuel droplets: a short review. Int J Heat Mass Transf 96:482–503

Bergsmark E, Simensen CJ, Kofstad P (1989) The oxidation of molten aluminum. Mater Sci Eng A 120:91–95

Bezmelnitsyn A, Thiruvengadathan R, Barizuddin S, Tappmeyer D, Apperson S, Gangopadhyay K, Gangopadhyay S, Redner P, Donadio M, Kapoor D, Nicolich S (2010) Modified nanoenergetic composites with tunable combustion characteristics for propellant applications. Propellants, Explos, Pyrotech 35(4):384–394

Campbell TJ, Aral G, Ogata S, Kalia RK, Nakano A, Vashishta P (2005) Oxidation of aluminum nanoclusters. Phys Rev B Condens Matter Mater Phys 71(20)

Chaalane A, Chemam R, Houabes M, Yahiaoui R, Metatla A, Ouari B, Metatla N, Mahi D, Dkhissi A, Esteve D (2015) A MEMS-based solid propellant microthruster array for space and military applications

Chakraborty P, Zachariah MR (2014) Do nanoenergetic particles remain nano-sized during combustion? Combust Flame 161(5):1408–1416

Chen B, Zheng H, Riehn M, Bok S, Gangopadhyay K, Maschmann MR, Gangopadhyay S (2018) In situ characterization of photothermal nanoenergetic combustion on a plasmonic microchip. ACS Appl Mater Interfaces 10(1):427–436

Cheng JL, Hng HH, Lee YW, Du SW, Thadhani NN (2010a) Kinetic study of thermal- and impact-initiated reactions in Al–Fe_2O_3 nanothermite. Combust Flame 157(12):2241–2249

Cheng JL, Hng HH, Ng HY, Soon PC, Lee YW (2010b) Synthesis and characterization of self-assembled nanoenergetic Al–Fe_2O_3 thermite system. J Phys Chem Solid 71(2):90–94
Chowdhury S, Sullivan K, Piekiel N, Zhou L, Zachariah MR (2010) Diffusive vs explosive reaction at the nanoscale. J Phys Chem C 114(20):9191–9195
Clark R, Wang W, Nomura KI, Kalia RK, Nakano A, Vashishta P (2011) Heat-initiated oxidation of an aluminum nanoparticle
Comet M, Martin C, Schnell F, Spitzer D (2017) Nanothermite foams: from nanopowder to object. Chem Eng J 316:807–812
Crouse CA, Pierce CJ, Spowart JE (2010) Influencing solvent miscibility and aqueous stability of aluminum nanoparticles through surface functionalization with acrylic monomers. ACS Appl Mater Interfaces 2(9):2560–2569
Dikici B, Dean SW, Pantoya ML, Levitas VI, Jouet RJ (2009) Influence of aluminum passivation on the reaction mechanism: flame propagation studies. Energy Fuels 23(9):4231–4235
Farley CW, Pantoya ML, Levitas VI (2014) A mechanistic perspective of atmospheric oxygen sensitivity on composite energetic material reactions. Combust Flame 161(4):1131–1134
Firmansyah DA, Sullivan K, Lee KS, Kim YH, Zahaf R, Zachariah MR, Lee D (2012) Microstructural behavior of the alumina shell and aluminum core before and after melting of aluminum nanoparticles. J Phys Chem C 116(1):404–411
Fu S, Zhu Y, Li D, Zhu P, Hu B, Ye Y, Shen R (2013) Deposition and characterization of highly energetic Al/MoOx multilayer nano-films. EPJ Appl Phys 64(3)
Geeson J, Staley C, Bok S, Thiruvengadathan R, Gangopadhyay K, Gangopadhyay S (2018) Graphene-based Al-Bi_2O_3 nanoenergetic films by electrophoretic deposition. In: 12th IEEE nanotechnology materials and devices conference, NMDC 2017, Institute of Electrical and Electronics Engineers Inc
Gibot P, Comet M, Eichhorn A, Schnell F, Muller O, Ciszek F, Boehrer Y, Spitzer D (2011) Highly insensitive/reactive thermite prepared from Cr_2O_3 nanoparticles. Propellants, Explos, Pyrotech 36(1):80–87
Gordeev VV, Kazutin MV, Kozyrev NV (2017) Effect of additives on CuO/Al nanothermite properties. In: All-Russian conference with international participation on modern problems of continuum mechanics and explosion physics: dedicated to the 60th anniversary of Lavrentyev Institute of Hydrodynamics SB RAS, MPCMEP 2017, Institute of Physics Publishing
Gottfried JL, Smith DK, Wu CC, Pantoya ML (2018) Improving the explosive performance of aluminum nanoparticles with aluminum iodate hexahydrate (AIH). Sci Rep 8(1)
Granier JJ, Pantoya ML (2004) Laser ignition of nanocomposite thermites. Combust Flame 138 (4):373–383
Grzelczak M, Vermant J, Furst EM, Liz-Marzán LM (2010) Directed self-assembly of nanoparticles. ACS Nano 4(7):3591–3605
He G, Yang Z, Zhou X, Zhang J, Pan L, Liu S (2016) Polymer bonded explosives (PBXs) with reduced thermal stress and sensitivity by thermal conductivity enhancement with graphene nanoplatelets. Compos Sci Technol 131:22–31
Henz BJ, Hawa T, Zachariah MR (2010) On the role of built-in electric fields on the ignition of oxide coated nanoaluminum: ion mobility versus Fickian diffusion. J Appl Phys 107(2)
Hong S, Van Duin ACT (2015) Molecular dynamics simulations of the oxidation of aluminum nanoparticles using the ReaxFF reactive force field. J Phys Chem C 119(31):17876–17886
Huber GW, Chheda JN, Barrett CJ, Dumesic JA (2005) Chemistry: production of liquid alkanes by aqueous-phase processing of biomass-derived carbohydrates. Science 308(5727):1446–1450
Jacob RJ, Wei B, Zachariah MR (2016) Quantifying the enhanced combustion characteristics of electrospray assembled aluminum mesoparticles. Combust Flame 167:472–480
Jacob RJ, Kline DJ, Zachariah MR (2018) High speed 2-dimensional temperature measurements of nanothermite composites: probing thermal vs. Gas generation effects. J Appl Phys 123(11)
Jeurgens LPH, Sloof WG, Tichelaar FD, Mittemeijer EJ (2002) Growth kinetics and mechanisms of aluminum-oxide films formed by thermal oxidation of aluminum. J Appl Phys 92(3):1649–1656

Jian G, Chowdhury S, Sullivan K, Zachariah MR (2013) Nanothermite reactions: is gas phase oxygen generation from the oxygen carrier an essential prerequisite to ignition? Combust Flame 160(2):432–437
Jouet RJ, Warren AD, Rosenberg DM, Bellitto VJ, Park K, Zachariah MR (2005) Surface passivation of bare aluminum nanoparticles using perfluoroalkyl carboxylic acids. Chem Mater 17(11):2987–2996
Jouet RJ, Granholm RH, Sandusky HW, Warren AD (2006) Preparation and shock reactivity analysis of novel perfluoroalkyl-coated aluminum nanocomposites
Kelly D, Beland P, Brousseau P, Petre CF (2017a) Electrostatic discharge sensitivity and resistivity measurements of Al nanothermites and their fuel and oxidant precursors. Centr Eur J Energ Mater 14(1):105–119
Kelly DG, Beland P, Brousseau P, Petre CF (2017b) Formation of additive-containing nanothermites and modifications to their friction sensitivity. J Energ Mater 35(3):331–345
Kim SH, Zachariah MR (2004) Enhancing the rate of energy release from nanoenergetic materials by electrostatically enhanced assembly. Adv Mater 16(20):1821–1825
Kim H, Hambir SA, Dlott DD (2002) Ultrafast high repetition rate absorption spectroscopy of polymer shock compression. Shock Waves 12(1):79–86
Kim JH, Cho MH, Kim KJ, Kim SH (2017) Laser ignition and controlled explosion of nanoenergetic materials: the role of multi-walled carbon nanotubes. Carbon 118:268–277
Lee BD, Thiruvengadathan R, Puttaswamy S, Smith BM, Gangopadhyay K, Gangopadhyay S, Sengupta S (2013) Ultra-rapid elimination of biofilms via the combustion of a nanoenergetic coating. BMC Biotechnol 13
Levitas VI (2009) Burn time of aluminum nanoparticles: strong effect of the heating rate and melt-dispersion mechanism. Combust Flame 156(2):543–546
Levitas VI, Asay BW, Son SF, Pantoya M (2006) Melt dispersion mechanism for fast reaction of nanothermites. Appl Phys Lett 89(7)
Levitas VI, Asay BW, Son SF, Pantoya M (2007) Mechanochemical mechanism for fast reaction of metastable intermolecular composites based on dispersion of liquid metal. J Appl Phys 101(8)
Levitas VI, Pantoya ML, Dikici B (2008) Melt dispersion versus diffusive oxidation mechanism for aluminum nanoparticles: critical experiments and controlling parameters. Appl Phys Lett 92 (1)
Levitas VI, Dikici B, Pantoya ML (2011) Toward design of the pre-stressed nano- and microscale aluminum particles covered by oxide shell. Combust Flame 158(7):1413–1417
Levitas VI, McCollum J, Pantoya M (2015) Pre-stressing micron-scale aluminum core-shell particles to improve reactivity. Sci Rep 5
Liu LM, Car R, Selloni A, Dabbs DM, Aksay IA, Yetter RA (2012) Enhanced thermal decomposition of nitromethane on functionalized graphene sheets: Ab initio molecular dynamics simulations. J Am Chem Soc 134(46):19011–19016
Mahadevan R, Lee D, Sakurai H, Zachariah MR (2002) Measurement of condensed-phase reaction kinetics in the aerosol phase using single particle mass spectrometry. J Phys Chem A 106 (46):11083–11092
Malchi JY, Foley TJ, Yetter RA (2009) Electrostatically self-assembled nanocomposite reactive microspheres. ACS Appl Mater Interfaces 1(11):2420–2423
Martirosyan KS (2011) Nanoenergetic gas-generators: principles and applications. J Mater Chem 21(26):9400–9405
Mukasyan AS, Rogachev AS (2016) Combustion behavior of nanocomposite energetic materials. In: Energetic nanomaterials: synthesis, characterization, and application. Elsevier Inc, pp 163–192
Muthiah R, Krishnamurthy VN, Gupta BR (1992) Rheology of HTPB propellant. I. Effect of solid loading, oxidizer particle size, and aluminum content. J Appl Polym Sci 44(11):2043–2052
Mutlu M, Kang JH, Raza S, Schoen D, Zheng X, Kik PG, Brongersma ML (2018) Thermoplasmonic ignition of metal nanoparticles. Nano Lett 18(3):1699–1706

Nixon E, Pantoya ML, Sivakumar G, Vijayasai A, Dallas T (2011) Effect of a superhydrophobic coating on the combustion of aluminium and iron oxide nanothermites. Surf Coat Technol 205 (21–22):5103–5108

Oh HU, Ha HW, Kim T, Lee JK (2016) Thermo-mechanical design for on-orbit verification of MEMS based solid propellant thruster array through STEP cube lab mission. Int J Aeronaut Space Sci 17(4):526–534

Ohkura Y, Rao PM, Zheng X (2011) Flash ignition of Al nanoparticles: mechanism and applications. Combust Flame 158(12):2544–2548

Pantoya ML, Levitas VI, Granier JJ, Henderson JB (2009) Effect of bulk density on reaction propagation in nanothermites and micron thermites. J Propul Power 25(2):465–470

Park K, Lee D, Rai A, Mukherjee D, Zachariah MR (2005) Size-resolved kinetic measurements of aluminum nanoparticle oxidation with single particle mass spectrometry. J Phys Chem B 109 (15):7290–7299

Patel VK, Ganguli A, Kant R, Bhattacharya S (2015) Micropatterning of nanoenergetic films of Bi_2O_3/Al for pyrotechnics. RSC Adv 5(20):14967–14973

Piekiel NW, Zhou L, Sullivan KT, Chowdhury S, Egan GC, Zachariah MR (2014) Initiation and reaction in Al/Bi_2O_3 nanothermites: evidence for the predominance of condensed phase chemistry. Combust Sci Technol 186(9):1209–1224

Plummer A, Kuznetsov V, Joyner T, Shapter J, Voelcker NH (2011) The burning rate of energetic films of nanostructured porous silicon. Small 7(23):3392–3398

Puszynski JA, Bulian CJ, Swiatkiewicz JJ, Kapoor D (2012) Formation of consolidated nanothermite materials using support substrates and/or binder materials. Int J Energ Mater Chem Propul 11(5):401–412

Rai A, Park K, Zhou L, Zachariah MR (2006) Understanding the mechanism of aluminium nanoparticle oxidation. Combust Theor Model 10(5):843–859

Rossi C (2014) Two decades of research on nano-energetic materials. Propellants, Explos, Pyrotech 39(3):323–327

Ru CB, Ye YH, Wang CL, Zhu P, Shen RQ, Hu Y, Wu LZ (2014) Design and fabrication of MEMS-based solid propellant microthrusters array. In: Applied mechanics and materials, vol. 490–491, pp. 1042–1046

Ru C, Dai J, Xu J, Ye Y, Zhu P, Shen R (2016a) Design and optimization of micro-semiconductor bridge used for solid propellant microthrusters array. EPJ Appl Phys 74(3)

Ru CB, Wang F, Xu JB, Dai J, Shen Y, Ye YH, Zhu P, Shen RQ (2016b) Micropropulsion characteristics of nanothermites prepared by electrospray. Hanneng Cailiao/Chin J Energ Mater 24(12):1136–1144

Sabourin JL, Dabbs DM, Yetter RA, Dryer FL, Aksay IA (2009) Functionalized graphene sheet colloids for enhanced fuel/propellant combustion. ACS Nano 3(12):3945–3954

Séverac F, Alphonse P, Estève A, Bancaud A, Rossi C (2012) High-energy Al/CuO nanocomposites obtained by DNA-directed assembly. Adv Func Mater 22(2):323–329

Shende R, Subramanian S, Hasan S, Apperson S, Thiruvengadathan R, Gangopadhyay K, Gangopadhyay S, Redner P, Kapoor D, Nicolich S, Balas W (2008) Nanoenergetic composites of CuO nanorods, nanowires, and Al-nanoparticles. Propellants, Explos, Pyrotech 33(2):122–130

Singh RP, Verma RD, Meshri DT, Shreeve JM (2006) Energetic nitrogen-rich salts and ionic liquids. Angew Chem Int Ed 45(22):3584–3601

Slocik JM, McKenzie R, Dennis PB, Naik RR (2017) Creation of energetic biothermite inks using ferritin liquid protein. Nat Commun 8

Smith DK, Bello MN, Unruh DK, Pantoya ML (2017a) Synthesis and reactive characterization of aluminum iodate hexahydrate crystals $[Al(H_2O)_6](IO_3)_3(HIO_3)_2$. Combust Flame 179:154–156

Smith DK, Unruh DK, Pantoya ML (2017b) Replacing the Al_2O_3 Shell on Al particles with an oxidizing salt, aluminum iodate hexahydrate. Part II: synthesis. J Phys Chem C 121 (41):23192–23199

Son SF, Mason BA (2010) An overview of nanoscale silicon reactive composites applied to microengergetics. 48th AIAA aerospace sciences meeting including the New Horizons Forum and Aerospace Exposition, Orlando, FL

Staley CS, Morris CJ, Thiruvengadathan R, Apperson SJ, Gangopadhyay K, Gangopadhyay S (2011) Silicon-based bridge wire micro-chip initiators for bismuth oxide-aluminum nanothermite. J Micromech Microeng 21(11)

Staley CS, Raymond KE, Thiruvengadathan R, Apperson SJ, Gangopadhyay K, Swaszek SM, Taylor RJ, Gangopadhyay S (2013) Fast-impulse nanothermite solid-propellant miniaturized thrusters. J Propul Power 29(6):1400–1409

Staley CS, Raymond KE, Thiruvengadathan R, Herbst JJ, Swaszek SM, Taylor RJ, Gangopadhyay K, Gangopadhyay S (2014) Effect of nitrocellulose gasifying binder on thrust performance and high-g launch tolerance of miniaturized nanothermite thrusters. Propellants, Explos, Pyrotech 39(3):374–382

Stamatis D, Jiang Z, Hoffmann VK, Schoenitz M, Dreizin EL (2009) Fully dense, aluminum-rich Al-CuO nanocomposite powders for energetic formulations. Combust Sci Technol 181(1):97–116

Sundaram D, Yang V, Yetter RA (2017) Metal-based nanoenergetic materials: synthesis, properties, and applications. Prog Energy Combust Sci 61:293–365

Tappan AS (2007) Microenergetics: combustion and detonation at sub-millimeter scales

Thiruvengadathan R, Bezmelnitsyn A, Apperson S, Staley C, Redner P, Balas W, Nicolich S, Kapoor D, Gangopadhyay K, Gangopadhyay S (2011) Combustion characteristics of novel hybrid nanoenergetic formulations. Combust Flame 158(5):964–978

Thiruvengadathan R, Korampally V, Ghosh A, Chanda N, Gangopadhyay K, Gangopadhyay S (2013) Nanomaterial processing using self-assembly-bottom-up chemical and biological approaches. Rep Prog Phys 76(6)

Thiruvengadathan R, Chung SW, Basuray S, Balasubramanian B, Staley CS, Gangopadhyay K, Gangopadhyay S (2014) A versatile self-assembly approach toward high performance nanoenergetic composite using functionalized graphene. Langmuir 30(22):6556–6564

Thiruvengadathan R, Staley C, Geeson JM, Chung S, Raymond KE, Gangopadhyay K, Gangopadhyay S (2015) Enhanced combustion characteristics of bismuth trioxide-aluminum nanocomposites prepared through graphene oxide directed self-assembly. Propellants, Explos, Pyrotech 40(5):729–734

Trunov MA, Schoenitz M, Dreizin EL (2005a) Ignition of aluminum powders under different experimental conditions. Propellants, Explos, Pyrotech 30(1):36–43

Trunov MA, Schoenitz M, Zhu X, Dreizin EL (2005b) Effect of polymorphic phase transformations in Al_2O_3 film on oxidation kinetics of aluminum powders. Combust Flame 140(4):310–318

Trunov MA, Schoenitz M, Dreizin EL (2006a) Effect of polymorphic phase transformations in alumina layer on ignition of aluminium particles. Combust Theor Model 10(4):603–623

Trunov MA, Umbrajkar SM, Schoenitz M, Mang JT, Dreizin EL (2006b) Oxidation and melting of aluminum nanopowders. J Phys Chem B 110(26):13094–13099

Umbrajkar SM, Seshadri S, Schoenitz M, Hoffmann VK, Dreizin EL (2008) Aluminum-rich Al-MoO_3 nanocomposite powders prepared by arrested reactive milling. J Propul Power 24 (2):192–198

Wang H, Jian G, Delisio JB, Zachariah MR (2014a) Microspheres composite of nano-Al and nanothermite: an approach to better utilization of nanomaterials. 52nd AIAA aerospace sciences meeting—AIAA science and technology forum and exposition, SciTech 2014, National Harbor, MD, American Institute of Aeronautics and Astronautics Inc

Wang H, Jian G, Egan GC, Zachariah MR (2014b) Assembly and reactive properties of Al/CuO based nanothermite microparticles. Combust Flame 161(8):2203–2208

Wang H, Zachariah MR, Xie L, Rao G (2015) Ignition and combustion characterization of nano-Al-AP and nano-Al-CuO-AP micro-sized composites produced by electrospray technique. In: 12th international conference on combustion and energy utilisation, ICCEU 2014, Elsevier Ltd

Wang J, Bassett WP, Dlott DD (2017) Shock initiation of nano-Al/Teflon: high dynamic range pyrometry measurements. J Appl Phys 121(8)
Wang A, Bok S, Thiruvengadathan R, Gangopadhyay K, McFarland JA, Maschmann MR, Gangopadhyay S (2018) Reactive nanoenergetic graphene aerogel synthesized by one-step chemical reduction. Combust Flame 196:400–406
Watson KW, Pantoya ML, Levitas VI (2008) Fast reactions with nano- and micrometer aluminum: a study on oxidation versus fluorination. Combust Flame 155(4):619–634
Wuillaume A, Beaucamp A, David-Quillot F, Eradès C (2014) Formulation and characterizations of nanoenergetic compositions with improved safety. Propellants, Explos, Pyrotech 39(3):390–396
Xue Y, Shi CJ, Ren XM, Liu L, Xie RZ (2014) Study of MEMS based micropyrotechnic igniter. In: Applied mechanics and materials, vol. 472, pp. 750–755
Yang Y, Wang PP, Zhang ZC, Liu HL, Zhang J, Zhuang J, Wang X (2013) Nanowire membrane-based nanothermite: towards processable and tunable interfacial diffusion for solid state reactions. Sci Rep 3
Yetter RA, Risha GA, Son SF (2009) Metal particle combustion and nanotechnology. Proc Combust Inst 32(II):1819–1838
Yu C, Zhang W, Gao Y, Ni D, Ye J, Zhu C, Ma K (2018) The super-hydrophobic thermite film of the Co_3O_4/Al core/shell nanowires for an underwater ignition with a favorable aging-resistance. Chem Eng J 338:99–106
Zakiyyan N, Wang A, Thiruvengadathan R, Staley C, Mathai J, Gangopadhyay K, Maschmann MR, Gangopadhyay S (2018) Combustion of aluminum nanoparticles and exfoliated 2D molybdenum trioxide composites. Combust Flame 187:1–10
Zarko VE (2016) Nanoenergetic materials: a new era in combustion and propulsion. In: Energetic nanomaterials: synthesis, characterization, and application. Elsevier Inc, pp 1–20
Zeng C, Wang J, He G, Huang C, Yang Z, Liu S, Gong F (2018) Enhanced water resistance and energy performance of core–shell aluminum nanoparticles via in situ grafting of energetic glycidyl azide polymer. J Mater Sci 53(17):12091–12102
Zhang W, Yin B, Shen R, Ye J, Thomas JA, Chao Y (2013) Significantly enhanced energy output from 3D ordered macroporous structured Fe_2O_3/Al nanothermite film. ACS Appl Mater Interfaces 5(2):239–242
Zhang D, Xiang Q, Li X (2016) Highly reactive Al-Cr_2O_3 coating for electric-explosion applications. RSC Adv 6(103):100790–100795
Zheng G, Zhang W, Shen R, Ye J, Qin Z, Chao Y (2016) Three-dimensionally ordered macroporous structure enabled nanothermite membrane of Mn_2O_3/Al. Sci Rep 6
Zheng Z, Zhang W, Yu C, Zheng G, Ma K, Qin Z, Ye J, Chao Y (2018) Integration of the 3DOM Al/Co_3O_4 nanothermite film with a semiconductor bridge to realize a high-output micro-energetic igniter. RSC Adv 8(5):2552–2560
Zhou W, Yu D (2013) Fabrication, thermal, and dielectric properties of self-passivated Al/epoxy nanocomposites. J Mater Sci 48(22):7960–7968
Zhou L, Piekiel N, Chowdhury S, Zachariah MR (2009) T-Jump/time-of-flight mass spectrometry for time-resolved analysis of energetic materials. Rapid Commun Mass Spectrom 23(1):194–202
Zhou L, Piekiel N, Chowdhury S, Zachariah MR (2010) Time-resolved mass spectrometry of the exothermic reaction between nanoaluminum and metal oxides: the role of oxygen release. J Phys Chem C 114(33):14269–14275
Zhou X, Torabi M, Lu J, Shen R, Zhang K (2014) Nanostructured energetic composites: synthesis, ignition/combustion modeling, and applications. ACS Appl Mater Interfaces 6(5):3058–3074

Chapter 3
Nanostructured Energetic Composites: An Emerging Paradigm

Hema Singh and Shaibal Banerjee

Abstract Nanotechnology has ushered a remarkable progress in the field of medicine, environment, ceramics, especially considering its applications in the defence sector. This progress has been inspired by the ordered assembly of molecular and nanoscale elements to develop multifunctional smart reactive materials for energetic applications. An important class of these materials is the nano-energetic materials or nanothermites, which are composed of nanometals and nano-oxidizers. A major drawback of classical micron-sized metal particles is that they ignite after a comparatively long delay. These micron-sized metal particles when combined with oxidizer such as metal oxides as in thermite result in metal delays which are usually associated with diffusion of oxidizer and/or fuel through the protective layer of metal oxides. The motive behind nano-energetic materials is to develop a new synthetic procedure, which could limit both the oxidizer and the fuel balance in the thermites. Development of assembly of energetic composite materials (by number of techniques like self-assembly, cold spraying, ball milling, sol–gel, gas-phase processes) is touching new horizons of research. In this chapter, emphasis is laid on the current research focusing on manipulation of individual atoms and molecules to produce organized and systematic structure of nanocomposites for applications in nanothermites. Nanothermites are comparatively a new class of energetic material that consist of metallic fuel and metal oxide-based oxidizer with critical dimensions on the nanoscale. The standard powder-mixing protocol has intrinsic constraints, particularly random distribution of fuel and oxidizer particles and unavoidable fuel pre-oxidation. The present research scenario deals with an alternative approach for nanostructured energetic composites by varied processes. The subsequent sections of this chapter will meticulously describe the strategies adopted for the preparation of such nanostructure assemblies. These hierarchical structures provide desirable performance in combustion, ignition and mechanical characteristics. In the end, some promising applications of nanostructured energetic composites incorporated into various systems ranging from

H. Singh · S. Banerjee (✉)
Organic Synthesis Laboratory, Department of Applied Chemistry, Defence Institute of Advanced Technology (DU), Girinagar, Pune 411025, India
e-mail: shaibal.b2001@gmail.com

S. Bhattacharya et al. (eds.), *Nano-Energetic Materials*, Energy, Environment and Sustainability, https://doi.org/10.1007/978-981-13-3269-2_3

microelectromechanical systems (MEMS) devices to rocket propellants to explosives that permit new functions to be performed are illustrated.

Keywords Nanothermites · Nano-energetic materials · Self-assembly

3.1 Introduction

Technological superiority is a necessary requirement in defence due to increasing threats from antisocial elements. Towards this, a significant progress has been made to develop novel materials with improved performance. Nanotechnology has played a seminal role in this direction. An important step in this regard is transition of particle size from micro- to nanoregime. The beauty of nanotechnology lies in the fact that it permits scientists to functionalize or tune the properties of material at molecular level. An innovation in materials, design and fabrication to exploit novel properties of nanomaterials offers fundamentally new approach of synthesizing nanostructures. An important class of nanostructure material is nano-energetic materials (nEMs). The motive behind exploring the nano-energetic materials is to develop new synthetic procedures, which could limit both the oxidizer and the fuel balance. On the basis of these advancements, the study on energetic materials is of great interest to several research groups. The uniqueness of nanoparticles is that they have a significantly high surface-to-volume ratio. Researchers have established that energetic materials that are in nanometre regime have shown remarkable improvements in their properties and performances, especially towards ignition and combustion behaviour, mechanical properties and heat release (Dreizin 2009). This ability of the nanostructures to convert chemical enthalpy to thermal enthalpy rightly gives them the name of nano-energetic material, and such materials are found to liberate more energy in comparison with conventional energetic materials. There are primarily two types of nano-energetic composites (also called nanothermites), namely the metastable intermolecular composites (MICs) which possess the aluminium nanoparticles and an oxidizer (Zhou et al. 2010) and the bimetallic energetic nanostructures (e.g. nanoparticles and nanolaminates) such as Ni/Ti, Co/Al, Ni/Al and Pt/Al (Picard et al. 2008).

The traditional methods of synthesis of energetic material involve either the physical mixing of solid oxidizers and fuels (Aumann et al. 1994) (like black powder), a process that has remained virtually unchanged for centuries or a process that involves creating a monomolecular energetic material, such as TNT ($C_7H_5N_3O_6$) (Clarkson et al. 2003) which incorporates oxidizer and fuel moieties in one molecule. In the MICs, the energy density is found to be much greater as compared to monomolecular (energy density for thermite 16,736 Jg^{-1} vs. 2094 Jg^{-1} for TNT) (Zarko and Gromov 2016) materials as the desired energy can be attained by varying the ratios of oxidizer and fuel. The metallic fuels such as Al, B, Mg, Ti, Zr used in the composites have higher enthalpies of combustion and energy densities as compared to the enthalpies of some selected monomolecular materials. However, the rate at which this energy is released is relatively slow in contrast to the release rate of monomolecular materials (Rossi et al. 2007). The main drawback of these composites is that the micron-sized metal particles ignite after a comparatively longer delay as monomolecular energetic compounds. Secondly, when these micron-sized metal particles are combined with oxidizers such as metal oxides as in thermites, the metal delays are usually coupled with diffusion of oxidizer and/or fuel through the protective layer of metal oxides (Rossi et al. 2007).

Reviews of the recent literature on energetic nanocomposites have revealed that thermite-like reactants (fuel/oxidizer) have burning rate ~ 1000 times higher (Danen and Martin 1993) than conventional energetic materials. The probable cause of this is the decrease in diffusion distance between the nanoparticles of fuel and the oxidizer which leads to a rise in the rate of reaction (Chen and Sachtler 1998). Figure 3.1 depicts the different oxidizer and oxophillic metals which combined to produce an energetic mixture.

A scheme of the reaction is illustrated below in Eq. 3.1.

$$\mathbf{M}^{\mathbf{1}}_{(s)} + \mathbf{M}^{\mathbf{2}}\mathbf{O}_{(s)} \rightarrow \mathbf{M}^{\mathbf{1}}\mathbf{O}_{(s)} + \mathbf{M}^{\mathbf{2}}_{(s)} + \mathbf{\Delta H} \quad (3.1)$$

where M^1 is fuel metal particle and typically Al and M^2O are the oxidizers.

The selection of fuel component in energetic composite materials is dependent on numerous properties such as high heat of formation, high density, low melting temperature, low toxicity and compatibility with other ingredients. After considering these properties, Al has emerged as the most preferred candidate for nanofuel.

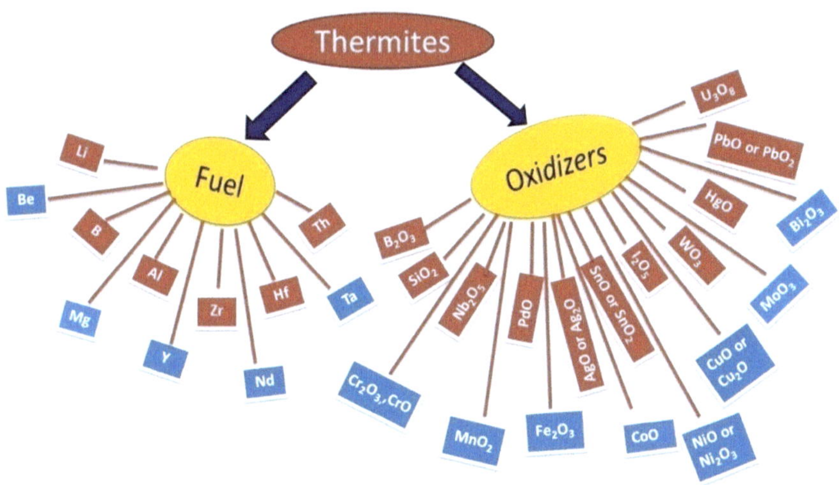

Fig. 3.1 Different types of fuels and oxidizers used in thermites

The reasons behind the use of aluminium nanoparticles (n-Al) as a fuel are summarized below:

- High enthalpy combustion value of ~1676 kJ mol^{-1}
- Low melting temperature (~660 °C) leading to low ignition temperature
- Propensity to form passive oxide prevents it from spontaneous combustion (Park et al. 2005)
- Enhances combustion velocity of the composite due to its high thermal conductivity
- Low vapour pressure.

In the thermite reaction, the oxygen is available by the oxidation of the oxidizers. This free oxygen atom binds more strongly with aluminium to produce aluminium oxide and thus releases large amounts of energy (Bernstein 2014). Ivanov and Tepper (1997) have reported that the burning rate could be augmented by a factor of 5–10 by adding nano-aluminium in the thermite reaction. The nanothermites have high rate of energy release (as shown in Fig. 3.2), as they are able to produce considerable blast due to their intermediate gas-phase reaction products and their ignition sensitivity.

With time, nano-energetic materials have gone through various modifications in their synthesis protocols, which can be briefly classified into different generations as summarized below:

- First generation
  - Synthesis of nano-Al (fuel) and oxidizers

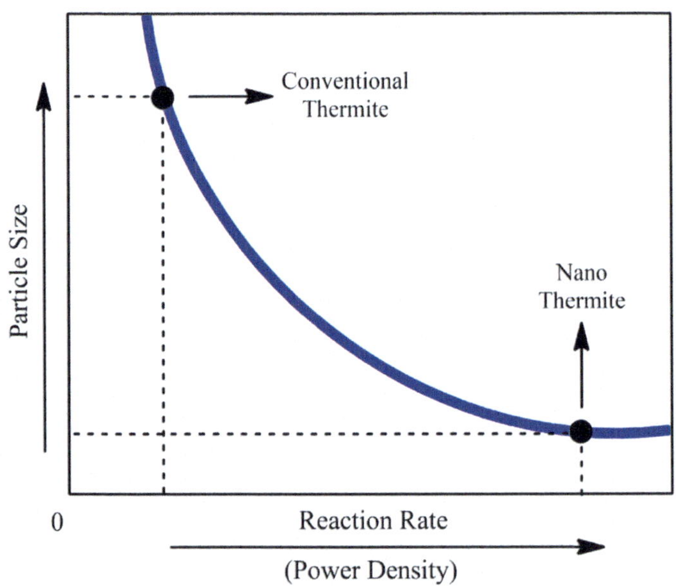

Fig. 3.2 Comparative account of reaction rate of conventional and nanothermites

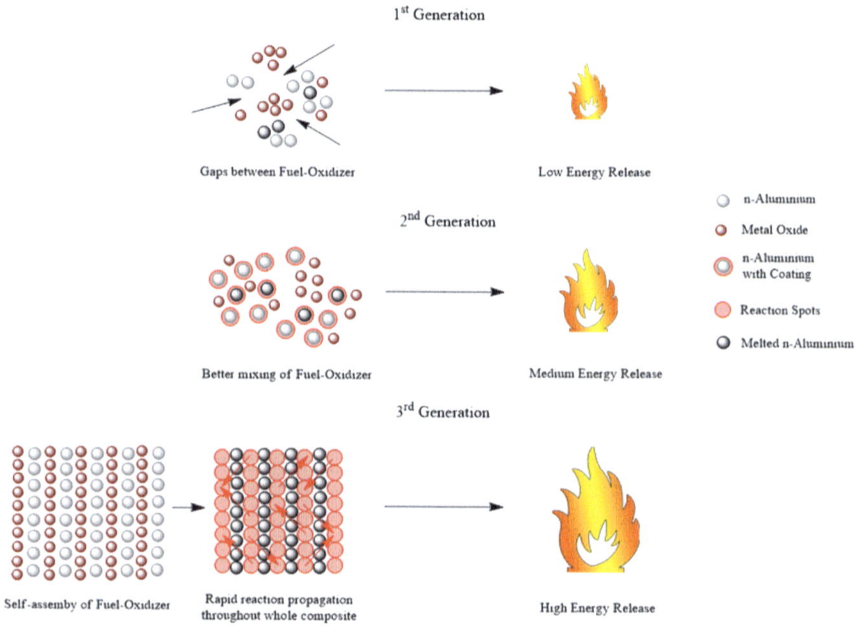

Fig. 3.3 Schematic diagram of timescale generations of nanothermites

- Second generation
  - Coated nanometre—sized particles
- Third generation
  - Three-dimensional nano-energetic materials
  - Ordered assembly of nano-energetic materials.

The three-dimensional nano-energetic materials have become hot research topic owing to its distinctive properties and better performances. Figure 3.3 depicts the third generation of the nanothermites consisting of self-assembled highly ordered structures, which has led to explosion rates much higher as compared to the conventional second- or first-generation nanothermites.

3.2 Synthesis of Nanothermites

Traditionally, thermites are prepared by conventional mixing of fuel and oxidizer components. However, synthesis of nanothermites has been classified into two main categories: bottom-up and top-down techniques. The 'bottom-up' technique uses atoms and molecules as building blocks for the nanomaterial, whereas in the 'top-down' approach the nanomaterials are produced because of the breakdown of bulk domains or relatively coarse powders. Figure 3.4 depicts a broad classification for nanothermite preparation techniques.

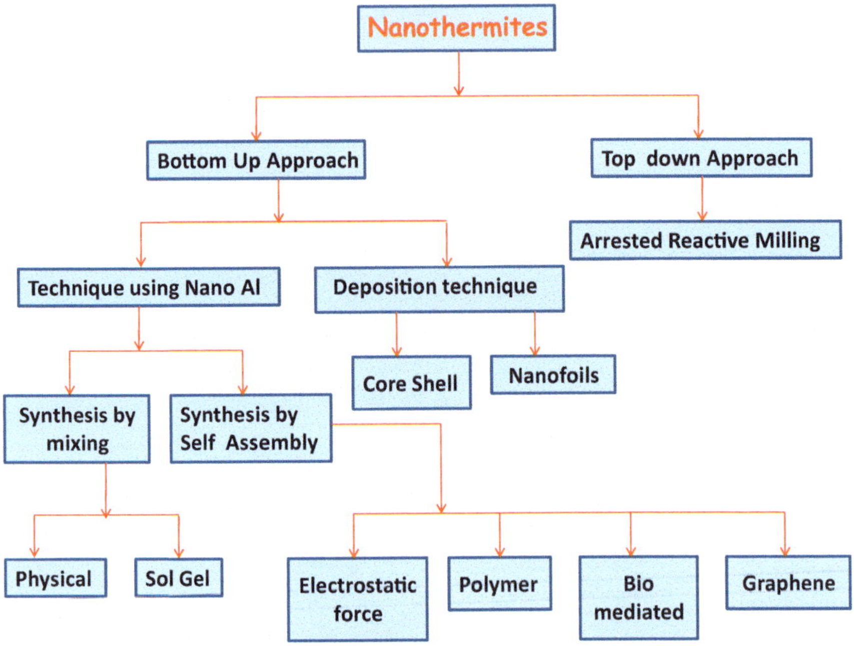

Fig. 3.4 Classification of nanothermites on the basis of their synthesis method

3.2.1 Nanofuel Synthesis Using Bottom-Up Approach

Bottom-up approach is further classified into two categories, firstly technique using nano-Al powder and secondly deposition technique of fuel and oxidizer.

3.2.1.1 Technique Using Nano-Al

In the case of technique using nano-Al powder, the use of nanometre-sized fuel and oxidizers is involved. The various methods for the synthesis of nanofuel and nano-oxidizers are discussed below.

Synthesis of Nanofuel

Aluminium nanoparticles being the most prominent nanofuel, we describe here various methods for their synthesis. Conventionally, Al metal has been formed electrochemically by the Hall–Heroult method via a carbon anode and an Al as the cathode at 1233 K temperature (Mandin et al. 2009). The major setback to this process is high-energy consumption, release of greenhouse gases such as CO_2, CO and CF_6 and restricted reaction conditions. Al nanoparticles have been synthesized by some physical methods, such as wire explosion, vaporization from molten poll into an inert gas under vacuum, induction and arc plasma methods.

Wire explosion is a top-down approach in which the particles are generated by evaporating a thin metal conductor at high current and in an inert atmosphere (Yavorovsky 1995). The first successful synthesis of aluminium nanopowder was performed by Tomsk. The well-known powder known as Alex was produced by fast heating of aluminium wire using electric current, which resulted in a wire explosion (Ivanov et al. 2003). Another group has also presented a new approach for the production of metal powders by wire explosion method. Al nanoparticles of 10–50 nm in size have also been synthesized via laser ablation method (Tepper 2000). Sarathi et al. (2007) have synthesized n-Al using aluminium wire of 150-mm length and 0.5 mm-wide, 3 μF capacitance, charging voltage of 25 kV under nitrogen, argon and helium atmosphere. It was observed that the wide-angle X-ray diffraction (WAXD) pattern depicted impurity peaks of aluminium nitride in case of nitrogen atmosphere, whereas others showed pure Al. The thermal analysis showed lowering of melting point in n-Al in comparison with micron-sized aluminium (m-Al) (654 °C).There are also reports for Al nanopowder synthesis by vapour condensation in a stream of inert gas under vacuum (Puszynski 2004). However, due to metal's inherent property of reacting with oxygen, an oxide layer of 2.3-nm thickness was found to cover the surface of Al nanopowder.

In 1988, Haber and Bulero (1998) carried out a reaction of lithium aluminium hydride [$LiAlH_4$] with aluminium chloride [$AlCl_3$] at 165 °C in 1,3,5-trimethyl benzene (Haber and Buhro 1998). They obtained aluminium nanomaterial having a size of $\sim 160 \pm 50$ nm. The by-product of this reaction, lithium chloride [LiCl], was removed by washing using methanol at temperature below 0 °C (Eq. 3.2).

$$3LiAlH_4 + AlCl_3 \xrightarrow[165°C,\ 24h]{1,3,5\text{-trimethyl benzene}} n\text{-}Al + 3LiCl + 6H_2 \quad (3.2)$$

In yet another method (Eq. 3.3), the authors have synthesized aluminium nanopowder by decomposition of intermediate compound H_3Al (NMe_2Et) under reflux in 1,3,5-trimethylbenzene at 160–164 °C, with (W) or without (W/O) catalyst[$Ti(OPr)_4$]. The average particle size of aluminium nanoparticles synthesized by this method was in the range of 40–180 nm (Cui et al. 2015).

$$H_3Al(NMe_2Et) \xrightarrow[\substack{\text{W (or) W/O Catalyst} \\ \text{Ti } i\text{-(OPr)}_4,\ < 164^{o}C}]{1,3,5\text{ trimethylbenzene}} n\text{-}Al + 3/2\ H_2 + NMe_2Et \quad (3.3)$$

In most of the chemical methods, Al ions are reduced by using another metal with a higher reduction potential, such as Na or Li. Mahendiran et al. (2009) have reported room-temperature synthesis of aluminium nanoparticles via pulsed sono-electrochemical method. In this method, an ultrasound horn acts both as the cathode and as the ultrasound emitter.

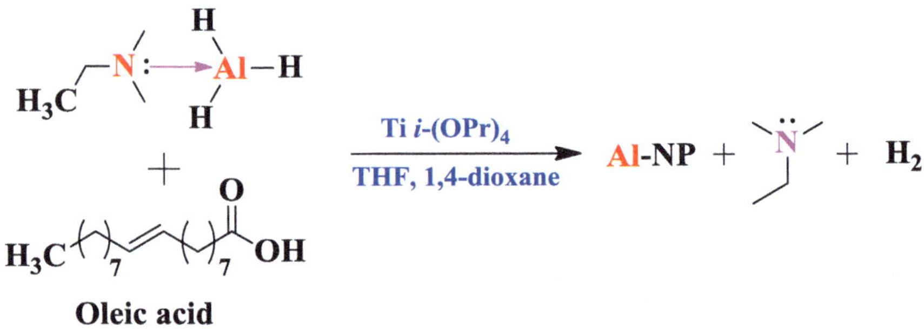

Scheme 3.1 Formation of aluminium nanoparticles using thermal decomposition method

Thermal decomposition of aluminium hydride in the presence of titanium catalyst was reported to produce n-Al. However, control of its shape and size is a major issue. McClain and co-workers have synthesized n-Al crystals over a range of size from 70 to 220 nm using oleic acid as a stabilizer (Scheme 3.1). A mixture of THF and dioxane was used as solvent, and it was found that the ratio of THF and dioxane played a vital role in controlling the size of the Al nanocrystals (McClain et al. 2015).

Al particles are passivated with an unreactive Al_2O_3, and as the surface area of the particles decreases, the Al_2O_3 layer increases which results into a considerable burden. The oxide layer lowers the energy density, slows down the combustion rate and prevents the complete Al consumption. The ignition and combustion velocity of nanothermites as a function of Al particle size are studied by various groups. Their studies showed reduction in ignition delay time with decrease in the size of Al nanoparticles. The combustion velocity increases significantly with the increasing n-Al percentage in the mixture (Moore et al. 2007). Recently, to reduce the surface oxidation of Al nanoparticles, attempts were made to stabilize the nanoparticles inside a polymeric matrix. Synthesized nano-aluminium particles from reduction of aluminium trichloride by lithium aluminium hydride in the presence of poly (vinylpyrrolidone) or poly(methyl methacrylate) are depicted in Eq. 3.4. This method was found successful as nanoparticles were in metallic form after successive months of synthesis indicated by XRD analysis (Ghanta and Muralidharan 2010).

$$\mathrm{AlCl_3 + 3LiAlH_4 + PVP\ or\ PMMA \xrightarrow[165^\circ C,\ 24h]{Mesitylene} 4Al/PVP\ or\ PMMA + 3LiCl + 6H_2} \tag{3.4}$$

Al nanoparticles were synthesized in solution by catalytic decomposition of H_3AlNMO_3 or $H_3AlN(Me)Pyr$ by $Ti(OPr)_4$ and coated with perfluoroalkyl carboxylic acid. In situ, coatings of the particles prevent it from oxidation and thus form oxygen-free nanoparticles (Jason et al. 2003).

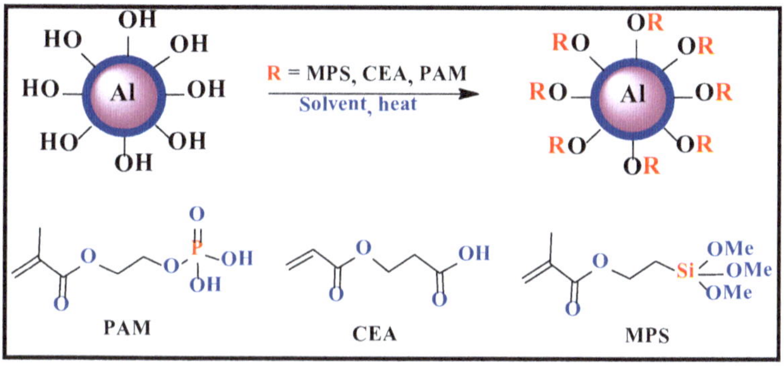

Scheme 3.2 Nano-aluminium functionalized using acrylic acids

In another work, n-Al has been functionalized with three acrylic monomers, 3-methacryloxypropyltrimethoxysilane (MPS), 2-carboxyethyl acrylate (CEA) and phosphonic acid 2-hydroxyethyl methacrylate ester (PMA) (Crouse et al. 2010) as per Scheme 3.2. Figure 3.5 describes the surface morphology by HRTEM of functionalized Al composite.

Researchers from NAVAIR, China, have reported another method for the wet synthesis of aluminium nanopowder from triethylamine adduct of alane [AlH_3•NEt_3] in heptanes (Foley et al. 2005) (Scheme 3.3). They passivated n-Al with transition metals such as gold, nickel, palladium and silver. It was observed that the nickel treatment increased the active metallic aluminium content in comparison with the untreated Al or when treated with other metals.

Aluminium nanoparticles have also been passivated with alkyl-substituted epoxides (Scheme 3.4) that are further polymerized to oxygen-rich layer, polyether. These organic shell Al composites have very low alumina layer. Epoxyisobutane capping on n-Al of 5:1 [Al/epoxide] ignites upon exposure to air (Chung et al. 2009).

Aluminium-based nano-energetic composites are formulated by using poly tetrafluoro ethylene (PTFE) or Teflon as an oxidizer (Zamkov et al. 2007). The use of PTFE was found to lower flame speeds and to increase peak pressures. Pantoya and Dean have observed (Watson et al. 2008) that the reduction in the particle size of Al delays the onset temperature of the fluoropolymer to higher temperature near to the ignition temperature of n-Al. Pre-ignition reaction (PIR) was noted with n-Al/Teflon mixture, and the reason suggested by them was fluorination of alumina (Al_2O_3) shell passivating the n-Al particles. Another group has examined that the initial combustion behaviour can change depending on the ignition method (Yarrington et al. 2011).

Synthesis of Nano-oxidizers

Various oxidizers that include Fe_2O_3, CuO, Bi_2O_3, WO_3, NiO have been synthesized with a range of morphologies which includes spherical, rod, wire, mesoporous and hollow.

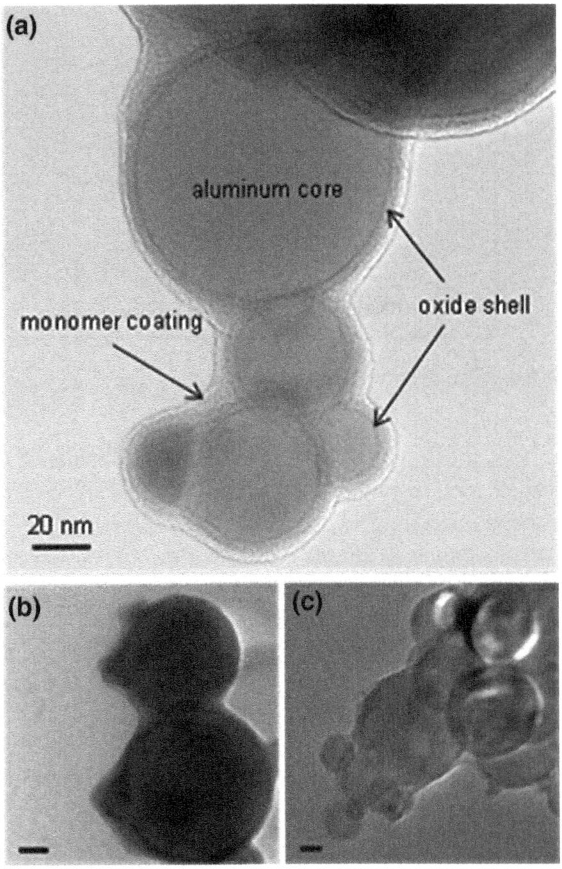

Fig. 3.5 HRTEM images of the functionalized Al composites with acrylic monomer coating. Reproduced with kind permission from Crouse et al. (2010)

Conventionally, iron oxide was prepared by alkaline precipitation from the aqueous solutions of iron(III) salts. Addition of base solution like ethanol has paved a path for the synthesis of Fe_2O_3 gels from such solutions (Gash et al. 2001a). In the literature, hydrolysis of Fe^{3+} at low pH has been reported. Scheme 3.5 shows the production of $[Fe(OH_2)_6]^{3+}$ complex from Fe(III) $(NO_3)_3$•$9H_2O$ in the presence of ethanol with the release of water molecules and nitrate ions. The $[Fe(OH_2)_6]^{3+}$ species are unstable and produce dimer with water. Further, propylene oxide was added as a scavenger, which produces α-FeOOH on hydrolysis. When two molecules of α-FeOOH combine, α-Fe_2O_3 is formed along with water molecule. The scheme of reaction is summarized below (Prakash et al. 2004).

The oxidizers are generally non-porous in nature, so an attempt was made to prepare mesoporous material to achieve even higher burning rate. In the porous materials, the energy transfer relies on convection mechanism, which is proclaimed

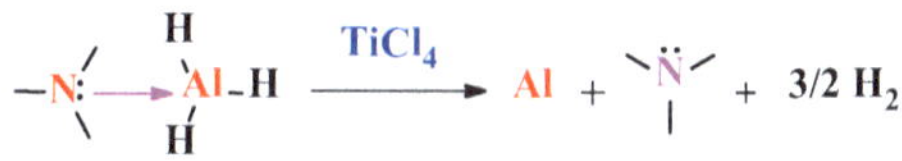

Scheme 3.3 Synthesis of aluminium nanopowder from alane

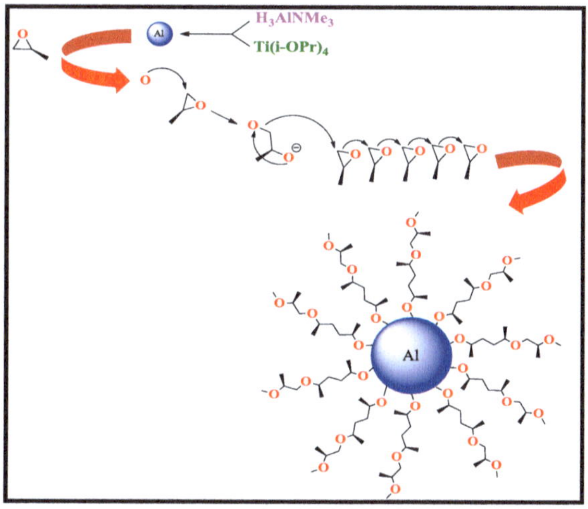

Scheme 3.4 Passivation of nano-aluminium with epoxides

to produce much higher burning rate. Mesoporous Fe_2O_3 has been synthesized using Fe(III)-ethoxide (Srivastava et al. 2002) with cetyltrimethylammonium bromide (CTAB) as a surfactant. Yet, another method involves the use of $FeCl_3$ as precursor and propylene oxide or epichlorohydrin as proton scavenger to attain mesoporous Fe_2O_3 having a pore size of 2–3 nm. However, ordered pores were not observed in the mesoporous Fe_2O_3 particles formed (Prakash et al. 2004).

Fe_2O_3 nanoporous materials with pore size 8–10 nm are obtained from Fe $(NO_3)_3$ precursor with propylene oxide as a proton scavenger and Brij 76 as a surfactant (Mehendale et al. 2006). The TEM results depicted that the pore size (8–10 nm) obtained using Brij 76 was almost double the pore size (2–3 nm) attained using cetyltrimethylammonium chloride (CTAC). The results showed that by addition of surfactant narrow size distribution was attained and distribution of pores was homogeneous throughout the sample. Secondly, on increment of the chain length of the surfactant, the pore size was also found to be increased.

Several methods have also been developed to synthesized CuO oxidizer in various morphologies. For example, CuO nanowires have been synthesized by thermal oxidation of Cu film of 1 μm in static air at different temperatures and time periods. The oxidation process of Cu included two steps as mentioned in Eqs. 3.5 and 3.6:

$$4Cu + O_2 \xrightarrow[5h]{360\ ^\circ C} 2Cu_2O \qquad (3.5)$$

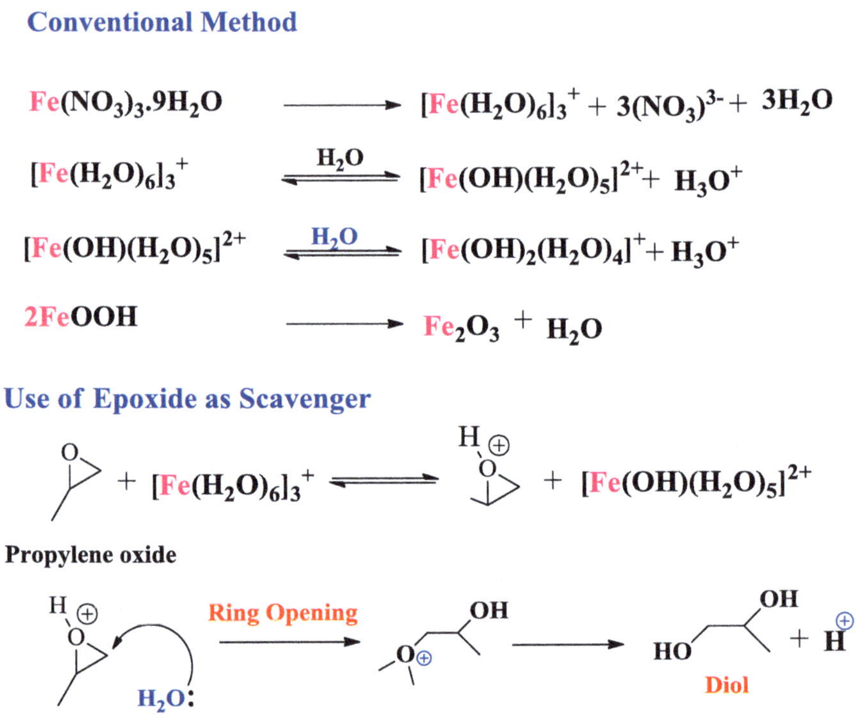

Scheme 3.5 Synthesis of iron oxide nanoparticles using epoxide scavenger

$$2Cu_2O + O_2 \xrightarrow{400\text{–}800\,^{\circ}C} 4CuO \tag{3.6}$$

The authors reported (Yang et al. 2012) that the presence of Cu_2O seed phase is essential for the growth of CuO nanowires.

In another method, CuO has been synthesized by using poly(ethylene glycol) (PEG) surfactant as a template. The function of PEG is to attain controlled and anisotropic growth of copper hydroxide. PEG-20000 surfactant is known to produce Cu(I) oxide nanowires, while PEG-400 surfactant has been used for the synthesis of both nanorods and nanowires (Shende et al. 2008), TEM images are presented in Fig. 3.6.

CuO nanowires have also been synthesized by electrospinning and subsequent calcination using $Cu(NO_3)_2$ and polyvinyl pyrrolidine (PVP) (Ahn et al. 2011a). Figure 3.7 describes the schematic diagram of CuO nanowires formed by electrospinning method.

Another approach has been developed to prepare hollow CuO spheres of size ~10 nm by a 'droplet-to-particle' aerosol spray pyrolysis protocol. The sucrose and hydrogen peroxide as in situ blowing agents were used. Figure 3.8 demonstrates the schematic representation of the technique (Jian et al. 2013).

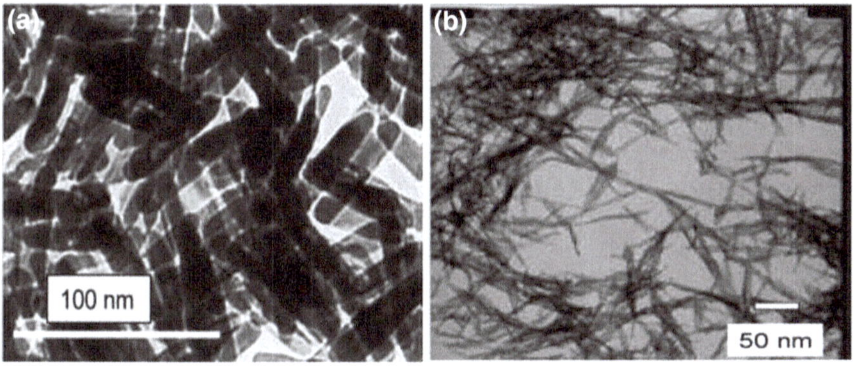

Fig. 3.6 CuO nanorods (**a**) and nanowires (**b**) prepared using PEG micelles. Reproduced with kind permission from Shende et al. (2008)

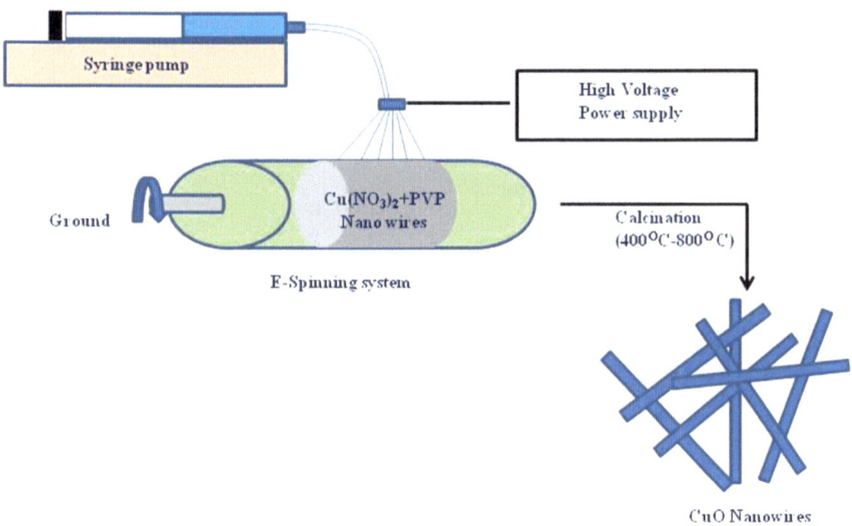

Fig. 3.7 Schematic diagram of CuO nanowires formed by electrospinning method

Tungsten trioxide (WO_3) is considered as a good oxidizer since it has a high material density (7.20 g/cm^3 as against 5.25 g/cm^3 for Fe_2O_3) and sublime at relatively low temperatures. The gas generated during sublimation could contribute to an increased reaction velocity (Perry et al. 2004).

Synthesis of WO_3 nanoparticles has been reported using crash precipitation method, wherein ammonium paratungstate ($(NH_4)_{10}W_{12}O_{41}$) mixed in acid was poured in distilled water to obtain a precipitate of tungstic acid which on heating in the presence of air at 200 °C or 400 °C forms cubic or monoclinic WO_3, respectively (Gash et al. 2004).

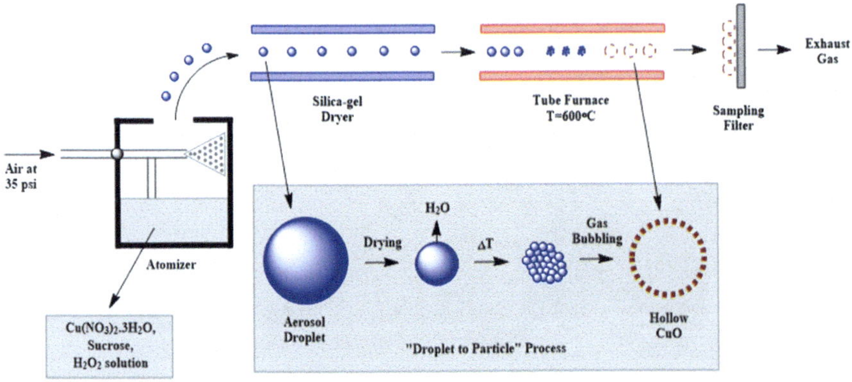

Fig. 3.8 Spray pyrolysis method to synthesize CuO nanospheres

Synthesis of Nanocomposite

Nanothermites synthesized by using aluminium nanoparticle powders can be divided into the following two categories: (a) mixing and (b) self-assembly method.

3.2.1.2 (a) Synthesis by Mixing

The mixing can further be divided into physical mixing and sono-chemical mixing.

i. **Physical mixing**

It is the easiest and widely used process towards synthesis of nanocomposites (Moore et al. 2004). In atypical reaction procedure, the powder was suspended in an inert liquid (to reduce static charges) such as hexane or isopropyl alcohol or any other solvent, which is volatile in nature, and they were then sonicated. The sonication process breaks the macro-sized particles to nanosize, and also, it ensures better mixing of fuel and oxidizer. The composite synthesized using mechanical mixing renders it a non-uniform structure. Moreover, the oxidizer and fuel particles are randomly distributed in the composites leading to lower interfacial contact area, and as a result lower combustion velocity is attained.

As opposed to the mechanical mixing of oxidizer and fuel components, a second synthesis method is based on sol–gel chemistry.

ii. **Sol–gel synthesis**

Researchers at Lawrence Livermore National Laboratory (Clapsaddle et al. 2003) were first to introduce sol–gel method to synthesize nano-energetic materials. The sol–gel reaction mechanism comprised of two steps: hydrolysis and condensation. The process starts when a metal salt, M_yX_z, is dissolved into its respective metal cations and anionic species in an aqueous solution. The positively charged metal cations in solution, M^{z+}, attract the partial negative charge of the oxygen species in surrounding water molecules to form a hydrated coordination sphere.

This is followed by hydrolysis of the hydrated metal complex, $[M(OH_2)_N]^{z+}$, generated by the coordination of the water molecules in the solution to the metal cations. The metal complex reacts with a base in solution, such as water, through a proton transfer reaction to form $[M(OH)(OH_2)_{N-1}]^{(z-1)+}$, a hydroxo ligand, and the conjugated acid, H_3O^+, as shown in Scheme 3.7 (Mabuchi et al. 2005):

$$\underset{\text{Acidic Hydrated Metal Complex}}{[M(OH_2)_N]^{Z+}} + \underset{\text{Base}}{H_2O} \rightleftharpoons \underset{\text{Hydroxo Ligand}}{[M(OH)(OH_2)_{N-1}]^{(Z-1)+}} + \underset{\text{Conjugated Acid}}{H_3O^+} \tag{3.7}$$

The hydroxo ligand reacts with another water molecule and undergoes deprotonation to form hydroxo ligand from an oxo ligand, $[MO(OH_2)_{N-1}]^{(z-2)+}$ and the conjugated acid, H_3O^+as depicted in Scheme 3.6.

The oxo ligands polymerize to form a three-dimensional network of metal hydroxide/oxyhydroxides via condensation step. Further, ageing and drying remove the residual hydroxyl species and yield the final stoichiometric metal oxide network, M_xO_z (Pierre 2013).

The gel so formed can be dried in air to form xerogels and on treatment with supercritical fluids removes liquid in the pores to form aerogels.

Tillotson et al. (2001) first synthesized both aerogel and xerogel monoliths of Al/Fe_2O_3 nano-energetic composite by sol–gel method and showed that high surface area metal oxides could be prepared by this method. In addition, aerogel composites could be easily ignited because of low thermal conductivity in comparison with xerogel (Brinker and Scherer 1990).

Further, the technique has been expanded to synthesize other oxides such as Cr_2O_3, Al_2O_3, In_2O_3, Ga_2O_3, SnO_2, ZrO_2, NbO_3 and WO_3 (Gash et al. 2001b). It was observed that the hydroxyl group is an inherent component in this technique. Therefore, Plantier et al. (2005) calcined the aerogel and xerogel powder at 410 °C to remove the hydroxyl impurity and found a noteworthy enhancement in the

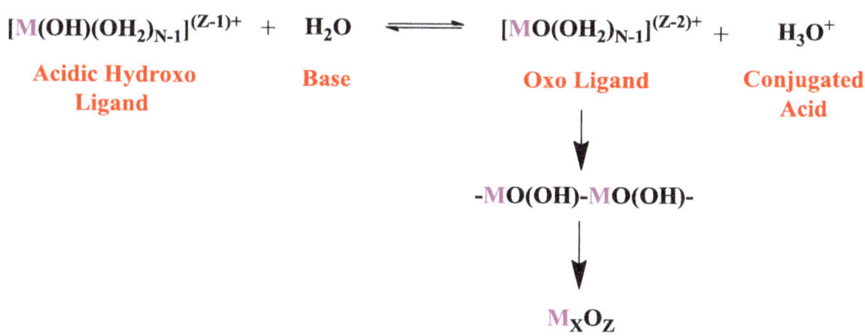

Scheme 3.6 Formation of oxo ligand

reaction velocity from ~10 to 900 ms^{-1}. Thus, it was demonstrated that sol–gel-synthesized composites had high reaction velocity over commercially available materials.

3.2.1.3 (b) Synthesis by Self-Assembly

In self-assembly method, fuel nanoparticles are orderly arranged around the oxidizer and as a result such composite possesses maximum hot spots and releases high rate of energy. The self-assembly of nanoparticles is synthesized by various methods such as electrostatic interaction, biomediated, polymer binding and graphene oxide functionalization.

i. **Electrostatic interaction-assisted self-assembly**

In the electrostatic interaction self-assembly approach, the diffusion between oxidizer and fuel aerosol particles is improved by increasing electrostatic forces on each of the components. In Brownian type of collision, fuel (Al) particles are held together in a linear chain by weak interaction with the oxidizer particles as depicted in Fig. 3.9a. While, in the case of electrostatic interaction, enhanced charged collision yields assembly of a larger number of Al particles around oxidizer nanoparticles which has better contact between fuel and oxidizer than Brownian assembly (Kim and Zachariah 2004) (Fig. 3.9b).

Another study has been performed, in which the surface of the n-Al was coated with alkanoic acid, $COOH(CH_2)_{10}NMe_3^+Cl^-$ and nCuO with alkanethiol,

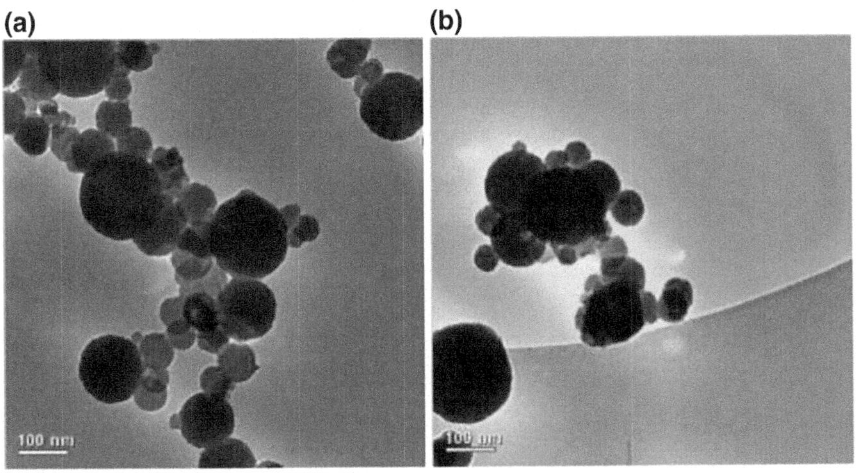

Fig. 3.9 TEM images of nanocomposite particles synthesized by: **a** Brownian coagulation and **b** bipolar coagulation, respectively. Reproduced with kind permission from Zachariah and Kim (2004)

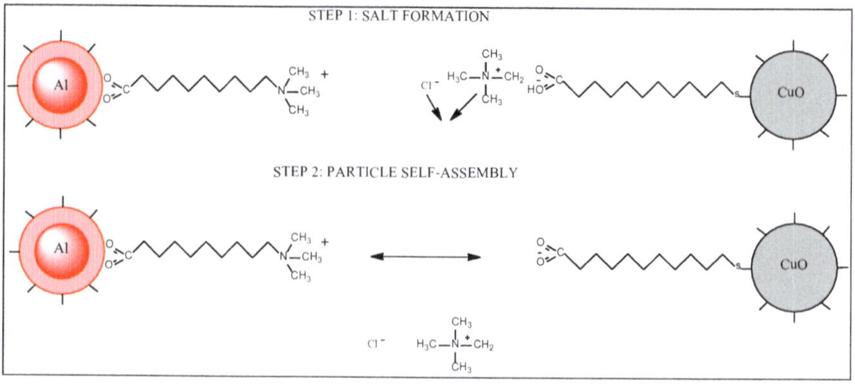

Fig. 3.10 Schematic of salt formation and electrostatic self-assembly of particles

$SH(CH_2)_{10}COO^-NMe_4^+$ as illustrated in Fig. 3.10. Customized tailoring of the reactive properties based on the nanoscale structure provides an insight into physics which governs the propagation (Malchi et al. 2009).

ii. **Polymer-assisted self-assembly**

Gangopadhyay et al. have reported the synthesis of nanocomposites comprising of CuO nanorods and nanowires, and Al nanoparticles using self-assembly and ultrasonic mixing process. In the self-assembly process, the poly(4-vinylpyridine) was coated on copper nanorods (20 × 100 nm) followed by another coating with 80-nm aluminium powder. TEM images are depicted in Fig. 3.11 (Shende et al. 2008).

iii. **Biologically assisted self-assembly**

DNA-directed energetic nanocomposite was prepared by assembling Al and CuO, which has remarkable increase in energetic performance as compared to the physically mixed Al/CuO counterparts. The thiol moieties which are organosulfur compounds (–SH) react with CuO surfaces to form Cu–S bonds and result into functionalized CuO with thiol-modified oligonucleotide (Folkers et al. 1995).

Another strategy was adopted to bind DNA with Al. In that, initially, NeutrAvidin (a tetrameric protein) which forms a strong non-covalent bond with biotin, a small vitamin which can be combined with any biomolecule (Diamandis and Christopoulos 1991), was absorbed on alumina present on n-Al and then was functionalized with biotin-modified oligonucleotide. Thereafter, CuO and Al were bridged by using a linker. The schematic diagram of DNA-directed energetic nanocomposite is presented in Fig. 3.12.

iv. **Graphene-assisted self-assembly**

Functionalization of graphene sheets has attracted attention in recent years as it could be easily modified to incorporate different ordered assembly approaches. The

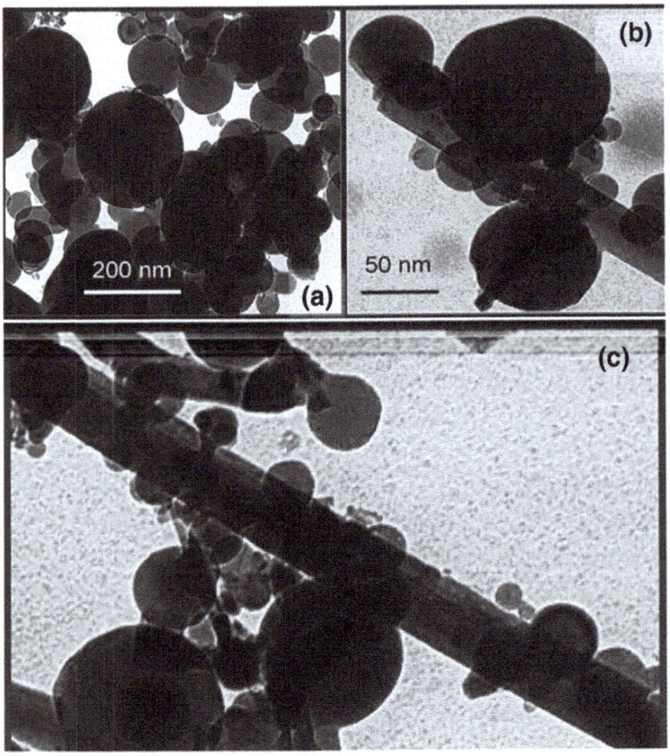

Fig. 3.11 TEM micrographs of **a** n-Al, **b** self-assembled n-Al around CuO nanorods and **c** nanostructured assembly. Reproduced with kind permission from Shende et al. (2008)

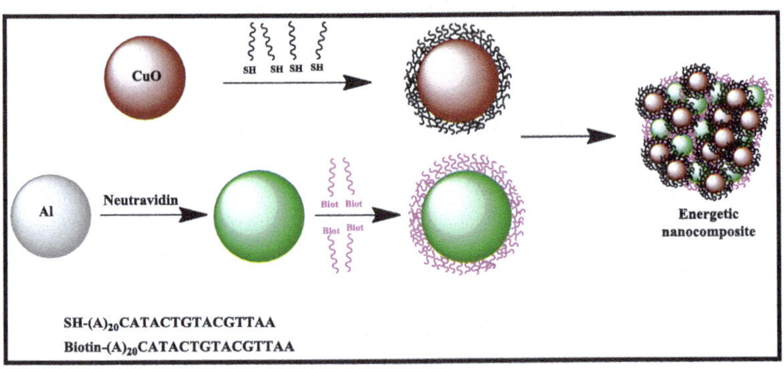

Fig. 3.12 Schematic diagram of DNA-directed energetic nanocomposite

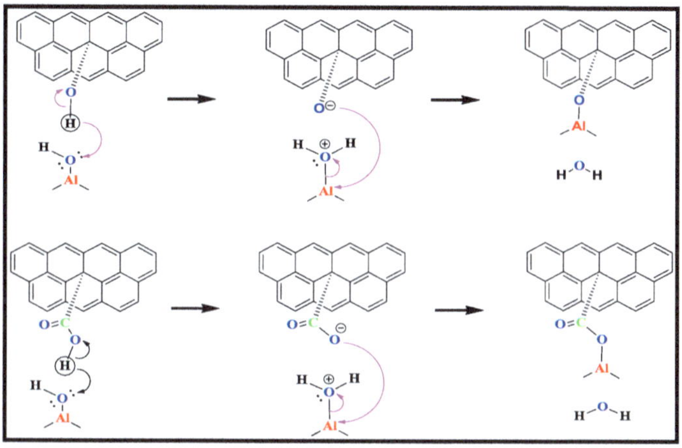

Scheme 3.7 Functionalization of graphene with aluminium

tailoring of the graphene at the molecular level with the energetic groups such as nitro (–NO_2) and amines (–NH_2) has boosted the combustion performance. In line with this work, Thiruvengadathan and team (2014) have utilized functionalized graphene sheets (FGSs) as an additive in nanothermites (Al as fuel and Bi_2O_3 as an oxidizer) to form an ordered assembly. Scheme 3.7 shows the self-assembly of GO/ Al/Bi_2O_3 composites.

3.2.1.4 Deposition Technique

i. **Nanofoils**

Highly structured nanofoils are prepared in argon environment by deposition of reactants onto a substrate (silicon wafer) by rotating them above metal and oxidizer targets, respectively. In the multilayer foil synthesis method, self-propagating exothermic reaction occurs between an oxidizer and the fuel (Wang et al. 2004). Elemental layers react and form a single intermetallic product. Figure 3.13 presents the typical layer-by-layer arrangement of fuel and oxidizer on silicon wafer.

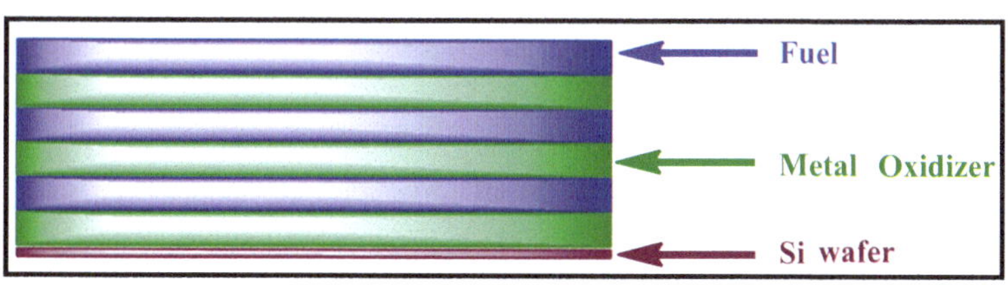

Fig. 3.13 Deposition of fuel and metal oxidizer on Si wafer

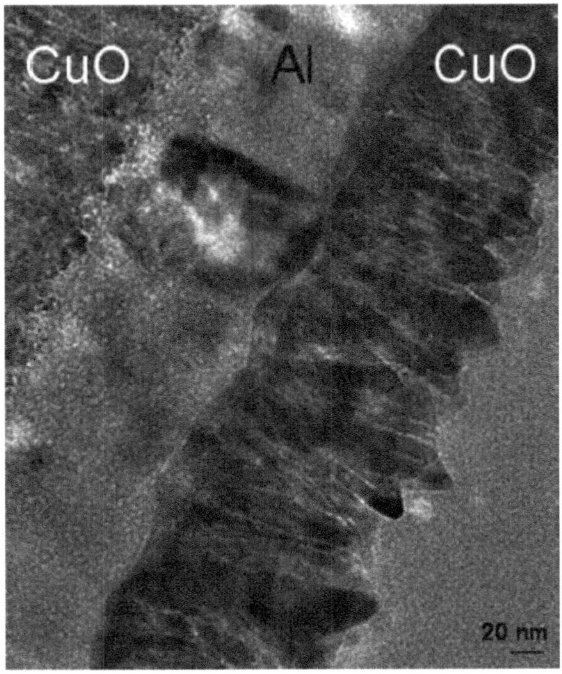

Fig. 3.14 TEM image depicting multilayer geometry of Al/CuO_x. Reproduced with kind permission from Blobaum et al. (2003)

Al/CuO_x multilayer foils as shown in Fig. 3.14 are deposited in a multilayer geometry using magnetron sputtering (Blobaum et al. 2003a). Sputter gun was shielded to minimize the premature reaction between Al and CuO_x preventing aluminium oxidation. In yet another report, differential thermal analysis was performed of these foils and it was found that the reaction proceeded through two-step exotherms ($\sim$850–950 and 975–1127 K). This suggests that the reaction was controlled by lateral growth of Al_2O_3 nuclei followed by diffusion of oxygen through Al_2O_3 layer growth (Blobaum et al. 2003b).

Petrantoni et al. (2010a) have deposited micro- and nanostructure of Al/CuO on oxidized silicon wafers as multilayered thin film by the reactive magnetron sputtering in argon–oxygen gas mixture. The study showed that the films are well suited for MEMS because the technique is versatile and produces low stress (<50 MPa) on the surface. Another layered structure is reported in the literature by growing CuO nanowires from Cu film followed by Al deposition through thermal evaporation (Zhang et al. 2007).

ii. **Core–Shell**

A core–shell method is another technique to synthesize nanothermites with ordered structure. In this method, an intimate contact is made between fuel and oxidizer by coating the oxidizer onto n-Al particles. Atomic layer deposition (ALD) has emerged as a promising protocol for applying a continuous and uniform film as a shell onto a core material. A continuous layer of ZnO and SnO_2 as an oxidizer is deposited on nanoparticles (Qin et al. 2013). The thickness of the oxidizer layer on to the Al surface can be controlled by tuning the number of ALD cycles. The reaction rate is enhanced in comparison with previously reported values. The enhancement can be ascribed to excellent conformity of the oxidizer film.

3.2.1.5 Arrested Reactive Milling (ARM)

ARM synthesis is a 'top-down' method of synthesis of nanocomposites. This method is a modification of reactive milling which works on the principle of high-energy mechanical milling of micron-scale starting powder which causes an exothermic reaction (Baláž et al. 2003). In ARM, the mechanically initiated exothermic reaction is avoided by ceasing the milling when the nanocomposite powder is formed. The mixing takes place at the nanoscale level, while the particle sizes are in the micron-range. Therefore, each micron-sized particle represents a nanocomposite structure of reactive components. ARM method has been applied for nano-energetic materials as well. Many articles in the literature account for the fabrication of Al–CuO and Al–MoO_3 nanocomposite by ARM method (Ermoline et al. 2011; Ermoline and Dreizin 2012). Various milling parameters were studied and the results indicated that the crystallite size was decreased with increase in the milling intensity, but undesirable partial reaction in the starting materials was observed. The milling media with densities in the range of 5–8 g cm^{-3} was found to be the ideal condition for highest structural refinement along with lowest degree of partial reaction. Steel was reported to be the optimum milling media for the synthesis of highly structured Al-MoO_3 nanocomposite.

The advantages and the disadvantages of the various methods of synthesis of nanothermites are listed in Table 3.1.

3.3 Properties of Nanothermites

Reaction of nanothermites is usually studied by igniting the powder and measuring the flame velocity and the pressurization rate. The two relative measurements bear correlation to each other where flame velocity is calculated using electric conductivity (Tasker et al. 2006), but imaging the flame with high-speed camera or photodiodes spaced a known distance apart is the most common technique. This has been done for various sample configurations, involving burning in an open channel (Kwon et al. 2003), in the micro-channels (Son et al. 2007) and in a cylindrical tube (Malchi et al. 2008). Monitoring of time-resolved emission can be used to find the kinetics of the reaction, along with the use of pressure sensors to collect simultaneous pressure data as the reaction propagates past a transducer (Vassiliou et al. 1993). The reaction mechanism during the self-propagating step can be well understood by the knowledge of the emission and pressure rise along with burning velocities. Combusting of fixed mass in a constant volume chamber and dynamic

Table 3.1 Pros and cons of various methods of synthesis of nanothermites

Mode of synthesis		Pros	Cons
Mixing	Physical	• Simple • Low cost • Scalable	• Handling is difficult • Homogeneous mixing is not possible • Impurity percentage is more • Compatibility with MEMS is less
	Sol–gel	• Low cost • Compatibility with MEMS is better • Tailoring of nanostructure is controllable	• Handling is difficult • Organic impurity is high • Self-sustaining reaction is inhibited
Self-assembly method		• Excellent controllability on tailoring of nanostructures • Homogeneity is more • Easily scalable	• Tedious and complicated process • Compatibility with MEMS is less • Impurity content is more
Nanofoils		• Controllable synthesis • Alumina shell is negligible • Highly pure • Self-sustaining reaction is possible • Integration with MEMS is convenient	• Expensive • Time consuming and difficult to scale up. • Undesired side products
ARM		• Inexpensive • Scalable	• Poor controllability on nanostructure • Require additional protective coating for safe handling • Total heat of reaction cannot be calculated • Integration with MEMS is difficult

pressurization estimations is among the other experimental techniques of studying the reactions of composite materials. As per studies of slow heat transfer time to the walls relative to the reaction timescale, flame temperature in these experiments can be speculated to be somewhat adiabatic in nature.

3.4 Types of Nanothermites

There are many possible thermodynamically stable fuel–oxidizer combinations. Some of them are listed below, and a few among them are discussed below:

- Aluminium-iron (II, III) oxide
- Aluminium-molybdenum (VI) oxide
- Aluminium-bismuth (III) oxide

- Aluminium-copper (II) oxide
- Other oxides such as aluminium-potassium permanganate, aluminium-tungsten (VI) oxide hydrate.

Aluminium/Iron (III) Oxide Nano-energetic Material

Historically, the thermites are referred to an aluminium and iron oxide reaction. The iron oxide in ferromagnetic phase (Fe_2O_3) is preferred as an oxidizing agent as it possesses superparamagnetic behaviour with decrease in size (Prakash et al. 2004). Herein, we describe the reports available related to Al/Fe_2O_3 nanothermites.

The burning rate of Al/Fe_2O_3 nanocomposites synthesized by sol–gel protocol was evaluated, and it revealed that the nanocomposites burn much faster and are more sensitive to thermal ignition than conventional thermites. The experimental value of heat of thermite reaction was found to be 1.5 kJ g^{-1}, which was significantly lower than the theoretical value (3.9 kJ g^{-1}). The declination of the total measured energy would be due to a native oxide coating on ultra-fine Al grain (Tillotson et al. 2001). Another report on Al/Fe_2O_3 nanocomposite synthesized by aero-sol–gel method showed vigorous burning with a propagation speed of 4 m/s (Son et al. 2002). The pressurization rate was reported to be 96 Pa s^{-1}, which was found to be lower than the traditional thermite. The low pressurization rate may be due to high density of iron oxide, which shows conductive burning rather than convective (low density material), which shows slow flame propagation (Shin et al. 2012). It was noted by another group that without heat treatment the Al/xerogel and Al/aerogel do not propagate within the confined tube arrangement. Heat treatment also leads to increase in the combustion wave propagation. Another observation made by them was that the aerogel has low density in comparison with xerogel and therefore is thrown ahead further at a much higher velocity. To improve homogeneity of the mixture, they have incorporated sodium dioctyl sulfosuccinate (SDS) and found a threefold improvement in the combustion wave speed (Kim and Zachariah 2004). In another work, Al was dispersed in bulk xerogel Fe_2O_3 by sol–gel process. It was noticed that sol–gel nanocomposites have shown high heat of enthalpy (order of four) than the nanocomposite by physical mixing. The enhancement in the enthalpy may be due to the increased intimacy between the fuel and the oxidizer due to direct dispersion of n-Al into Fe_2O_3 gel (Bhattacharya et al. 2006).

In another article, authors have reported the synthesis of mesoporous Fe_2O_3 using a cationic cetyltrimethylammonium chloride (CTAC) and non-ionic co-polymer Briz 76 (Mehendale et al. 2006). The composites of ordered Fe_2O_3 and Al nanoparticles were found to have higher burning rate than Fe_2O_3 with no ordering of mesopores (Fig. 3.15). The increase in the burning rate may be attributed to enhancement in the interfacial contact between oxidizer and fuel due to the use of the surfactant.

Thus, the use of the surfactant increases ordering of the mesoporous due to self-assembly of micelles and thereby decreases the pore size distribution in an oxidizer. Recently, a bottom-up approach, combining soft template self-assembly with sol–gel process, has been applied in which Brij S10 micelle was employed as

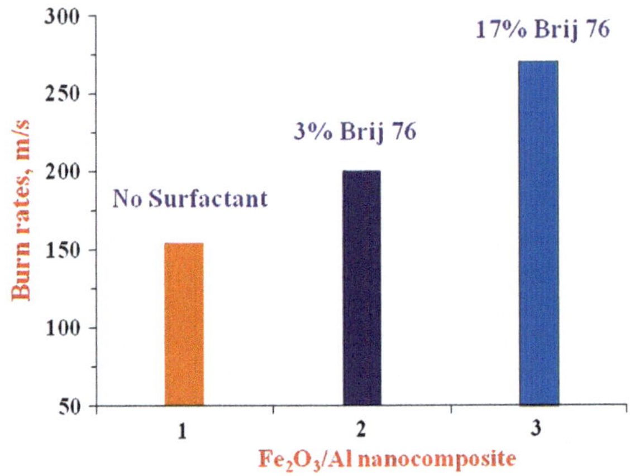

Fig. 3.15 Comparison of burning rate of Al/Fe_2O_3 nanocomposite with and without surfactant

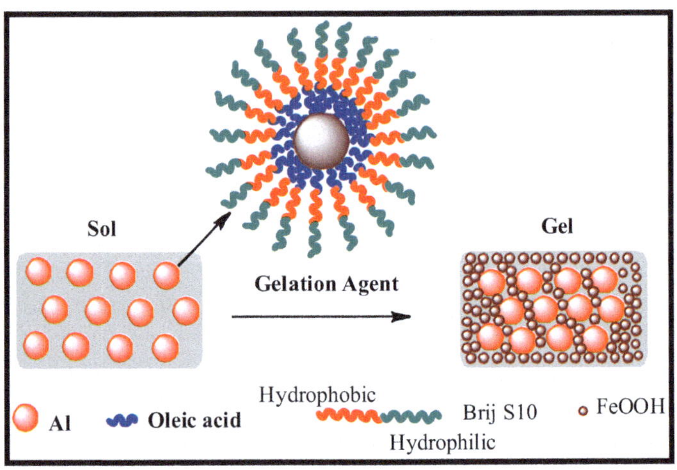

Fig. 3.16 Schematic diagram of synthesis of Al/Fe_2O_3 nanothermites using self-assembly

passivating n-Al, depicted in Fig. 3.16 (Zhang et al. 2013). A comparative study of physical mixing, sol–gel and self-assembly was also made. The heat releases by the samples synthesized by the aforementioned methods are 1097, 1689 and 2088 J g^{-1}, respectively. Hence, ordering of the composites has resulted in the betterment of the energetic content. When this oxidizer is mixed with the fuel nanoparticles and it is ignited, the hot spot density in the combustion wavefront increases leading to higher burning rates.

Table 3.2 Products obtained by different formulations

Thermite	Method	Temperature (°C)	Detected product
n-Al/xero-Fe_2O_3	Sol–gel	660 1020	Fe, $FeAl_2O_4$, γ-Fe_2O_3, Fe_3O_4 Fe, $FeAl_2O_4$, γ-Fe_2O_3, Fe_3O_4
n-Al/micro-Fe_2O_3	Sol–gel	660 1020	$FeAl_2O_4$, FeO $FeAl_2O_4$, FeO
Micro-Al/xero-Fe_2O_3	Mixing	660 1020	γ-Fe_2O_3, Al, Fe, $FeAl_2O_4$, Fe_3O_4, γ-Fe_2O_3, $Al_{2.667}O_4$
Micro-Al/ micro-Fe_2O_3	Mixing	660 1020	α-Fe_2O_3, Al, Fe, $FeAl_2O_4$, Fe_3O_4 γ-Fe_2O_3, α-Fe_2O_3, $Al_{2.667}O_4$, Al
Micro-Al/xero-Fe_2O_3	Mixing	660 1020	γ-Fe_2O_3, Al α- Fe_2O_3, α-Al_2O_3
n-Al/xero-Fe_2O_3	Mixing	660 1020	α-Fe_2O_3, $FeAl_2O_4$, Al α-Fe_2O_3, α-Al_2O_3, Fe

Three-dimensionally (3D) ordered macroporous Al/Fe_2O_3 nanothermite films with a polystyrene spheres as temperate were employed. A remarkable enhancement in the energy content 12.83 kJ g^{-1} has been reported by the authors (Cheng et al. 2010).

The mechanistic study of thermites was performed by combining Al and iron oxide to form nanocomposites using of sol–gel and physical mixing method (Durães et al. 2007). Two nanocomposites (nano-Al/xero-Fe_2O_3 and micro-Al/xero-Fe_2O_3) and four simple mixed (nano-Al + xero-Fe_2O_3, nano-Al + micro-Fe_2O_3, micro-Al + xero-Fe_2O_3 and micro-Al + micro-Fe_2O_3) samples were heated up to 1020 °C. The X-ray diffraction of the product obtained at 600 and 1020 °C was studied, and the products obtained are given in Table 3.2. These results signify only a theoretical existence of this equation $2Al + Fe_2O_3 \rightarrow Al_2O_3 + 2Fe$.

Hübner et al. (2017) have studied Al-based iron oxide/hydroxide nanothermites and reported that non-magnetic iron hydroxide exhibited higher flame propagation in comparison with magnetic iron oxide. Apart from this, non-magnetic nanothermite requires 45.8 mJ of spark energy to ignite, while the magnetic nanothermite ignites at 1212 J energy. They concluded that the non-magnetic nanothermite could be an alternative for lead or mercury for initiation of secondary explosives. Another work on iron oxide includes deposition of Al/Fe_2O_3 core–shell structure on grapheme sheet. The authors claim that these nanocomposites exhibited enhanced energy released and remarkable reduction in electrostatic ignition hazard (Yan et al. 2017).

Aluminium/Molybdenum Trioxide Nano-energetic Material

Today, nanothermites are not centred to an aluminium and iron oxide reaction as described earlier, but it implies to any exothermic reaction, which involves a metal reacting with a metal oxide to form a stable oxide. Molybdenum trioxide particles are often used as an oxidizer in nanocomposite energetic materials due to its high energy density (Hammons et al. 2008). Plantier et al. (2005) has reported the standard deviation of Al/Fe_2O_3 to be as high as 15% of the mean combustion velocity, whereas as on the other hand Granier and Pantoya (2004) found that in case of Al/MoO_3 nanocomposite the combustion speed is greater than 20% of the mean values. This grabbed the attention of researchers, and hence, the reports related to Al/MoO_3 are described below.

Eckert et al. (1993) prepared the composite of Al and MoO_3 by mechanically mixing using sonic waves and found that the burning rate increases from 4 to 12 m/s as the particle size of Al decreases from 200 to 50 nm. Further, decrease in the particle size below 50 nm showed lowering of the burning rate. Al and molybdenum trioxide (MoO_3) nanocomposite were examined as a function of Al particle size (Bockmon et al. 2005). The average speed increased from 750 to 950 m s^{-1} when the particle size was reduced from 121 to 80 nm which was remarkably higher than micrometre scale Al/MoO_3 (10 mm s^{-1}). Pantoya and Granier (2005) reported that the burning rate of n-Al/MoO_3 (micron size MoO_3) was augmented by tenfold when the size of Al decreases from 20,000 to 50 nm. The effect of environmental factors such as UV radiation, fluorescent and humidity on combustion properties of Al/MoO_3 formulations was also studied by a group (Moore and Pantoya 2006). The result demonstrates that no remarkable change in the burning rate on exposure to UV and fluorescent light was observed, whereas in humid environment a decrease in deterioration performance was seen. Valliappen et al. (2005) and Sun and Simon (2007) group have prepared nanosheets of MoO_3 with specific area of 76 and 42 m^2 g^{-1}, respectively, and found the combustion speed for Al/MoO_3 to be 362 and 410 m s^{-1}, respectively, indicating that increase in the surface area increases the combustion speed. In an another work, Dutro et al. (2009) synthesized nanosheets of MoO_3 of 30 × 200 nm dimension and observed a significant improvement in the combustion speed (1000 m s^{-1}). Recently, Zakiyyan et al. (2018) have prepared which exfoliated two-dimensional MoO_3 sheets which has resulted into reduction in the mass diffusion length and thereby provided large surface of interaction. A noteworthy increment in the combustion speed of Al/MoO_3 was reported in comparison with micron size composite from 51.3 to 1730 m s^{-1}.

Aluminium/Bismuth Oxide Nano-energetic Material

Another metal oxide of particular interest in nanothermite formulations is bismuth trioxide, which has a relatively low vapour pressure. The Al/Bi_2O_3 nanothermites are reported to generate the highest pressure pulse and thus may be used as a nano-energetic gas generator.

In this context, Puszynski and co-workers (2007) studied the thermodynamic analysis and revealed that Al-Bi_2O_3 system is more reactive than other systems (Al–CuO, Al–MoO_3). In addition to that, they studied the effect of coating on Al/Bi_2O_3 and obtained that the combustion front velocity of Al-Bi_2O_3 system for aluminium nanopowder coated with 5 wt% oleic acid was found to be higher (757 m s^{-1}) in comparison with uncoated sample (617 m s^{-1}) for the same fuel-to-oxide ratio. Martirosyan et al. (2009) have prepared Bi_2O_3 by a modified aqua-combustion protocol using bismuth nitrate and glycine, and time-depended mixing of fuel and oxidize was studied. Their finding demonstrated that a maximum front velocity of ~2500 m s^{-1} was obtained for a sample mixed for 6 h while, a very low front velocity i.e. ~760 m s^{-1} was obtained for 0.5 h. On the other hand, when the time was increased to 30 h, the front velocity was found to decrease to 1030 m s^{-1}. To explore the use of grapheme oxide in the area of thermites, Thiruvengadathan et al. (2014) attempted to synthesize grapheme oxide-functionalized Al/Bi_2O_3 self-assembled nanocomposite and reported a noteworthy increase in energy release from 739 to 1421 J g^{-1}. Same author in another work compared the self-assembled system with randomly mixed nanocomposite and found an increment in the pressure generation from 60 to 200 MPa, pressurization rate from 3 to 16 MPa μs^{-1} and burning rate from 1.15 to 1.55 km s^{-1} (Thiruvengadathan et al. 2015).

Nanothermites are attracting much attention, but it suffers from high electrostatic discharge (ESD) issues so in this regard. Pichot et al. (2015) have succeeded in desensitizing Al/Bi_2O_3 nanothermites by adding nanodiamond but with the increase in nanodiamond (0–2.65 wt%) which has lead to the reduce of combustion rate from 500 to 100 ms^{-1}. Yet et al. (2018) adhered nitrocellulose on to the surface of Al and Bi_2O_3 nanoparticles by a facile electrospray method with the aim to reduce ESD. They suggested that outside the nitrocellulose the covering dissipates input energy and does not allow the temperature to arise inside the Al/Bi_2O_3 matrix before ignition. The possible mechanism of use of nitrocellulose is that it serves as a low-temperature gas generator to release gas products and does not lead the nanoparticles to sinter into larger particles before ignition; as a result, a violent combustion occurs after ignition.

Aluminium/Copper Oxide Nano-energetic Material

Among all nanothermites, Al/CuO is commonly used nano-energetic material due to its highly exothermic nature and thus more explored.

Bhattacharya et al. (2006) determined the burning rate of nanosize Al/CuO to be 440 ms^{-1}. Later, different morphologies of the oxidizers (CuO) like nanowells, nanowires, nanorods have been studied for their varying combustion wave speeds with Al forming a composite material using self-assembly and ultrasonic method (Table 3.3) (Ahn et al. 2011a). In case of composite of CuO nanorods and Al nanoparticles, the combustion wave speed was found to be 1650 ms^{-1}, which

Table 3.3 Burning rate of different compositions of nanothermites

S. no.	Different compositions of nanothermites	Flame speed (m/s)
1.	CuO nanoparticles/n-Al	550–780
2.	Copper oxide (CuO) nanowells/n-Al	2100–2400
3.	CuO nanorods/n-Al	1500–1800
4.	CuO nanowire/n-Al	1900
5.	CuO nanorods mixed with Al nanoparticles and polymer by mixing	1800–1900
6.	CuO nanorods self-assembled/n-Al	1800–2200

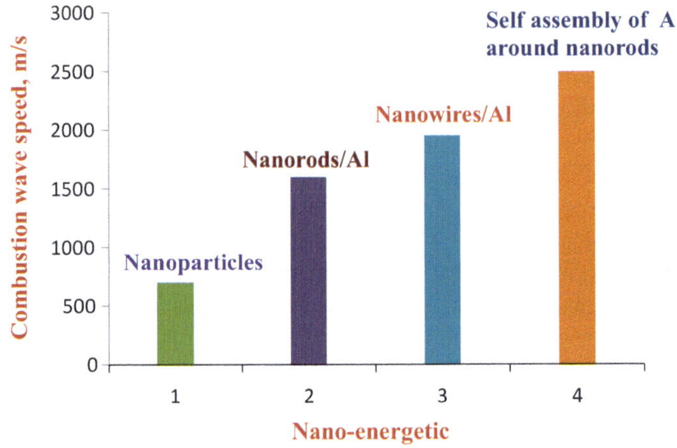

Fig. 3.17 Comparison of combustion wave speed of the self-assembled composite with the other nano-energetic composites prepared using physical mixing

increased to 1900 m s^{-1} for the composite containing nanowires and Al nanoparticles. Higher combustion wave speed of composite containing nanowires can be accredited to higher surface area creating higher hot spot density as compared to nanorods. The maximum combustion wave speed of 2400 m s^{-1} was noted for composite of CuO nanorod and 80-nm Al nanoparticles prepared by self-assembly method as demonstrated in Fig. 3.17. The increased combustion velocity in the self-assembled composite was due to the availability of lone pair of electron on the nitrogen of pyridyl group in the polymer to form a covalent bond with the metal and also interact with oxidizer. Another group recently worked on different morphologies of CuO (branch, plate and sphere) (He et al. 2015) mixed with ~70-nm Al particles. The authors have observed the first exothermic peak at ~587 °C which was lower than the micron-sized CuO/Al composites (1040 °C). The highest heat release was seen in case of hollow sphere-like morphology of CuO.

Zhang et al. (2013) have synthesized nano-Al/CuO nanowire on silicon substrates and determined that the n-Al react with CuO nanowire at about 500 °C prior to the melting of Al as revealed by DSC and DTA studies. Further, the unreacted material reacts with the Cu_2O film underneath CuO nanowire and reported that this will open a new venture in silicon-based functionalized nanodevices (Heat of reaction was estimated to be 2950 J g^{-1}). Yet, in another work, author prepared a core–shell nanothermite which incorporates an exothermic peak to be 100 °C lower than the melting point of n-Al and no endothermic peak associated with residual Al as observed in conventional thermites prepared by mixing. These observations suggested that core–shell structure has excellent spatial uniformity within core (Ohkura et al. 2011). In another approach, CuO nanowires were grown on the surface of copper wire which played a dual role as a base material for growing CuO and as a nanothermite shell. The heat release was reported to be 3108 J g^{-1} and burning velocity ~43.8 cm s^{-1} (Chiang and Wu 2017). Yin et al. prepared tubular structure nanothermite by depositing nanotubes/nanorod CuO with 43.8 cm, size ~ 100–200 nm in diameter and 5–7 μm in length on to n-Al using electrophoretic deposition. They have mentioned in their work that more energy release and combustion performance were observed in nanotube-structured nanothermite due to larger surface area which resulted in the enhancement of intimate contact and mass transmission between fuel and oxidizer (Yin et al. 2017a). Marín et al. (2015) reported the deposition of copper layer (~5 nm) thin between Al and CuO layers which resulted into enhancement in the reactivity. The increase in the activity is due to the formation of Al:Cu alloy which decreased the M.P of Al, thereby increasing the reactivity.

In order to improve the energetic properties of the nanothermite, Wang's group have demonstrated the route to assemble Al and CuO nanoparticles into micron composites via electrospray technique with a small percentage of nitrocellulose as an energetic binder leading to higher reactivity compared with nanothermite prepared through conventional physical mixing (Wang et al. 2014). Thermites formed by electrospray technique fuel and oxidizer were well-dispersed nanoparticles with intimate, while segregation was seen in sample prepared via physical mixing. The pressurization rate was augmented nine times with the increase in the percentage of binder (up to 5%) in the sample, whereas a reverse effect was observed in the physically mixed sample. Another energetic material CL-20 (2,4,6,8,10,12-hexanitrohexaazaisowurtzitane) was added to Al/CuO onto a SiO_2/Cr/Pt/Au microheater chip (Shen et al. 2014).The combustion of Al/CuO/50% CL-20 generated a strong light spot in contact with Al/CuO nanocomposite. The pressurization rate was also measured to be higher in case of Al/CuO/CL-20 reason being the compensation of pressure loss at microscale by the addition of CL-20. Energetic materials are often used in aqueous environment, such as underwater propulsion, welding, cutting. However, the use of energetic materials in underwater applications is limited as water perturbs the reactants and quenches the combustion on ignition. To address this challenge, Kim et al. (2015) added an optimum amount of sea urchin-like carbon nanotube (SUCNT) as the optical igniter into Al/CuO nanothermite matrix. The SUCNT absorbs the irradiated flash energy, converts into thermal energy and thereby leads to flash ignition and underwater explosion.

The burning behaviour of aluminium/CuO (spherical-shaped nano-aluminium 80 nm and cylindrical CuO—21 nm × 100 nm) was studied by Malchi et al. (1993) by incorporating 40-nm Al_2O_3 as an additive. They claimed that there was significant reduction in gas and flame production with the increase in the Al_2O_3 content from 0 to 20%. The burning velocity was found to be same on adding both 0 and 5% Al_2O_3 additives.

Zhou et al. (2017) prepared core–shell Al/CuO nanothermites on silicon substrate having an adhesive thin film of Cr and Cu. The thermal analysis revealed a low-temperature exothermic reaction at temperature ranging from 210 to 400 °C. The method resulted into crack-free core–shell Al/CuO which is useful in preparing energetic chips that could be exploited in the area of micro-ignition, micropropulsion, etc. Recently, nanothermites composed of aluminium and three-dimensionally ordered macroporous CuO are reported which claim to solve the problem of homogeneity. They also suggested that thermite reaction is augmented where oxidation occurs via convective propagation (Kim et al. 2018). Three-dimensional porous Al/CuO films were fabricated by Yin et al. using electrophoretic deposition (EPD) at a different pH ranging from 1 to 4. The films prepared at pH = 2 exhibited remarkable enhancement in surface area, combustion performance and energy output (Yin et al. 2017b).

Recently, Deng et al. (2018) reported the drawbacks of using n-Al reported in the literature which includes its relative thick native oxide layer, safety issues related to its sensitivity to electrostatic discharge, impact friction, etc., costly routes involved in the synthesis of n-Al and further its agglomeration leads to reduction in surface area. To overcome these challenges, the authors have suggested the use of micron-sized Al particles (m-Al) and prepared Al/CuO thermites by two simple methods, i.e. precipitation and displacement. They claimed that by using these methods they were successful in preparing thermites with reduced agglomeration and shorter diffusion distance between oxidizer and fuel.

Other Types of Nano-energetic Material

Tungsten oxides (WO_3) are commonly used in environmentally friendly electric matches and have gained premises by maintaining good performance without the use of lead. WO_3 has a high density (7.20 g cm^{-3} compared with 5.25 g cm^{-3} for Fe_2O_3). Combustion wave speeds of tungsten trioxide-based thermites were studied using two types of oxidizers (Prentice et al. 2006). First aerogel was dried at 120 °C (aerogel 120) and another at 400 °C (aerogel 400). Combustion wave speed was measured for the loose powder and the pellet of both the thermites. Aerogel 120 depicted high velocity when loose powder was adopted for measurement. On the contrary, aerogel 400 showed higher velocity in pellet form. Observed behaviour may be credited to small-size particles in the aerogel 120, and thus, energy transport via convection phenomenon is predominant. In pellet form, the aerogel 400 was comparatively free from hydroxyl impurities owing to more energy transfer through conduction mechanism. Another researcher studied hydrated tungsten oxide ($WO_3 \cdot H_2O$) thermites and compared its oxide formulations (Lee Perry et al. 2007). The results of the hydrated formulation were remarkably superior than dehydrated sample in terms of energy release and pressurization rate. The conclusion from their finding was that the water participates in adding energy and produces hydrogen gas.

Another type of Al/NiO nanothermites was synthesized using NiO nanowires by sono-chemical method (Wen et al. 2013). The energy release increased drastically when the NiO amount was increased in the sample. Al/NiO thermites are theoretically reported to generate less gas and had a lower onset temperature than the Al/CuO thermite. Zhang and Li (2015), prepared a Al/NiO thermite film using electrophoretic deposition (EPD) method. The heat release calculated from DSC was 931.1 J g^{-1} which was comparable with the thermite synthesized using ultrasonic method. Al/NiO thermites with copper additive are also studied which showed a better material joining quality than pure thermites (Bohlouli-Zanjani et al. 2013).

The theoretical value of heat of reaction for Al/Co_3O_4 nanothermites was reported to be ~4232 J g^{-1} which is higher than the other thermites attracted researcher to work on it.

The thermal studies of Al/Co_3O_4 prepared via thermal evaporation method (Wang et al. 2015a) reported an exothermic peak at 520 °C, which signifies that it can rapidly react and release heat below the melting point of Al (660 °C) which is lower than the other thermites (Al/Fe_2O_3 at 588 °C and Al/CuO at 550 °C).

Prakash et al. (2005) have prepared nanothermite using strong oxidizing metal salt ($KMnO_4$), the pressurization rate of $Al/KMnO_4$ to be 2 MPa μs^{-1} which is found to be 40 times higher than that of Al/CuO measured under the same condition. However, there are chemical stability issues with the permanganate salt as it reacts with organic substances.

Another system, Al/I_2O_5, has been used in this area as it has highest reaction enthalpy per volume (25.7 kJ cm^{-3}) in comparison with other nano-energetic formulations. Hobosyan et al. (2012) have reported the synthesis of I_2O_5 nanorods by using ball milling process with controlled energy dosage. The pressure discharge of nanocomposite was found to be twofold higher than the composite prepared using commercial available I_2O_5 particles.

Sullivan et al. (2010) have studied another system, $Al/AgIO_3$, and reported that the pressurization rate is ~0.4 MPa μs^{-1} which was found to notably higher than 0.062 and 0.0001 MPa μs^{-1} for Al/CuO and Al/Fe_2O_3, respectively. The propagation rate of $Al/AgIO_3$ was found to be 630 m s^{-1} which was significantly more than Al/CuO (340 m s^{-1}).

Al/Bimetallic Oxide Nanoparticles Nano-energetic Material

Ahn et al. (2011) have investigated the effect of the mixed two oxidizers, namely CuO and Fe_2O_3. The schematic diagram of the synthesis of mixed oxide-based thermites is presented in Fig. 3.18. Explosive reactivity of nano-energetic materials augmented linearly as the mass fraction of CuO in the CuO–Fe_2O_3 (bimetallic oxide) increased as per results obtained from DSC and pressure cell tester (PCT) systems. In their findings, they demonstrated that energy release and pressurization rate of nano-energetic materials can be tuned by designing the elemental compositions using a strong oxidizer (CuO) and a mid-oxidizer (Fe_2O_3) in the bimetallic oxide. The tailoring of energetic outputs of Al/Fe_2O_3-based nano-energetic materials was performed by adding a third reactant, SiO_2 (Clapsaddle et al. 2005), in the composite, but the addition of SiO_2 was shown to decrease the property of thermite by 2–3 reaction velocities.

Sulphates-Based Nanothermites

Sulphates-based nanothermites are also expanding the horizon of nanothermites as a 'green' alternative for primary explosives. In this regard, Comet et al. (2015) have reported that the hydrated salts of $MgSO_4$ have augmented the heat of reaction (4–6 MJ Kg^{-1}) in comparison with traditional nanothermites (1.5–4.8 MJ Kg^{-1}). They have mentioned in their report that ignition of such salts produces hydrogen along with the metal oxide which leads to high reaction heat. It has also reduced the safety issues related to high electrostatic discharge pertaining to the use of oxide-based

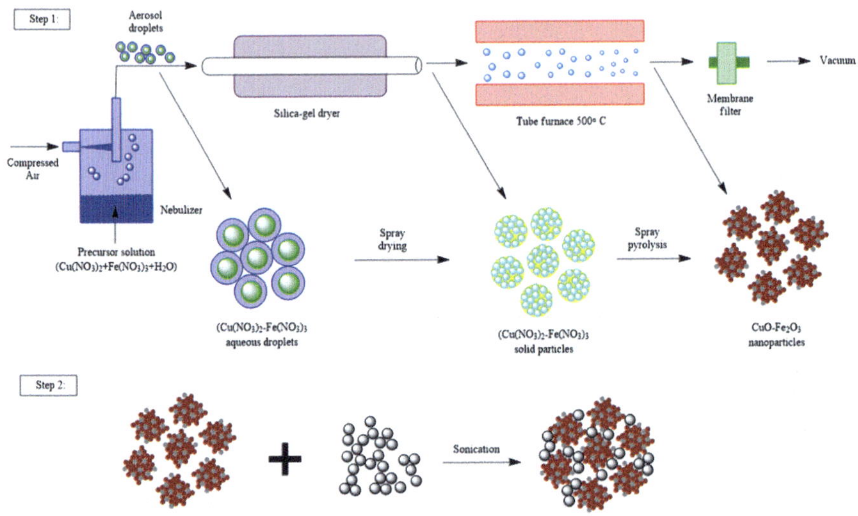

Fig. 3.18 Schematic diagram of the synthesis of mixed oxide-based thermites

nanothermites. In another work, authors (Yi et al. 2018) have synthesized Al/$CuSO_4$•$5H_2O$ nanothermite using electrostatic spray protocol. The peak pressure and pressurization rate of Al/$CuSO_4$•$5H_2O$ nanothermite were found to be ten- and 20-folds, respectively, higher than Al/CuO nanothermite. They also proclaimed that it could be a substitute for lead containing primary explosive.

3.5 Applications of Nanothermites

As we have studied until now that thermite is a mixture of a metal and an oxide of another metal resulting in an exothermic reaction, these are being used in a number of applications. Chemistry tells us that smaller the particle size, faster is the reaction. Nanothermites have thus improved the overall exothermic characteristic of the reaction, which proved useful for applications where thermites were being used already as well as in new applications. Some of the applications where nanothermites are being presently used are micropropulsion technology in rockets, micro-initiators for actuators, biomedical shock wave therapy and joining and welding works. Thermites are also used in the area of explosives and propellants owing to its high volumetric energy density of thermites; for example, Al/Fe_2O_3 has 3.9 kcal cm^{-3}, whereas TNT has 1.0 kcal cm^{-3}. Some of these applications are elaborated as below.

Micropropulsion

The key to future of spacecraft lies in its miniaturization. More and more effort is being devoted today to develop smaller spacecraft. A spacecraft with the weight range of 20–100 kg is termed as microspacecraft and those with the weight less than 20 kg fall into the category of nanospacecraft. This miniaturization offers many advantages like overall mission and launch cost reduction due to reduced weight, launching microspacecraft cluster consisting of many microspacecrafts at once instead of a single large spacecraft and reduced mission risk. However, this all comes at the price of significantly reducing the size of all the subsystems of the spacecraft. Thankfully, the microelectromechanical systems (MEMS) technologies provide us with a solution to this problem. Microthrusters are micropropulsion methods used to provide numerous applications to the microspacecraft like orbit adjustment, controlling altitude, drag compensation and delta-v manoeuvres.

Apperson et al. (2009) have studied the combustion properties of Al/CuO nanothermite composites and found their potential application in microthrusters. Using two types of thrust motors, i.e. one with no nozzle and another with a convergent–divergent nozzle, they tested thrust-generating characteristics using various quantities of mixture. Depending upon the material packing density, distinct impulse characteristics were observed. For example, thrust forces of ≈75 N with duration < 50 μs full-width at half-maximum (FWHM) and 3–5 N with a duration of 1.5–3 ms were recorded at low packing pressure and high packing density, respectively. Specific impulse value of 20–25 s in both cases and short thrust duration makes this material promising for space-limited micropropulsions.

In a similar concept (Staley et al. 2013), applications of Al/Bi_2O_3 nanoparticle mixture and cupric oxide/aluminium nanoparticle mixture for micropropulsion have been studied. Fabrication of steel-based miniaturized engine was done without converging–diverging nozzle. The various parameters related to the nanothermite materials in micropropulsion applications are summarized in Table 3.4. Further, it was observed that if a small amount of nitrocellulose was added to these solid propellants the value of specific impulse and volumetric impulse increased to 59.4 s and 2.3 mN s mm^{-3}, respectively. This ensured prolonged burning duration and

Table 3.4 Various nanothermite materials in micropropulsion application

S. no.	Nanothermite material	Average thrust force (N)	Specific impulse (s)	Duration (ms)	Remarks	Refs
1.	Al/CuO	≈75	20–25	<50 μs FWHM	Low packing density	Apperson et al. (2009)
2.	Al/CuO	3–5	20–25	1.5–3 ms	High packing density	Apperson et al. (2009)
3.	Al/Bi_2O_3	46.1	41.4	1.7 ms	–	Staley et al. (2013)
4.	Al/Cu_2O	4.6	20.2	5.1 ms	–	Staley et al. (2013)

controllably reduced average thrusts. A maximum energy conversion efficiency of 0.19% was recorded for thrusters using rotary-arm measurement.

In a separate study, the same group (Staley et al. 2014) studied the effect of nitrocellulose doping in Al/Bi_2O_3-miniaturized nanothermite thrusters in detail. Nanothermites can be extremely sensitive to ignition stimuli and susceptible to phase separation. Hence, in the study, nitrocellulose was used as a gasifying and nanothermite desensitizing binder. $Al/Bi_2O_3/NC$ nanothermites through thrust measurements were found to offer specific impulse values of up to 63.2 s and greater combustion stability compared to nanothermites in their pure form.

Initiators

Nanothermite material also has applications in miniaturized actuation technologies (Staley et al. 2011). By fabricating a silicon-based bridge wire microchip initiator using micromanufacturing process, the initiator was able to produce chemical energy in joules as output when just microjoules of electrical energy was provided as input. Nanothermite energetic composite material bismuth oxide–aluminium was integrated with the microchip initiator, which was assembled with an open material reservoir and utilized a novel 47 °C melting point solder alloy bonding.

Nanoporous silicon bed was used to maintain the structural integrity of the device as well as enhance the electrothermal conversion efficiency. This is done by preventing the thermal coupling of the bridge wire and bulk silicon substrate. On studying the electrical behaviour of the ignition elements, it was found that the minimum input power and the minimum input energy had the values of 382.4 mW and 26.51 μJ, respectively. These values are independent of whether the Al/Bi_2O_3 nanothermite composition was injected in the device or otherwise. Full success rate was achieved over a 30–80 μJ firing energy range of the programmed combustion of Al/Bi_2O_3 nanothermite, which was placed in the fabricated initiator. The ignition response times were found to be less than 2 μs when the operation was carried out at high input power.

In another study (Taton et al. 2013), successful demonstration of the non-contact propellant ignition has been achieved using a new nanothermite-based polymeric electrothermal initiator. The reactive nanothermite Al/CuO multilayer was insulated from the silicon bulk substrate by placing it on a 100-μm-thick SU-8/PET membrane (polyethyleneterephtalate). On supplying current, sparkles are produced up to a distance of several millimetres due to the chemical reaction Al and CuO. The team further investigated the characteristics of the fabricated initiator by comparing it to two bulk electrothermal initiators, of which one had a substrate of silicon and another of Pyrex glass. On analysis, it was found that PET devices achieved 100% Al/CuO ignition success at electric current of >250 mA, whereas the glass-based initiator achieved the same success rate for electric current of >500 mA. The silicon-based reactive initiators were not able to initiate even at electric current as high as 4 A. It was also observed that the initiation time at low currents (<1 A) was 100 times faster in PET-based initiator as against Pyrex-based initiator. In addition, the Al/CuO thermite film of PET membrane reacted within 1 ms, whereas the sparkles duration for the Pyrex initiator was around 4 ms. The intensity of the

Table 3.5 Different parameters for various types of initiators

Parameter	PET-based initiator	Pyrex-based initiator	Silicon-based initiator
100% Al/CuO ignition success rate	Electric current of >250 mA	Electric current of >500 mA	Did not initiate even at electric current of 4 A
Initiation time at low currents (<1 A)	Least among the three	100× more than PET-based initiator	–
Sparkles duration	1 ms	4 ms	–
Thermite reaction intensity	40× more than Pyrex-based initiator	Very less as compared to PET-based initiator	–

thermite reaction was around 40 times more in PET-based initiator than in the Pyrex-based initiator. All of these parameters have also been summarized in Table 3.5.

Petrantoni et al. (2010b) also developed a micro-initiator based on a similar principal of nano-energetic applications of nanothermite materials. The team proposed the potential use of their micro-initiator in safe arm and fire systems. The Al/CuO nanothermite was ignited by the Joule effect, which required supplying a few Watts of energy initially to start ignition process. This ignited nanothermite material is further used to ignite a propellant in close proximity. The propellant can be ignited by being in contact with the nanothermite material or by non-contact mode through air. The maximum distance through which the non-contact ignition of the propellant worked without failing was calculated to be around 270 μm.

In Biomedical Field

Even after decades of innovative research and eradication of a number of deadly diseases, some diseases still today cannot be treated by conventional approaches of medical science. For curing such diseases, some more advanced mechanisms like delivering molecular and genetic material directly into cells are required. Though there are a number of reported methods to do so, recently a novel method using shock waves is generated by nanothermite material (Korampally et al. 2012). Current drug carrier vehicles are broadly categorized as viral and non-viral. Viral vehicles like adeno- and retroviruses though have a stable gene expression and high efficiency of delivery, but they also possess a number of disadvantages like poor target specificity, therapeutic gene size restrictions and undesirable immunogenic responses. Hence, there is a need to look at their non-viral counterparts. One way the non-viral vehicle makes foreign molecular transport possible is by using pressure waves, rendering the cell membrane permeable temporarily. The chemical energy liberated by the burning of nanothermite material like bismuth trioxide–aluminium (Bi_2O_3/Al) can provide this type of pressure waves. These transient pressures create shock wave fronts, which are generated by the combination of very

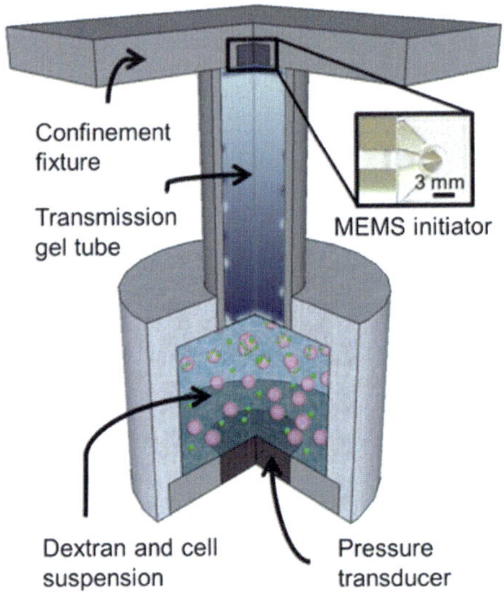

Fig. 3.19 Cross-sectional illustration of MEMS-based nanothermite. Reproduced with kind permission from Korampally et al. (2012)

Table 3.6 Efficiency parameters of FITC-dex

Parameter	Efficiency (%)
Cytoplasmic delivery efficiency	>90
Tailored intranuclear delivery efficiency	18–84
24 h post-delivery cell viability	50–95

high reaction propagation velocities (in the range of 550–2600 m/s) and gaseous product evolution of the nanothermite material.

The team was successfully able to demonstrate the generation of pressure pulse in their MEMS-based actuator (Fig. 3.19), using which they delivered 59–77 kDa of fluoresceinisothiocyanate dextran (FITC-dex) as fluorescent probe into chicken cardiomyocytes and obtained the following parameters (Table 3.6).

The threatening due to bioterrorism leads to use of iodine containing nanothermites such as Al/I_2O_5, $Al/Bi(IO_3)_3$, metal iodate/C_2I_4 as biocidal reactive materials (Wang et al. 2015b).

Joining/Welding

Bohlouli-Zanjani (2013) in his study presents an elaborate account of the welding and joining applications of the thermite and nanothermite reactions. Bulk thermite reactions have shown their application in a number of chemical processes like material synthesis, preparation of metals and their alloys, centrifugal coating for a long time now. The nanoscale thermites have proved to be of great advantage in the similar direction, further improving upon their properties in such applications meanwhile adding some new ones in the basket as we already discussed previously like in microthrusters, biomedical domain, propellants.

Table 3.7 Shear strength values of nanothermites

Material	Weld strength (MPa)
Mo–TiB_2–Mo	20–40
Mo–TiC–Mo	≈10
Al 1100 alloy sheets using Al–CuO–Ni	27

Nanothermites specifically find application in the area of welding where conditions demand a molten state of materials and the release of very high amount of energy for the joining process to take place. Temperatures as high as 3000 K resulting from nanothermite reactions make it possible for the metals to melt and thus join in the required fashion.

The lower activation energies of nanothermites cause them to ignite easily as compared to their bulk counterparts, which make them a more suitable choice for welding of both similar and dissimilar materials. Ceramic-metal welds like that of Mo–TiB_2–Mo and Mo–TiC–Mo can be obtained by using pressurized combustion reaction. TiB_2 and TiC welds can be prepared when pressed against Mo interfaces by first forming a mixture of Ti powder with B or C, and then sandwiching them between two Mo surfaces while subjecting them to electrical ignition via exothermic reaction. Weld strength of 20–40 MPa was obtained for Mo–TiB_2–Mo and ≈10 MPa for Mo–TiC–Mo. Al–Ni mixture through the process of self-propagating high-temperature synthesis (SHS) is used for welding superalloy substrates. Al and Ni mixture when heated to 920 K form AlNi. This also marks the initiation of an exothermic reaction, which raises the temperature up to 1950 K by the evolving heat of AlNi exothermic reaction. AlNi melts at 1950 K, causing the substrate surface to melt and forming an Al-rich Ni-based superalloy at the interfaces. In another study, Al 1100 alloy sheets can be joined by micron-sized particle mixture of Al–CuO–Ni using a uniaxial pressure of 9 MPa (Kim et al. 2011). Observations show that the mean shear strength of 27 MPa was obtained by this method as against 8.05 MPa for normal diffusion-welded samples. Shear strength values as high as 480 MPa have also been reported using Al–Ni multilayer foil to weld metallic glass using exothermic reaction of the self-propagating nature. An electrical potential of 30 V was applied to the foil to initiate the reaction. It was further found that both the thickness of the foil and the pressure applied can affect the shear strength value of the final weld (Table 3.7).

3.6 Conclusion

Nanothermites have emerged as a 'green energetic' contribution to the civil and defence applications with the evolution of time. The present work summarizes the various methodologies adopted towards the synthesis of different classes of nanothermites which includes basic form to the three-dimensionally ordered assembly of nano-energetic materials. Further, two approaches, namely bottom-up and top-down, are taken into account. It is interesting to note that researchers have played with the synthetic procedure to artistically tune the energetic parameters of different nanothermites. Additionally, the literature reveals that additions of explosive materials resulting in the formation of hybrid mixtures have greatly enhanced the properties of nanothermites. The mechanistic aspect of the nanothermites is also discussed in the literature which explains the transition considering conduction, radiation, convection and hydrodynamic mechanism of heat transfer. The conduction mechanism involves layer-by-layer combustion process which grows with the ambient pressure, whereas in convection fast energy transfer occurs from hot reaction products to cold zone. Researchers have reported high burning rates of nanothermites (2000–2500 m s^{-1}) which meets the requirement of the energetic materials. However, these materials are very sensitive to electrostatic discharge, so attempts are made to reduce its sensitivity but it has resulted into reduction in its performance and reactivity. Thus, this aspect has still to be explored. Further, aluminium oxide layer on n-Al surface is a critical issue; approaches should be adopted to enrich the performance of nanothermites. The attempts until now focus on enhancing the intimate contact between fuel and oxidizer and its uniform dispersion. The nanothermites with decreased length scale and uniformly dispersed structures have a limited usage, so incorporation of additives to absorb oxygen molecules released during the reaction can possibly be explored more. Moreover, hybrid materials which include nanothermite along with energetic materials like RDX or inert gas generating agents are being explored on a broad scale. This can solve both the perspective energetic materials as well as triggering the performance with acceptable sensitivity threshold. The design of three-dimensionally ordered structure perhaps would be a strategy for improving the homogeneous mixing at controllable length scale.

References

Ahn JY, Kim WD, Cho K, Lee D, Kim SH (2011a) Effect of metal oxide nanostructures on the explosive property of metastable intermolecular composite particles. Powder Tech 211:65–71

Ahn JY, Kim WD, Kim JH, Kim JH, Lee JK, Kim JM, Kim SH (2011b) Gas-phase synthesis of bimetallic oxide nanoparticles with designed elemental compositions for controlling the explosive reactivity of nanoenergetic materials. J Nanomater 42

Apperson SJ, Bezmelnitsyn AV, Thiruvengadathan R, Gangopadhyay K, Gangopadhyay S, Balas WA, Anderson PE, Nicolich SM (2009) Characterization of nanothermite material for solid-fuel microthruster applications. J Propul Power 25:1086–1091

Aumann CE, Murray AS, Skofronick GL, Martin JA (1994) Metastable interstitial composites: super thermite powders. In Proceedings insensitive munitions technology symposium, Williamsburg, VA, USA, pp 6–9

Baláž P, Takacs L, Boldižárová E, Godočíková E (2003) Mechanochemical transformations and reactivity in copper sulphides. J Phys Chem Solids 64:1413–1417

Bernstein ER (2014) On the release of stored energy from energetic materials. In: Advances in quantum chemistry, vol 69. Academic Press, pp 31-69

Bhattacharya S, Gao Y, Apperson S, Subramaniam S, Talantsev E, Shende RV, Gangopadhyay S (2006) A novel on-chip diagnosis method to detect flame velocity of nanoscale thermites. J Energ Mater 24:1–5

Blobaum KJ, Reiss ME, Plitzko JM, Weihs TP (2003a) Deposition and characterization of a self-propagating CuOx/Al thermite reaction in a multilayer foil geometry. J Appl Phy 94:2915–2922

Blobaum KJ, Wagner AJ, Plitzko JM, Heerden DV, Fairbrother DH, Weihs TP (2003b) Investigating the reaction path and growth kinetics in CuOx/Al multilayer foils. J Appl Phy 94:2923–2929

Bockmon BS, Pantoya ML, Son SF, Asay BW, Mang JT (2005) Combustion velocities and propagation mechanisms of metastable interstitial composites. J Appl Phys 98:64903

Bohlouli Zanjani G (2013) Synthesis, characterization, and application of nanothermites for joining. Master's thesis, University of Waterloo

Bohlouli-Zanjani G, Wen JZ, Hu A, Persic J, Ringuette S, Zhou YN (2013) Thermo-chemical characterization of a Al nanoparticle and NiO nanowire composite modified by Cu powder. Thermochim Acta 572:51–58

Brinker CJ, Scherer GW (1990) Sol-gel science. Academic Press. San Diego, p 2

Chen HY, Sachtler WMH (1998) Activity and durability of Fe/ZSM-5 catalysts for lean burn NOx reduction in the presence of water vapor. Catal Today 42:73–83

Cheng JL, Hng HH, Lee YW, Du SW, Thadhani NN (2010) Kinetic study of thermal-and impact-initiated reactions in Al–Fe_2O_3 nanothermite. Combust Flame 157:2241–2249

Chiang YC, Wu MH (2017) Assembly and reaction characterization of a novel thermite consisting aluminum nanoparticles and CuO nanowires. Proc Combust Inst 36:4201–4208

Chung SW, Guliants EA, Bunker CE, Hammerstroem DW, Deng Y, Burgers MA, Jelliss PA, Buckner SW (2009) Capping and passivation of aluminum nanoparticles using alkyl-substituted epoxides. Langmuir 25:8883–8887

Clapsaddle BJ, Gash AE, Satcher JH, Simpson RL (2003) Silicon oxide in an iron (III) oxide matrix: the sol–gel synthesis and characterization of Fe–Si mixed oxide nanocomposites that contain iron oxide as the major phase. J Non-Cryst Solids 331:190–201

Clapsaddle BJ, Zhao L, Prentice D, Pantoya ML, Gash AE, Satcher Jr JH, Shea KJ, Simpson RL (2005) Formulation and performance of novel energetic nanocomposites and gas generators prepared by sol–gel methods. In: Proceedings of 36th international annual conference of ICT, Karlsruhe, Germany, p 39

Clarkson J, Smith WE, Batchelder DN, Smith DA, Coats AM (2003) A theoretical study of the structure and vibrations of 2, 4, 6-trinitrotolune. J MolStruct 648:203–214

Comet M, Martin C, Klaumünzer M, Schnell F, Spitzer D (2015) Energetic nanocomposites for detonation initiation in high explosives without primary explosives. Appl Phys Lett 107:113–119

Crouse CA, Pierce CJ, Spowart JE (2010) Influencing solvent miscibility and aqueous stability of aluminium nanoparticles through surface functionalization with acrylic monomers. ACS Appl Mater Interfaces 2:2560–2569

Cui Y, Huang D, Li Y, Huang W, Liang Z, Xu Z, Zhao S (2015) Aluminium nanoparticles synthesized by a novel wet chemical method and used to enhance the performance of polymer solar cells by the plasmonic effect. J Mater Chem C 3:4099–4103

Dai J, Xu J, Wang F, Tai Y, Shen Y, Shen R, Ye Y (2018) Facile formation of nitrocellulose-coated Al/Bi_2O_3 nanothermites with excellent energy output and improved electrostatic discharge safety. Mater Des 143:93–103

Danen WC, Martin JA (1993) Energetic composites. U.S. Patent 5,266,132, issued Nov 30

Deng S, Jiang Y, Huang S, Shi X, Zhao J, Zheng X (2018) Tuning the morphological, ignition and combustion properties of micron-Al/CuO thermites through different synthesis approaches. Combus Flame

Diamandis EP, Christopoulos TK (1991) The biotin-(strept) avidin system: principles and applications in biotechnology. Clin Chem 37:625–636

Dreizin EL (2009) Metal-based reactive nanomaterials. Prog Energy Combust Sci 35:141–167

Durães L, Costa BF, Santos R, Correia A, Campos J, Portugal A (2007) Fe_2O_3/aluminum thermite reaction intermediate and final products characterization. Mater Sci Eng A 465:199–210

Dutro GM, Yetter RA, Risha GA, Son SF (2009) The effect of stoichiometry on the combustion behavior of a nanoscale Al/MoO_3 thermite. Proc Combust Inst 32(II): 1921–1928

Eckert J, Holzer JC, Ahn CC, Fu Z, Johnson WL (1993) Melting behavior of nanocrystalline aluminum powders. Nanostruct Mater 2:407–413

Ermoline A, Schoenitz M, Dreizin EL (2011) Reactions leading to ignition in fully dense nanocomposite Al-oxide systems. Combust Flame 158:1076–1083

Ermoline A, Stamatis D, Dreizin EL (2012) Low-temperature exothermic reactions in fully dense Al–CuO nanocomposite powders. Thermochim Acta 527:52–58

Foley TJ, Johnson CE, Higa KT (2005) Inhibition of oxide formation on aluminum nanoparticles by transition metal coating. Chem Mater 17:4086–4091

Folkers JP, Gorman CB, Laibinis PE, Buchholz S, Whitesides GM, Nuzzo RG (1995) Self-assembled monolayers of long-chain hydroxamic acids on the native oxide of metals. Langmuir 11:813–824

Gash AE, Tillotson TM, Satcher JH, Poco JF, Hrubesh LW, Simpson RL (2001a) Use of epoxides in the sol-gel synthesis of porous iron (III) oxide monoliths from Fe (III) salts. Chem Mater 13:999–1007

Gash AE, Tillotson TM, Satcher Jr JH, Hrubesh LW, Simpson RL (2001b) New sol–gel synthetic route to transition and main-group metal oxide aerogels using inorganic salt precursors. J Non-Cryst Solids 285:22–28

Gash AE, Satcher Jr JH, Simpson RL (2004) Behaviour of sol Gel derived nanostructured iron (III) oxide. In: Proceedings of 31st international pyrotechnic seminar, Fort Collins, Colorado, USA

Ghanta SR, Muralidharan K (2010) Solution phase chemical synthesis of nano aluminium particles stabilized in poly (vinylpyrrolidone) and poly (methylmethacrylate) matrices. Nanoscale 2:976–980

Granier JJ, Pantoya ML (2004) Laser ignition of nanocomposite thermites. Combust Flame 138:373–383

Haber JA, Buhro WE (1998) Kinetic instability of nanocrystalline aluminum prepared by chemical synthesis; facile room-temperature grain growth. J Am Chem Soc 120:10847–10855

Hammons JA, Wang W, Ilavsky J, Pantoya ML, Weeks BL, Vaughn MW (2008) Small angle X-ray scattering analysis of the effect of cold compaction of Al/MoO_3 thermite composites. Phys Chem Chem Phys 10:193–199

He S, Chen J, Yang G, Qiao Z, Li J (2015) Controlled synthesis and application of nano-energetic materials based on the copper oxide/Al system. Cent Eur J Energ Mater 12:129–144

Hobosyan M, Kazansky A, Martirosyan KS (2012) Nanoenergetic composite based on I_2O_5/Al for biological agent defeat. In: Technical proceeding of the 2012 NSTI nanotechnology conference and expo, pp 599–602

Hübner J, Klaumünzer M, Comet M, Martin C, Vidal L, Schäfer M, Kryschi C, Spitzer D (2017) Insights into combustion mechanisms of variable aluminum-based iron oxide/-hydroxide nanothermites. Combust Flame 184:186–194

Ivanov GV, Tepper F (1997) Activated aluminum as a stored energy source for propellants. Int J Energetic Mater Chem Propul 4:1–6

Ivanov YF, Osmonoliev MN, Sedoi VS, Arkhipov VA, Bondarchuk SS, Vorozhtsov AB, Korotkikh AG, Kuznetsov VT (2003) Productions of ultra-fine powders and their use in high energetic compositions. Propell Explos Pyrot 28:319–333

Jason JR, Waren AD, Rosenberg DM, Bellitto UJ (2003) Surface passivation of base Al nanoparticles using perfluroalkyl carboxylic acids. In: Proceeding of materials research society symposium, vol 800, pp 67–78

Jian G, Liu L, Zachariah MR (2013) Facile aerosol route to hollow CuO spheres and its superior performance as an oxidizer in nanoenergetic gas generators. Adv Funct Mater 23:1341–1346
Kim SH, Zachariah MR (2004) Enhancing the rate of energy release from nanoenergetic materials by electrostatically enhanced assembly. Adv Mater 16:1821–1825
Kim DK, Bae JH, Kang MK, Kim HJ (2011) Analysis on thermite reactions of CuO nanowires and nanopowders coated with Al. Curr App Phy 11:1067–1070
Kim JH, Kim SB, Choi MG, Kim DH, Kim KT, Lee HM, Lee HW, Kim JM, Kim SH (2015) Flash-ignitable nanoenergetic materials with tunable underwater explosion reactivity: the role of sea urchin-like carbon nanotubes. Combus Flame 162:1448–1454
Kim WD, Lee S, Lee DC (2018) Nanothermite of Al nanoparticles and three-dimensionally ordered macroporous CuO: mechanistic insight into oxidation during thermite reaction. Combus Flame 189:87–91
Korampally M, Apperson SJ, Staley CS, Castorena JA, Thiruvengadathan R, Gangopadhyay K, Mohan RR, Ghosh A, Polo-Parada L, Gangopadhyay S (2012) Transient pressure mediated intranuclear delivery of FITC-Dextran into chicken cardiomyocytes by MEMS-based nanothermite reaction actuator. Sens Actuators B Chem 171:1292–1296
Kwon YS, Gromov AA, Ilyin AP, Popenko EM, Rim GH (2003) The mechanism of combustion of superfine aluminum powders. Combust Flame133:385–391
Lee Perry W, Tappan BC, Reardon BL, Sanders VE, Son SF (2007) Energy release characteristics of the nanoscale aluminum-tungsten oxide hydrate metastable intermolecular composite. J App Phy 101:064313
Mabuchi T, Nishikiori H, Tanaka N, Fujii T (2005) Relationships between Fluorescence properties of benzoquinolines and physicochemical changes in the Sol–Gel–xerogel transitions of silicon alkoxide systems. J Sol-Gel Sci Technol 33:333–340
Mahendiran C, Ganesan R, Gedanken A (2009) Sonoelectrochemical synthesis of metallic aluminium nanoparticles. Eur J Inorg Chem 14:2050–2053
Malchi JY, Yetter RA, Foley TJ, Son SF (2008) The effect of added Al_2O_3 on the propagation behavior of an Al/CuO nanoscale thermite. Combust Sci Technol 180:1278–1294
Malchi JY, Foley TJ, Yetter RA (2009) Electrostatically self-assembled nanocomposite reactive microspheres. ACS Appl Mater Interfaces 1:2420–2423
Mandin P, Wüthrich R, Roustan H (2009) Industrial Aluminium Production: the Hall-Heroult process modelling. ECS Trans 19:1–10
Marín L, Nanayakkara CE, Veyan JF, Warot-Fonrose B, Joulie S, Estève A, Tenailleau C, Chabal YJ, Rossi C (2015) Enhancing the reactivity of Al/CuO nanolaminates by Cu incorporation at the interfaces. ACS Appl Mater Interfaces 7:11713–11718
Martirosyan KS, Wang L, Vicent A, Luss D (2009) Synthesis and performance of bismuth trioxide nanoparticles for high energy gas generator use. Nanotechnol 20:405609
McClain MJ, Schlather AE, Ringe E, King NS, Liu L, Manjavacas A, Knight MW et al (2015) Aluminium nanocrystals. Nano Lett 15:2751–2755
Mehendale B, Shende R, Subramanian S, Gangopadhyay S, Redner P, Kapoor D, Nicolich S (2006) Nanoenergetic composite of mesoporous iron oxide and aluminum nanoparticles. J Energ Mater 24:341–360
Moore K, Pantoya ML (2006) Combustion of environmentally altered molybdenum trioxide nanocomposites. Propell Explos Pyrot 31:182–187
Moore DS, Son SF, Asay BW (2004) Time-resolved spectral emission of deflagrating nano-Al and nano-MoO3 metastable interstitial composites. Propell Explos Pyrot 29:106–111
Moore K, Pantoya ML, Son SF (2007) Combustion behaviors resulting from bimodal aluminium size distributions in thermites. J Propul Power 23:181–185
Ohkura Y, Liu SY, Rao PM, Zheng X (2011) Synthesis and ignition of energetic CuO/Al core/shell nanowires. Proc Combust Inst 3:1909–1915
Pantoya ML, Granier JJ (2005) Combustion behavior of highly energetic thermites: nano versus micron composites. Propell Explos Pyrotech 30:53–62

Park K, Lee D, Rai A, Mukherjee D, Zachariah MR (2005) Size-resolved kinetic measurements of aluminium nanoparticle oxidation with single particle mass spectrometry. J Phys Chem B 109:7290–7299

Perry WL, Smith BL, Bulian CJ, Busse JR, Macomber CS, Dye RC, Son SF (2004) Nano-scale tungsten oxides for metastable intermolecular composites. Propell Explos Pyrot 29:99–105

Petrantoni M, Rossi C, Salvagnac L, Conédéra V, Estève A, Tenailleau C, Alphonse P, Chabal YJ (2010a) Multilayered Al/CuO thermite formation by reactive magnetron sputtering: nano versus micro. J Appl Phy 108:084323

Petrantoni M, Bahrami M, Salvagnac L, Conédéra V, Rossi C, Alphonse P, Tenailleau C (2010b) Nanoenergetics on a chip: technology and application for micro ignition in safe arm and fire systems. In: Proceedings of power MEMS, vol 39

Picard YN, Joel PMD, Friedmann TA, Steven MY, David PA (2008) Nanosecond laser induced ignition thresholds and reaction velocities of energetic bimetallic nanolaminates. Appl Phys Lett 93:104104

Pichot V, Comet M, Miesch J, Spitzer D (2015) Nanodiamond for tuning the properties of energetic composites. J Hazard Mater 300:194–201

Pierre AC (2013) Introduction to sol-gel processing, vol 1. Springer Science & Business Media

Plantier B, Pantoya ML, Gash AE (2005) Combustion wave speeds of nanocomposite Al/Fe_2O_3: the effects of Fe_2O_3 particle synthesis technique. Combust Flame 140:299–309

Prakash A, McCormick AV, Zachariah MR (2004) Aero-sol-gel synthesis of nanoporous iron-oxide particles: a potential oxidizer for nanoenergetic materials. Chem Mater 16:1466–1471

Prakash A, McCormick AV, Zachariah MR (2005) Tuning the reactivity of energetic nanoparticles by creation of a core-shell nanostructure. Nano Lett 5:1357–1360

Prentice D, Pantoya ML, Gash AE (2006) Combustion wave speeds of sol-gel-synthesized tungsten trioxide and nano-aluminum: the effect of impurities on flame propagation. Energ Fuels 20:2370–2376

Puszynski JA (2004) Recent advances and initiatives in the field of nanotechnology. In: Proceedings of 31st international pyrotechnic seminar, Fort Collins, Colorado, USA, pp 233–240

Puszynski JA, Bulian CJ, Swiatkiewicz JJ (2007) Processing and ignition characteristics of aluminium-bismuth trioxide nanothermite system. J Propul Power 23:698–706

Qin L, Gong T, Hao H, Wang K, Feng H (2013) Core–shell-structured nanothermites synthesized by atomic layer deposition. J Nanopart Res 15:1–15

Rossi C, Zhang K, Esteve D, Alphonse P, Tailhades P, Vahlas C (2007) Nanoenergetic materials for MEMS: a review. IEEE/ASME J Microelectromech Syst 16:919–931

Sarathi R, Sindhu TK, Chakravarthy SR (2007) Generation of nanoaluminium powder through wire explosion process and its characterization. Mater Charact 58:148–155

Shen J, Qiao Z, Wang J, Zhang K, Li R, Nie F, Yang G (2014) Pressure loss and compensation in the combustion process of Al–CuO nanoenergetics on a microheater chip. Combust Flame 161:2975–2981

Shende R, Subramanian S, Hasan S, Apperson S, Thiruvengadathan R, Gangopadhyay K, Gangopadhyay S (2008) Nanoenergetic composites of CuO nanorods, nanowires, and al-nanoparticles. Propell Explos Pyrot 33:122–130

Shin MS, Kim JK, Kim W, Moraes CAM, Kim HS, Koo KK (2012) Reaction characteristics of Al/Fe_2O_3 nanocomposites. J IndEngChem18:1768–1773

Son SF, Busse JR, Asay BW, Peterson PD, Mang JT, Bockmon B, Pantoya M (2002) Propagation studies of metastable intermolecular composites (MIC). No. LA-UR-02–2954. Los Alamos National Laboratory

Son SF, Asay BW, Foley TJ, Yetter RA, Wu MH, Rish GA (2007) Combustion of nanoscale Al/MoO_3 thermite in microchannels. J Propul Power 23:715–721

Srivastava DN, Perkas N, Gedanken A, Felner I (2002) Sonochemical synthesis of mesoporous iron oxide and accounts of its magnetic and catalytic properties. J Phys Chem B 106:1878–1883

Staley CS, Morris CJ, Thiruvengadathan R, Apperson SJ, Gangopadhyay K, Gangopadhyay S (2011) Silicon-based bridge wire micro-chip initiators for bismuth oxide–aluminum nanothermite. J Micromech Microeng 21:115015

Staley CS, Raymond KE, Thiruvengadathan R, Apperson SJ, Gangopadhyay K, Swaszek SM, Taylor RJ, Gangopadhyay S (2013) Fast-impulse nanothermite solid-propellant miniaturized thrusters. J Propul Power 29:1400–1409

Staley CS, Raymond KE, Thiruvengadathan R, Apperson SJ, Gangopadhyay K, Swaszek SM, Taylor RJ, Gangopadhyay S (2014) Effect of nitrocellulose gasifying binder on thrust performance and high-g launch tolerance of miniaturized nanothermite thrusters. Propell Explos Pyrot 39:374–382

Sullivan KT, Piekiel NW, Chowdhury S et al (2010) Ignition and combustion characteristics of nanoscale Al/AgIO3: a potential energetic biocidal system. Combust Sci Technol 183:285–302

Sun J, Simon SL (2007) The melting behavior of aluminum nanoparticles. Thermochim Acta 463:32–40

Tasker DG, Asay BW, King JC, Sanders VE, Son SF (2006) Dynamic measurements of electrical conductivity in metastable intermolecular composites. J Appl Phy 99:023705

Taton G, Lagrange D, Conedera V, Renaud L, Rossi C (2013) Micro-chip initiator realized by integrating Al/CuO multilayer nanothermite on polymeric membrane. J Micromech Microeng 23:105009

Tepper F (2000) Nanosize powders produced by electro-explosion of wire and their potential applications. Powder Metall 43:320–322

Thiruvengadathan R, Chung SW, Basuray S, Balasubramanian B, Staley CS, Gangopadhyay K, Gangopadhyay S (2014) A versatile self-assembly approach toward high performance nanoenergetic composite using functionalized graphene. Langmuir 30:6556–6564

Thiruvengadathan R, Staley C, Geeson JM, Chung S, Raymond KE, Gangopadhyay K, Gangopadhyay S (2015) Enhanced combustion characteristics of bismuth trioxide-aluminum nanocomposites prepared through graphene oxide directed self-assembly. Propell Explos Pyrot 40(5):729–734

Tillotson TM, Gash AE, Simpson RL, Hrubesh LW, Satcher JH, Poco JF (2001) Nanostructured energetic materials using sol–gel methodologies. J Non-Cryst Solids 285:338–345

Valliappan S, Swiatkiewicz J, Puszynski JA (2005) Reactivity of aluminum nanopowders with metal oxides. Powder Technol 156:164–169

Vassiliou JK, Mehrotra V, Russell MW, Giannelis EP, McMichael RD, Shull RD, Ziolo RF (1993) Magnetic and optical properties of γ-Fe2O3 nanocrystals. J Appl Phys 73:5109–5116

Wang J, Besnoin E, Duckham A, Spey SJ, Reiss ME, Knio OM, Weihs TP (2004) Joining of stainless-steel specimens with nanostructured Al/Ni foils. J App Phy 95:248–256

Wang H, Jian G, Egan GC, Zachariah MR (2014) Assembly and reactive properties of Al/CuO based nanothermite microparticle. Combust Flame 161:2203–2208

Wang J, Qiao Z, Shen J, Li R, Yang Y, Yang G (2015a) Large-scale synthesis of a porous Co3O4 nanostructure and its application in metastable intermolecular composites. Propell Explos Pyrot 40:514–517

Wang H, Jian G, Zhou W, De Lisio JB, Lee VT, Zachariah MR (2015b) Metal iodate-based energetic composites and their combustion and biocidal performance. ACS Appl Mater Interfaces 7:17363–17370

Watson KW, Pantoya ML, Levitas VI (2008) Fast reactions with nano-and micrometer aluminum: a study on oxidation versus fluorination. Combust Flame 155:619–634

Wen JZ, Ringuette S, Bohlouli-Zanjani G, Hu A, Nguyen NH, Persic J, Petre CF, Zhou YN (2013) Characterization of thermochemical properties of Al nanoparticle and NiO nanowire composites. Nanoscale Res Lett 8:1–9

Yan N, Qin L, Hao H, Hui L, Zhao F, Feng H (2017) Iron oxide/aluminum/graphene energetic nanocomposites synthesized by atomic layer deposition: enhanced energy release and reduced electrostatic ignition hazard. Appl Surf Sci 408:51–59

Yang Y, Xu D, Zhang K (2012) Effect of nanostructures on the exothermic reaction and ignition of Al/CuOx based energetic materials. J Mater Sci 47:1296–1305

Yarrington CD, Son SF, Foley TJ, Obrey SJ, Pacheco AN (2011) Nano aluminum energetics: the effect of synthesis method on morphology and combustion performance. Propell Explos Pyrot 36:551–557

Yavorovsky NA (1995) Method of production of highly dispersed powders of inorganic materials. Patent of Russian Federation 2048277

Yi Z, Ang Q, Li N, Shan C, Li Y, Zhang L, Zhu S (2018) Sulfate-based nanothermite: a "green" substitute of primary explosive containing lead. ACS Sustain Chem Eng (accepted)

Yin Y, Li X, Shu Y, Guo X, Zhu Y, Huang X, Bao H, Xu K (2017a) Highly-reactive Al/CuO nanoenergetic materials with a tubular structure. Mater Des 5:104–110

Yin Y, Li X, Shu Y, Guo X, Bao H, Li W, Zhu Y, Li Y, Huang X (2017b) Fabrication of electrophoretically deposited, self-assembled three-dimensional porous Al/CuO nanothermite films for highly enhanced energy output. Mater Chem Phy 194:182–187

Zakiyyan N, Wang A, Thiruvengadathan R, Staley C, Mathai J, Gangopadhyay K, Maschmann MR, Gangopadhyay S (2018) Combustion of aluminum nanoparticles and exfoliated 2D molybdenum trioxide composites. Combust Flame 187:1–10

Zamkov MA, Conner RW, Dlott DD (2007) Ultrafast chemistry of nanoenergetic materials studied by time-resolved infrared spectroscopy: aluminum nanoparticles in Teflon. J Phys Chem C 111:10278–10284

Zarko VE, Gromov AA (eds) (2016) Energetic nanomaterials: synthesis, characterization, and application. Elsevier, Amsterdam

Zhang D, Li X (2015) Fabrication and Kinetics Study of Nano-Al/NiO Thermite Film by Electrophoretic Deposition. J Phys Chem A 119:4688–4694

Zhang K, Rossi C, Rodriguez GAA, Tenailleau C, Alphonse P (2007) Development of a nano-Al/CuO based energetic material on silicon substrate. Appl Phys Lett 91:3117

Zhang W, Yin B, Shen R, Ye J, Thomas JA, Chao Y (2013) Significantly enhanced energy output from 3D ordered macroporous structured Fe_2O_3/Al nanothermite film. ACS Appl Mater Interfaces 5:239–242

Zhou L, Piekiel N, Chowdhury S, Zachariah MR (2010) Time-resolved mass spectrometry of the xothermic reaction between nanoaluminum and metal oxides: the role of oxygen release. J Phys Chem 114:14269–14275

Zhou X, Wang Y, Cheng Z, Ke X, Jiang W (2017) Facile preparation and energetic characteristics of core-shell Al/CuO metastable intermolecular composite thin film on a silicon substrate. Chem Eng J 328:585–590

Chapter 4
Nano-energetic Materials for Defense Application

Sudarsana Jena and Ankur Gupta

Abstract Energetic materials are the reactive materials containing fuels and oxidizers that can liberate chemical energy preserved in their molecular structure. Nano-energetic materials have found to be the potential sources for extremely high heat release rates, tailored burning rate, extraordinary combustion efficiency, and reduced sensitivity. These materials play a vital role in defense applications as a recent advancement in emerging areas such as manufacturing of explosives, solid and liquid propellants, rocket propelling, advanced gun propellant materials. Considering the immense scope of these functional materials, this chapter focuses to cover the fundamental aspect of energetic materials, description of contemporary reported literature on design and synthesis of nano-energetic materials and their significance for microscale applications in the defense sector.

Keywords Nano-energetic materials · Tailored explosives · High surface area Microreaction technology

4.1 Introduction to Energetic Materials

Energetic materials release chemical energy stored in their molecular structures which are reactive by nature. These materials will emit energy upon external stimulations, such as heat, shock, or electrical current for a short duration of time. Thus, such materials contain both fuel and oxidizers. Based on how fuel and oxidizers are linked whether chemically or physically, energetic material is classified as homogeneous and heterogeneous. Energetic materials are basically used in defense application, but now a day's energetic materials are used more in fields of

S. Jena · A. Gupta (✉)
School of Mechanical Sciences, Indian Institute of Technology Bhubaneswar, Bhubaneswar 752050, Odisha, India
e-mail: ankurgupta@iitbbs.ac.in

S. Jena
Defence Research, and Development Organization (DRDO), Bhubaneswar, India

S. Bhattacharya et al. (eds.), *Nano-Energetic Materials*, Energy, Environment and Sustainability, https://doi.org/10.1007/978-981-13-3269-2_4

civil engineering and space exploration as well, such as construction, mining, and rocket propelling. Though there is a continuous development of such materials, still people are seeking a safer, more powerful, compact, and more cost-effective energetic materials. The prime focus to improve the engineering approach of energetic materials is to understand the chemistry involved in it. However, the fast reaction rate and extreme reaction conditions make the direct experimental measurement difficult and unsafe. Therefore, developing the knowledge through computer simulation provides a safer and convenient way to study the chemistry of energetic material.

Energetic materials are broadly classified based on their applications, like explosives, propellants, and pyrotechnics.

Explosives To meet the high energy requirement, explosives are extensively used which can enlarge very much in volume to produce force in the timescale of microseconds. Rapid reaction is required to perform to achieve high power output as known as detonate. Based on the propagation rate and their rate-determining step, different phenomena like deflagration, detonation, and usual fuel combustion are notable. In case of usual fuel combustion, the rate of reaction is limited by diffusion of reactive species; this process is comparatively slow and leading to a small propagation rate. The fuel and oxidizer are pre-mixed in the deflagration process, so the reactive species is no longer the rate-determining step during diffusion process. In fact, heat transfer plays an important role to control the propagation of reaction rate which results in its quicker rate than usual fuel combustion. A shockwave propagates at the detonation of energetic materials. The materials at wavefront are highly compressed, and the temperature is raised, which causes an exothermic chemical reaction and propagates after the wavefront. Because of this exothermic reaction, the pressure and the temperature are raised above the point, that is, the condition before the passage of shock wave. As a result, energy is generated to sustain the propagation of shock wave. Therefore, the detonation is in the speed of shock wave is supersonic. In contrast to the cases of deflagration and regular fuel combustion, which are subsonic, several parameters can be used to characterize explosives, as described below (Millar 2011; Singh et al. 2006; Tillotson et al. 2001, 1998; Proud 2014).

i. **Sensitivity**: This property tells, how easily it can be set off by the external stimuli, such as heat, mechanical shock, spark, friction, and impact. Based on sensitivity, explosives are classified as primary and secondary explosives. Primary explosives are highly sensitive which can easily undergo the deflagration to detonation transition (DDT). Secondary explosives are called high explosives which are less sensitive, but usually more powerful than primary explosives. This will be more effective, if both the explosives are to be placed adjacent to each other, i.e., a small amount of primary explosive placed adjacent to a large amount of secondary explosive. This is called three explosives trains. The fast deflagration to detonation transition (DDT) of primary explosive helps to enhance the early non-explosive impulse to shockwave, and this will further detonate the secondary explosive.

ii. **The heat of explosion (*Q*)**: The heat discharged from the decomposition of explosive at the time of the explosion is represented by heat of the explosion. This amount can be well estimated as the difference of the heat generated from the combustion fuels and explosive itself. For obtaining higher explosive power, large heat formation is preferred for explosives. This can be found out by multiplying the product of heat formation and the volume of gas used.
iii. **The velocity of detonation (V_D)**: This parameter tells how fast the detonation wave propagates, by which the rate of energy produced from explosives can be controlled. The detonation velocity is proportional to the density of explosives which can be correlated with the detonation pressure.

To maximize the shattering power for certain applications like rock cleaving and grenade, it is required for explosives to arrive at its peak pressure as fast as possible and detonation velocity should also be high. The shattering power can be determined by brisance, and it can be found out by multiplying three parameters, i.e., specific energy, detonation velocity, and loading density. The maximum detonation pressure is produced through this explosion power multiplied by the volume of detonation gases. The practicality of explosives is determined by many factors. Firstly, it is required to perform a quick deflagration to detonation transition in order to produce shock wave to start the secondary explosive detonation. Though explosives are very high sensitivities, still the chemical and thermal stability of primary explosives is important in order to have a longer shelf life. For example, heavy metal salts used as a primary explosive in form of mercury fulminate, lead azide, and lead styphnate. Combustion products of such primary explosives are very hazardous to health and cause the environment pollution. This leads to exploring for metal-free primary explosives. But in the case of secondary explosives, it is very imperative to have low sensitivity and long-term stability. As a result, it will be easier to store and handle these explosives in huge amounts. Performance requirements such as high detonation velocity and large heat of explosion have to be insured separately. The cost of the production is one of the important aspects to be looked into while determining the practicality of explosives.

Propellants Propellants are not used to detonate like explosives do, but these are used as a combust in a controlled way. Specific impulse (I_{sp}) plays an important role to determine the performance factor of propellants. Specific impulse is defined as the impulse generated by the consumption of one unit mass of propellants (impulse = force × time, or mass × velocity) (Klapötke 2017). I_{sp} is calculated as per unit mass. As the thrust comes from the only exhaust gas, it does not depend on the burning rate of propellants. So I_{sp} is a material-related parameter.

Propellants are mainly classified as a liquid and solid propellant. Solid propellants are produced by mixture of reductant powder like C and Al with oxidant like nitrate and perchlorate salts. In case of the explosives such as RDX or HMX, if there is no shock wave generated during the combustion process to begin detonation, then these explosives can also be used as propellants. Solid propellants are used in rocket motors and inter-continental ballistic missile (ICBM) in order to have a very high propellant fraction in weight. As there is no liquid pump or cryogenic tanks are used in rocket or ICBM missile, hence solid propellants are more reliable to operate in such cases. The only issue is in control at the starting of the combustion of solid propellants in practical usage.

Liquid propellants are classified into two categories; one is called monopropellants, and other is bipropellants. Hydrogen peroxide and hydrazine are the most common types of monopropellants. These monopropellants are able to decompose catalytically to produce heat and gas products. The specific impulse (I_{sp}) of such monopropellants is not more as compared to bipropellants, and this is because they have very less change in heat (ΔH). Therefore, monopropellants are used in very limited areas where small loading is required. In the case of bipropellants, oxidizer and fuel are injected and mixed in a combustion chamber. The most common type of bipropellants is hypergolic propellants. Such type of liquid propellants forms pairs of oxidizer and fuel. It ignites impulsively if both fuel and oxidizer are mixed. This is used for simplification of the ignition system in rocket thrusters, where a variable and intermittent thrust is required in its propulsion system for easy controlling. Ignition delay is the most essential parameter of hypergolic propellant apart from its I_{sp}. Ignition delay is defined as the time interval between two liquid surfaces in contact with a flame formation. If the ignition delay is shorter, it implies faster response and motion can be controlled easily. Normally monomethylhydrazine (MMH) and unsymmetrical dimethylhydrazine (UDMH) are commonly used as hydrazine fuels, and these are mixed with an oxidizer like nitrogen tetroxide (NTO) or HNO_3. Fuels like MMH or UDMH are derivatives of hydrazine which are carcinogens, and this can be replaced by alkyl amines. These alkyl amines groups of fuels are safer and suitable for such applications. Energy propagation and molecule activation are two important processes in the initiation of energetic material at its atomistic level, which create a positive feedback loop. For a cold and unreacted material, molecules are driven by external stimuli in a ground state over barriers. Initially, the external stimuli are not strong, due to which, only low-lying reaction channels are activated. Once the reaction occurs, the local temperature increased due to some of them are exothermic by the reaction. This heat energy slowly propagates to adjacent molecules due to the repetition of this process. In this process, there are mainly the following three factors that affect the rate of the propagation of reaction;

i. The height of the barrier of reactions: This controls the accessibility of reaction channels,
ii. Exothermicity of reactions: This determines the amount of heat to be produced to increase the local temperature.
iii. The efficiency of energy transfer to adjacent molecules.

The energy can be transferred between two neighboring molecules by means of two ways; one is the coupling of vibrations between neighboring molecules, and another one is the change of momentum carried out by ballistic gas due to exothermic reactions. The above three factors decide the sensitivity of energetic materials. The sensitivity of energetic materials will be highly exothermic channels

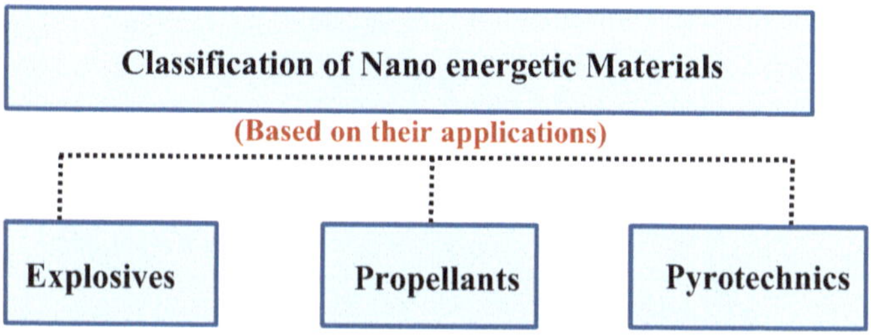

Fig. 4.1 Classification of nano-energetic materials based on their applications

with low barrier height. Multi-paradigm multi-scale simulations can be used to characterize the overall combustion process of energetic materials. The reaction methods such as exothermicity and height of the barrier are started from the unreacted molecule and can be constructed with first principle based methods. A study on the multi-molecular process such as energy transfer between molecules and reactions occurring in the condensed phase can be done through simulation of molecular dynamics. Based on this simulation results, a combustion model containing exothermicity reactions for all species can be constructed. When this model combines with continuum fluid dynamics (CFD), then it describes the time evolution of all species and complete engine operation can be simulated with its rate of diffusion and heat transfer.

Pyrotechnics This type of energetic materials is capable of enduring self-sustained and self-contained exothermic chemical reactions to produce heat, light, sound, gas, and smoke. This includes not only fireworks manufacturing but different items like explosive bolts and fasteners, components of the automotive air bag, quarrying, and demolition, gas pressure blasting in mining, safety matches, and oxygen candles. (Suceska 2012; Klapötke 2017; Tichapondwa et al. 2012; Schönhuber et al. 2011) (Fig. 4.1).

4.2 Types of Various Energetic Materials and Their Syntheses

Energetic heterocyclic compound and its synthesis are becoming familiar day by day. Generally, heterocyclic compounds have larger density and oxygen balance which is an imperative parameter for the improvement of performance of energetic material than that of carbocyclic analogues. The nitrogen content in heterocycles also has a propensity to have elevated heats of reaction. Below is the list of various energetic materials explored for microscale applications.

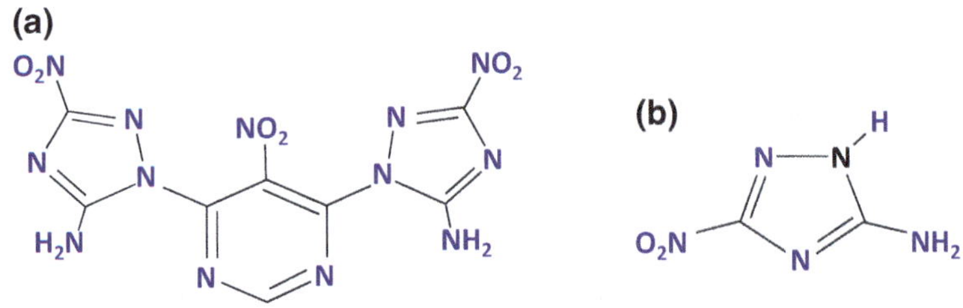

Fig. 4.2 **a** DANTNP structure and **b** ANTA structure

4.2.1 Nitrotriazoles

Nitrotriazole as one of the energetic materials has become interesting material and is able to draw attention from the past one decade (Ou et al. 1994). DANTNP [IUPAC nomenclature: derivative of amino-nitro-triazolyl-nitro-pyrimidine] is the most studied nitrotriazole explosive and was stated by Wartenberg et al. (1995). The synthesis of DANTNP is one of the novel emergences of energetic material. Method stated by Delpuech et al. (1981) is used to predict the sensitivity of the materials. DANTNP is not considered as a sensitive explosive. It has a crystal density = 1.865 g/cc, change in heat = 431 kJ/mol, melting point = 603 K, (see Fig. 4.2a).

First time Pevzner et al. (1979) prepared the ANTA of composition [IUPAC nomenclature: amino-nitro-triazole]. Subsequently, an advanced synthesis of ANTA is proposed by Lee and his team (1991), and also its synthesis is developed by Simpson et al. (1994) with little modification of scale-up as shown in Fig. 4.2b.

4.2.2 Ammonium Dinitramide

The synthesis of ammonium dinitramide is first reported in 1997 by Bottaro et al. (1997). This oxidizer is very strong which has a prospective use as an ingredient in cationic phase transfer agent and rocket propellant as well. A considerable amount of dinitramide anion salts has been synthesized, together with the hydroxylammonium ($HONH_2$), alkali salts, guanidinium (CH_5N_3 group), aminoguanidinium, cubane-1, biguanidinium salts, etc.

When pKã—5 is added with ADN, then it becomes very strong acid, and it is stable between pH value 0–15, but in concentrated acid, it slowly decomposes. The specific density of ADN is 1.801 g/cc, and the melting point is 365 K (Martin et al. 1997).

4.2.3 Pyrazoles

Pyrazole is one of the most familiar energetic materials for last few years. Synthesis of pyrazole ring containing energetic materials is studied by molecular simulations model and thermochemical codes. Modern tools are used to design the molecules of pyrazoles, and further, it is synthesized and characterized as well in additional stages. Normally, ring systems and caged molecular compounds have higher densities than that of single ring analogues. MOLPAK program is used to predict the density of the substances, whereas Hartree–Fock calculation is used to predict the heat formation rate with necessary corrections that is used to be calibrated separately for both energetic and non-energetic materials.

Densities of different ring systems have been compared, and it is found that five members of two fused rings have higher density than that of two fused four-member rings or six-member rings. Heterocyclic ring molecules are also chosen to improve the energy possess by molecule. Ring system has un-soaked bonding which can be improved by controlling the energy possess and density parameter of energetic materials.

Lawrence Livermore National Laboratory has developed the pyrazoles (Fig. 4.3). Chapman-Jouguet (C-J) state that release energy is more reliable for the performance of detonation than the velocity of detonation. For rapid metal acceleration, the energy release is said to be good if the relative volume of volume expansion is 2.2. Based on experimental results, the metal acceleration energy of LLM-119 compound is predicted to be 1.1 times more than HMX; similarly, a compound like LLM-121 and DNPP were also predicted to be 1.4 and 0.95 times more than HMX, respectively. Crystal density of X-ray crystallography is determined as 1.865 g/cc (Dremin and Shvedov 1964).

DNPP is considered as a thermally stable energetic material but comparatively insensitive to shock which is DNPP as a smart alternative to different other compounds such as TATB, RDX, and TNT. Acidic protons of DNPP with pKa = 6 and 3 may have some problems in some of the applications.

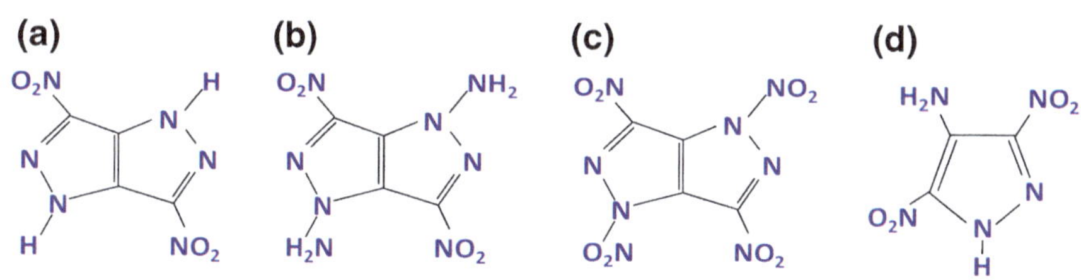

Fig. 4.3 Molecular structure of **a** DNPP, **b** LLM-119, **c** LLM-121, **d** LLM-116 compound

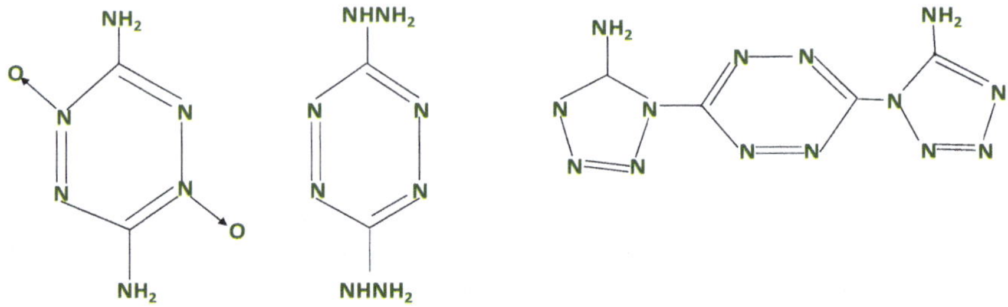

LAX112 DHT 2

Fig. 4.4 Molecular structure of LAX-112, DHT, and compound 2

4.2.4 *Tetrazines*

The synthesis of LAX 112 is first reported by Coburn et al. (1993). LAX-112 is an energetic material of a cycloaromatic type which does not contain oxidizer of the nitro group. LAX-112 compound has a greater value of specific density and also some detonation velocity. This sensitive energetic material is made up by the use of more insistent oxidation step as a composition of [3-amino-6- nitro-1,2,4, 5-tetrazine-2,4-dioxide] that decomposes at about a temperature of 383 K (Coburn et al. 1993) (see Fig. 4.4).

The research over this energetic material is further extended by Chavez and Hiskey (1999). He stated that the explosive based on 1,2,4,5-tetrazine synthesis and can be used as an interesting ingredient for pyrotechnic and propellant due to its lower carbon content. For example, DHT of composition 3,6-dihydrazino-1,2,4,5-tetrazine (Chavez and Hiskey 1998) is an energetic fuel having a density of 1.61 g/cc, and delta heat value (heat of combustion) is 536 kJ/ mol. DHT can be again structured as 3,6-bis-1,2,4,5-tetrazine, with a melting point of 537 K and change in heat (Δh) as measured is 883 kJ/mol.

4.2.5 *Furazans*

Coburn (1968) reported that the synthesis of DAF (structured as 3,4-Diaminofurazan) is a very important contribution to produce a series of energetic materials containing furazan, which is used as a vital ingredient in propellant and explosives making. Oxidation characteristics of DAF are further analyzed by Solodyuk et al. (1981). As per their analysis, the oxidation of DAF can yield ANF [amino-nitrofurazone]; DAAF [diamino–azoxyfurazan]; and also DAAzF [-diamino-azofurazan]. Again, synthesis of DAAF is scaled up by Chavez et al. (2000), and different measurements were performed to find out its explosive

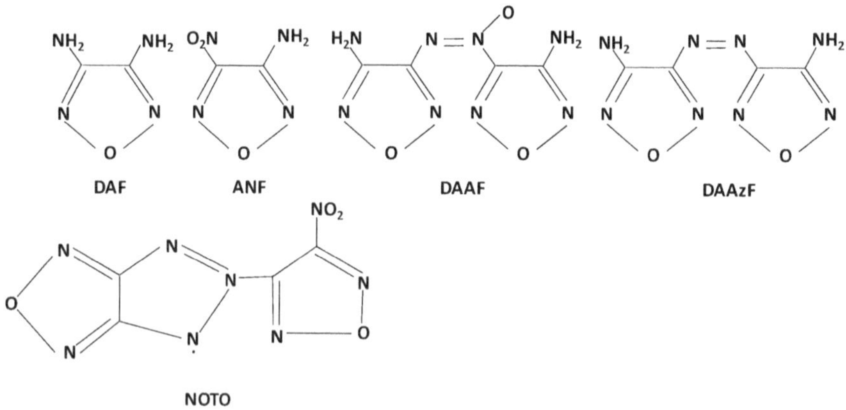

Fig. 4.5 Molecular structure of DAF, ANF, DAAF, DAAzF, and NOTO

properties. DAAF has more energy contained than TATB [triamino-trinitrobenzene]. The crystal density of DAAF was found to be 1.747 g/cc, and its change in heat (Δh) is 444 kJ/mol. DAAF is very insensitive to shock. Reactivity rate of nitro groups of DNAzF, DNAF, and DNF to nucleophiles was exploited by Sheremetev and his team (1999), and its synthesis is characterized of a high value of 3-substituted-4-nitrofurazone derivatives (see Fig. 4.5).

4.2.6 Pyridines and Pyrazines

The complexity involved in the synthesis of nitrated heteroaromatic systems depends on the electron deficiency. Such case of fabricating electrophilic aromatic substitution is difficult. Nitration may continue more rapidly if the substituent that donating free electrons is added to the heteroaromatic ring. Some of the examples are stated in next paragraphs, where pyrazine and pyridine predecessors are nitrated to give up the required alternative heterocycles. The idea to enhance not only the density but also the thermal stability of such compounds by the employ of an alternative amino and nitro groups array is also illustrated in following examples.

Synthesis of LLM-105 [diamino-dinitropyrazine-oxide] as reported by Pagoria et al. (1998) is obtained by the process oxidation of ANPZ [diamino-dinitropyrazine] (Fried et al. 2001). The density of LLM-105 is 1.913 g/cc (Pagoria et al. 2002), and the temperature of decomposition is 627 K. To enhance the oxygen balance and density of heterocyclic system, the exchange of tertiary amines at its subsequent N-oxides is also another method used frequently. In case of N-oxide, N–O bond is comparatively strong and contains considerable dual-bond character due to ¼ rear bonding by the single pair of electrons in oxygen (Albini 1991). Heterocyclic N-oxide formation leads to amplify in aromaticity of

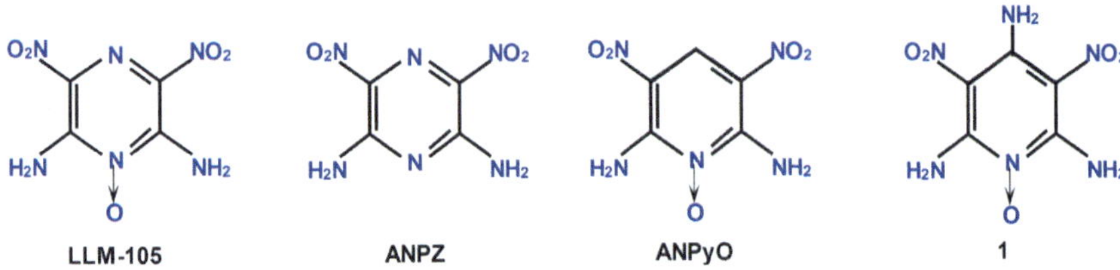

Fig. 4.6 Molecular structure of LLM-105, ANPZ, ANPyO, and compound 1

the heterocyclic ring and also the distribution of charge of the heterocyclic ring alters, which may stabilize the ring system (Albini 1991). Crystal density of ANPZ and LLM-105 is about 1.84 g/cc (Pagoria et al. 2002) and 1.913 g/cc, respectively. Therefore, the formation of N-oxide not only alters the number of oxygen molecules but also it allocates better crystal packing.

The synthesis of ANPyO [diamino-dinitropyridine-oxide] is reported by Ritter and Licht (1995). The specific density of ANPyO is 1.878 g/cc, and its melting point is more than 613 K at decomposition stage. This work has further extended by Hollins et al. (1995, 1996). Compound-1, which is not a sensitive energetic material (density = 1.876 g/cc) shows a melting point temperature of about 581 K with decomposition (see Fig. 4.6).

4.3 Methods of Producing Energetic Material

The conventional energetic materials such as composite or monomolecular substances are produced by the simple mixing process. Method of the stoichiometry of chemical reactants is used to estimate their performances. Basically, such energetic material is produced by the following methods;

i. The monomolecular energetic material is produced by combining oxidizer and fuel components into a single molecule, for example, nitroglycerine, nitrocellulose, and trinitrotoluene (Sanders et al. 2007).
ii. The composite energetic material is produced by mixing different oxidizer powders such as ammonium nitrate or potassium perchlorate with the fuel powders such as sulfur, carbon. Black powder is an example of composite energetic material (Sanders et al. 2007).

Composite energetic materials exhibit higher energy density as compared to the monomolecular energetic material, but the composite materials have slower energy release rates than that of monomolecular materials due to limited mass transport rate by the reactant granulometry. Since last few decades, these materials are used in the various fields such as defense, mining operation, and demolition applications purpose as these materials are the most useful medium of inbuilt energy to produce power, heat, and gas. Civil industry and scientific community are also using such materials since last two decades. Energetic materials have a major impact on micro-/nano-energetic field. The major application of microenergetic materials includes microthrust (Rossi et al. 2006; Youngner et al. 2000), micro initiation (Troianello 2001; Laucht et al. 2004), for actuation purpose using gas or moving fluid injection (Rossi and Esteve 1997; Rossi et al. 1999), gases used in chemical reaction (Ding et al. 1987; Vasylkiv et al. 2006), gases used in welding and heating purpose (Stewart 2005), and also for switching (Pennarun et al. 2005). Research has been made in various ways to adopt energetic materials to useful applications such as to produce gas, heat, or chemical compound to a useful form. However, the actual challenge lies to address the compatibility issues while emerging energetic materials for MEMS applications. Deposition of a thin film made of energetic materials has to be done below 250 °C. Freestanding microstructure technique is also used in some cases to deposit energetic materials film on a substrate, where stress needs to be controlled or reduced without using heat treatment process. There are some familiar processing techniques such as screen printing (Ismail et al. 2001), PVD with liftoff process (Li et al. 2011; Gao et al. 2011), or chemical reactions (Mattox 2010) which are frequently used to deposit energetic materials on a substrate of silicon. One of the most challenging issues is to minimize the heat loss in MEMS-based applications.

Although there are major improvements which have been made using chemical formulation by combining different substances such as chemical molecules or compounds (Pierson 1999) in the traditional composite and monomolecular energetic materials. But for microscale application, this technique is comparatively slow reaction rate. For example, in the case of HMX deflagration, it has to quench in steel tubes of different sizes of diameters at atmospheric pressure. Another example is the combustion of GAP-AP, which can be quenched into a glass tube of diameter of 1.4 mm (Sathiyanathan et al. 2011) at atmospheric pressure. To fulfill the present energy demands in this militarization trends, reactive nano-energetic materials are more effective than the traditional ones.

Since last few decades, there are major hard works which have been made to bring in metal powder with nano- and microsized elements into conventional energetic materials with faster combustion velocities. However, there is some difficulty to handle this metal powder particles and their amalgamation into obtainable formulations as stated by some researcher. Apart from conventional energetic materials, a combination of inorganic energetic materials, composite, and metal oxidizer produces an effective ingredient called metastable intermolecular composite (MIC) that are mostly used. Such materials experience a rapid solid-state redox reaction with very exothermic nature nearly about twice that of monomolecular energetic materials. The output of such research stated mainly empirically that the properties of combustion and initiation of energetic materials are sturdily affected by their microscopical properties. The particle size reduction to a nanoscale-sized may lead to a diminution of mass transport rate. Ultimately, it would enhance the rate of burning because of which nanoscale energetic materials are more useful and became an alternative to monomolecular structures (Badgujar et al. 2008; Rossi and Estève 2005; Tanaka et al. 2003; Rugunanan and Brown 1993; Chung et al. 2009; Son and Asay 2001).

4.4 Importance of Nano-energetic Materials

The present advances in nano-science are providing the capabilities to fabricate a material with a complex molecular structure. In a comparison with microparticles, there is a significantly higher surface area to volume ratio in the case of nanoscale particles, where a mixture of solid particles is closely contacted (Martirosyan et al. 2013). In general, conventional energetic materials are prepared of particles between 1 and 100 μm, which is limited to the boundary on maximum reaction rates due to its spatial scale and therefore has limited applications. Energetic materials with homogeneity by nature have reactants on the same molecule. Nanoscale energetic materials are made to enhance the energy release rate with better control manner, reliability, reduced sensitivity, safety, etc. The nanotechnology has precisely defined the scale for energetic materials on the order of 100 nm or even smaller. It is observed that the change in scale has a significant impact on the ignition or reactive properties. Though there is not much change observe in melting point at this scale; there is a dramatic change in the specific surface area of ignition and combustion behavior (Son et al. 2007).

The nano-energetic material provides greater energy density fuel elements to propellants, fuels, and explosives with greater concentrations. It is also used as modifiers to burning rate with low concentrations. Gelling agents can be easily applied to MEMS-based products and other applications in order to reduce the hazard effects. The nano-energetic material has unique properties like it enhanced the specific surface area, reactivity, catalytic activity, catalytic activity, and heats of reaction. Similarly, it decreased the melting temperatures and heat of fusion (Eichhorn et al. 2012).

4.5 Nano-energetic Materials (NEMs) for Microscale Application

The initial approach is adopted using traditional energetic materials to produce combustion in microscale level by introducing Al nanoparticles into conventional explosives or propellants. The process of doping is used to mix up composite propellant with Al nanoparticles to produce microscale combustion. With the progress of nanotechnology, researcher found a way to mix and synthesize the inorganic nano-energetic materials made of metal oxide as an oxidizer and Al as fuel particles to produce metastable intermolecular composite (MIC). There are basically two types of materials used which are discussed below (Kondo et al. 2004; Yarrington et al. 2010).

i. *Propellant made of aluminum nanoparticle doping process.*

Al is mostly used metal in industries, and also, cost is relatively low. Al powder is mostly used as a doping agent due to the following properties;

(a) In Al powder, each particle is covered with a thin oxide layer to prevent spontaneous combustion
(b) Nanoparticles of size 50–120 nm can be easily produced form, Al. However, nano-Al is commercially available.
(c) Due to the high thermal conductivity of Al, the material's reactive power has been improved by increasing the velocity of combustion.

The effect of introducing microsized Sb particles mixed with $KMnO_4$ was investigated by Brown et al. (1998), and he found that the burning rate will be increased by four times if the particle size is reduced from 14 to 2 μm. Adding aluminum nanoparticles in a propellant also increases the burning rate of 5–10 times, as reported by Ivanov and Tepper (1997). It was also demonstrated by Mench et al. (1998) that burning rate will be increased by 70% in case of solid propellant based on HTPB, doped with 20% of Al nanoparticles. If the size of the Al particles is reduced from 10 μm to 100 nm, then the burning rate will increase from 1 to more than 100 mm/s. One more benefits of using Al nanoparticles as a doping agent is that it reduces the ignition time of the propellant (Mench et al. 1998). However, use of nano-metallic powder is not proven effective in case of explosives (Li et al. 2005).

ii. *Metastable Intermolecular Composite (MIC)—Nano-thermite or superthermite Materials*

In case of the thermite reaction, a metal reacted with a nonmetallic oxide, which is a highly exothermic reaction. By this reaction, a stable oxide and respective metal and nonmetal reactant are formed (Wang et al. 1993). The reaction output as given below is a reduction reaction process of oxidation;

$$\mathrm{M} + \mathrm{XO} \rightarrow \mathrm{MO} + \mathrm{X} + \Delta h$$

where

M = alloy or metal
XO and MO = corresponding oxides
X = can be metal or a nonmetal
Δh = change in heat due to thermite reaction

Faster reaction rates have been observed in the case of a thermite reaction, use of such reaction is very energy efficient.

Al nano-power is mostly used as a fuel because of the following additional advantages for thermite reactions.

(a) Unlike calcium or magnesium, Al has a low vapor pressure; therefore, it does not require any specific pressure vessel to perform the reaction.
(b) Aluminum has a low ignition temperature as it's melting temperature is low (660 °C). Low ignition temperature helps to detonate first and burn quickly at the spark of fire.

In 1995, MoO_3/Al metastable intermolecular composite (MIC) of diameter varies from 20 to 50 nm which was produced by Aumann et al. (1995). By use of the stoichiometric mixture, the energy density can rise to 16 kJ/cm^3. And also, the burning rate of this mixture is 1000 times more than the macroscale materials produced by thermite reaction. It is observed, at a reduction size of Al particles from 121 to 44 nm, the average velocity of combustion increases from 685 to 990 m approximately. It is experimentally found that the burning rate of micrometer size of MoO_3/Al is 10 mm/s (Rossi et al. 2007). It is also patient to note that the reaction rate will not change below the critical diameter of the aluminum particle and is found as 40 nm diameter as examined by Bockmon et al. (2005). This could be further explained that by decreasing the diameter of aluminum particles, Al_2O_3 proportion over aluminum increases. As a result, the volume of active material is reduced, and there is a chance of inhibition in thermite reaction.

4.6 Synthesis of Nano-energetic Materials (nEMs) for Microscale Applications

Synthesizing of thermite composite of nano-size is addressed by different approaches, such as conventional approaches, MEMS-based compatible approaches, bottom-up approaches, and molecular science approaches. In the case of traditional approaches, fuel and oxidizer nanoparticles are mixed properly after synthesizing it. This process is called powder mixing process. In the MEMS-based approach, the nanoparticles of oxidizer and fuel are synthesized and then mixed together in a specific process like vapor deposition or solgel solution techniques. In case of molecular science approach, the fuel and oxidizer molecules/atoms are synthesized properly and then mixed up via a polymer chain or using molecular engineering in solution. The bottom-up approaches describe the basic requirement and understanding the different properties of energetic materials and its ingredients intimacy. Following paragraphs present the different approaches adopted to synthesized of MIC;

i. *Powder Mixing*: One of the easiest methods of producing nano-energetic material is powder mixing. Ultrasonic waves are used to mix aluminum nanoparticle and its oxidizer powder together (Simonenko and Zarko 1999; Prakash et al. 2005; Durães et al. 2006). Normally, the aluminum nanoparticle and oxidizer are easily dissolved in solvents {i.e., hexane (C_6H_{14})} and mix together in presence of ultrasonic wave. A process called sonication (>20 kHz) is used to disintegrate agglomerates and blend both constituents. After this

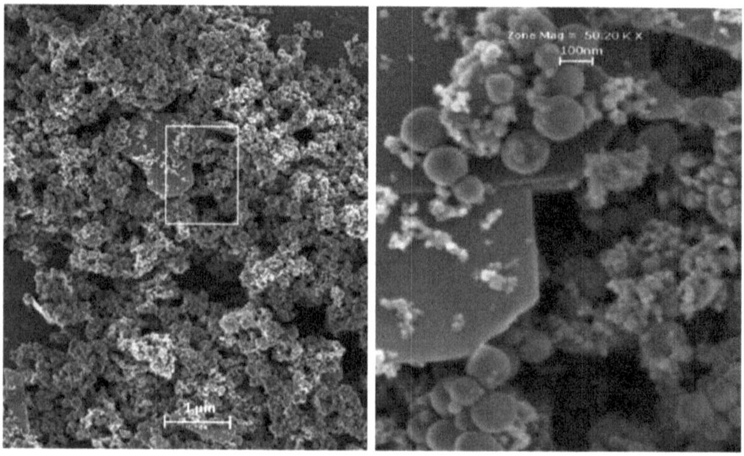

Fig. 4.7 Al particles of size 80 nm mixed with MoO_3 particles. Reprinted with kind permission from Granier and Pantoya (2004)

process, the solvent is evaporated by heating of complete mixture. During this evaporation process, the above mixture is continuously allowed to pass into a fine mesh to blend into agglomerates and produce powder so-called sub-micronic size. MoO_3/Al MIC mixture's SEM image is represented in Fig. 4.7 (Granier and Pantoya 2004). Mixing of fuel with an oxidizer is one of the easiest methods that is frequently used. However, it has some limitations as described below:

a. Intimate mixing of ultrafine power is a difficulty.
b. Homogeneous distribution of nanoparticle of fuel and oxidizer mixture is difficult to get.
c. It is dangerous to manipulate some powder components, and also, thin-film deposition for microsystems application is difficult to achieve.

ii. *Solgel/Aerogel Process*: This technique is widely used for synthesizing the glass materials or ceramic glass monoliths and its powder of nano-structure. This technique has some advantages such as processing temperature is low, and chemical homogeneity is high. Lawrence Livermore National Laboratory has first commenced the technique to use solgel solution to synthesize the nano-energetic materials (nEMs) (Gash et al. 2001; Kim and Zachariah 2004). This process entails the reactions in result to generate diffusion of nanoparticles in form of a liquid phase, so-called sol. When it condensates, a 3D solid network is created from sol, known as gel, with the open holes which are covered by means of the solvent. This solvent is removed by means of the evaporation process and generates pore-free structure by collapsing the gel material to get a xerogel. But in case of drying process (at the supercritical stage), solvents are removed without collapsing the gel structure. In

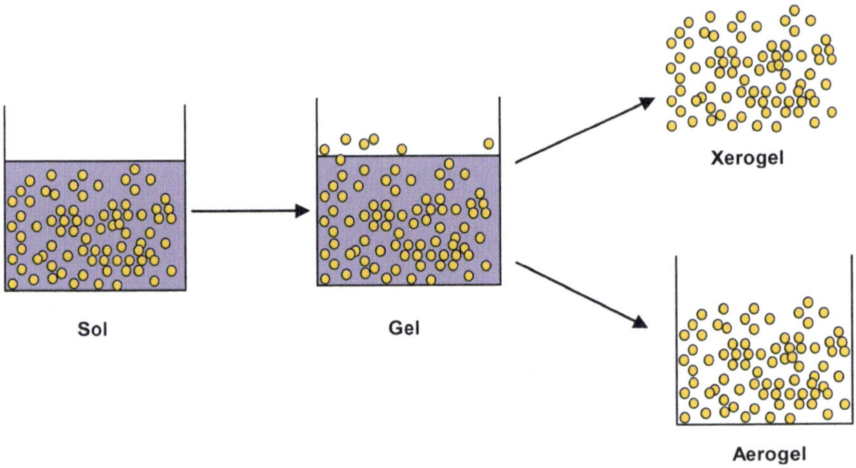

Fig. 4.8 Schematic of solgel methodology

supercritical drying process, material becomes very lightweight and porous called "aerogel." This gives excellent particle uniformity structure, and nanoscale-sized pores are achieved. The process of solgel is shown in Fig. 4.8. The synthesis of Fe_2O_3/Al MIC for solgel material where the size of Fe_2O_3 is varied from 3 to 10 nm in diameter and size of Al is 30 nm in diameter as reported by Tillotson et al. (2001).

The size of the specific surface of the nano-porous Fe_2O_3 composites is achieved about 300–400 m^2/g. As per the ignition and thermal analysis, heat produced in the above process is about 1.5 kJ/g, whereas the hypothetical value is 3.9 kJ/gas reported by some researcher. The reason for such decrease in total measured energy may due to the substantial passivation of aluminum powder particles coated with oxide and also due to the organic impurities present in the matrix (Miziolek 2002).

Advantages of using the solgel process over mixing methods are:

(a) Compatible to MEMS-based application
(b) Safe to use
(c) Low cost
(d) Potentially produce nano-energetic materials because of its high porosity properties.
(e) Easy to control the process.

There are also some disadvantages to this process:

(a) Since particles are randomly distributed in this process, therefore, fuel and oxidizer are separated from each other which may cause the inhibition of self-sustaining reaction;
(b) Reaction performance is significantly reduced due to the organic impurities present in the solgel mixture.

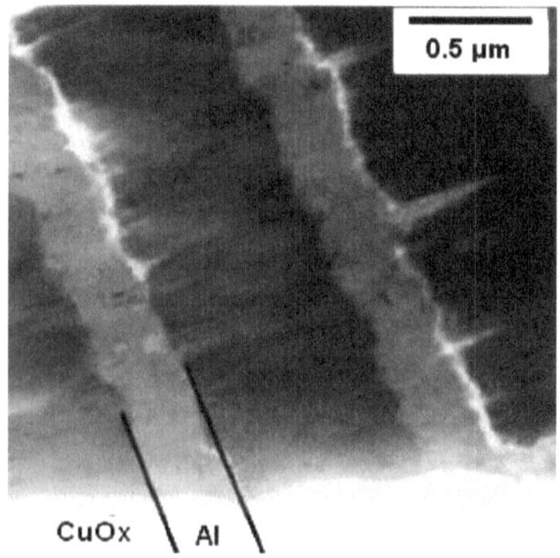

Fig. 4.9 Al/CuO_x multilayer nano-fossils. Reprinted with kind permission from Blobaum et al. (2003)

iii. *Vapor deposition process*: This is an alternative method to solgel or aerogel process. In vapor deposition or sputtering, a nano-layer of oxidizer and fuel is deposited over the substrate under vacuum condition. Vapor deposition is either a chemical or physical process of depositing a thin film of different materials on a substrate. This process is frequently used in chemical industries. In such process, the substrate is sometimes exposed to more than one volatile precursors. The desired deposition is produced by reaction and decomposition of such volatile material over the substrate surface. In this process, the thickness of the thin layer normally varies from 20 nm to 2 μm size. Sputtering is also a physical vapor deposition process in which atoms or molecules are injected in a solid target material in the gaseous state due to the bombardment of material by energetic ions. This process is very familiar with thin-film deposition and also widely used in many chemical industries. To avoid interdiffusion of metal, it is required to cool down the silicon wafer during the deposition process. In this deposition process, an Al layered structure of thickness 0.3 μm and Cu_4O_3 of thickness 0.7 μm is obtained as shown in Fig. 4.9 (Blobaum et al. 2003).

In this process, the diffusion distances between each layer of oxidizer or fuel are reduced by 10–1000 times as compared to powder mixing process. Since this process of deposition is performed in a high vacuum condition, the possibility of impurities present and the chance of aluminum oxidation is very low as compared to other processes like powder mixing or solgel process.

iv. *Nano-structuring*

Deposition using nano-structuring method includes mainly two processes which are presented in the following paragraphs;

a. *Atomic Layer Deposition (ALD)*: In this process, the thin film deposited in the substrate with a high uniformity manner and also the thickness of the layer are precisely controlled. Different gas pulses are utilized sequentially to deposit one layer of thin film at a time. By using an atomic layer deposition technique, many metal oxides such as NiO (Baek and An 2011), Co_3O_4 (Mukae et al. 1977), WO_3 (Kumar and Rao 2017) are deposited on the substrate material. The first time this process is used by Ferguson et al. (2005) for deposition of SnO_2 oxidizer layer on aluminum nanoparticles. ALD process is utilized to prepare the Al/SnO_2MIC mixture as illustrated in Fig. 4.10 (Ferguson et al. 2005). It is observed by experiments on samples of aluminum and SnO_2 that, a faster, as well as violent reaction, occurs despite the low molar ratio of Al/O and low mass percentage of Sn particles.
b. *Nano-porous Silicon With Oxidizer*: Reaction due to combustion of porous silicon (PSi) while submerged in nitric acid is first discovered by McCord et al. (1992). He found PSi can be used as a reactive material. The liquid oxidizer is used to fill up PSi pores of size 2–10 nm. For such properties of PSi pores, PSi-based nano-energetic material is widely used in industrial applications such as air bag initiator components (Tägtström et al. 1999; Bartuch et al. 2004).

As shown in Fig. 4.11 (Currano and Churaman 2009), the electrochemical etching process can be used to produce the porous silicon (PSi) from bulk silicon by using fluoride (HF) and H_2O_2 (Currano and Churaman 2009) solution. The structure size, as well as porosity of PSi, can be achieved from 2 to 1000 nm by suitably selecting the concentration proportion of different ingredients such as the concentration of fluoride, treatment duration, starting material, and current density (Bartuch et al. 2004). Then, a liquid oxidizer may be filled with the nano-pores. Apart from this process, CVD and PVD are also used to fill up PSi pores with liquid

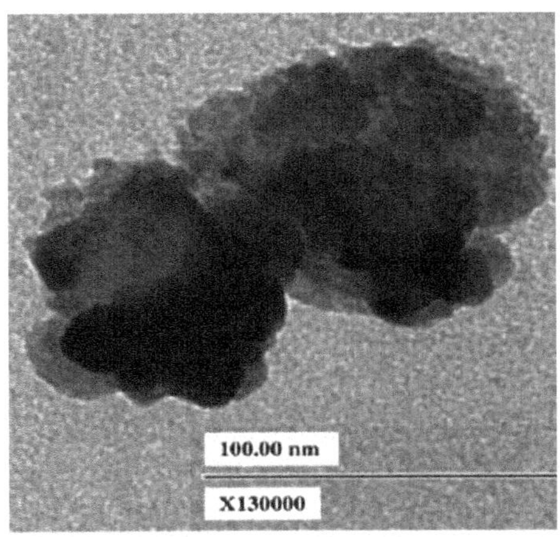

Fig. 4.10 Al/SnO_2 MIC prepared by ALD process. Reprinted with kind permission from Ferguson et al. (2005)

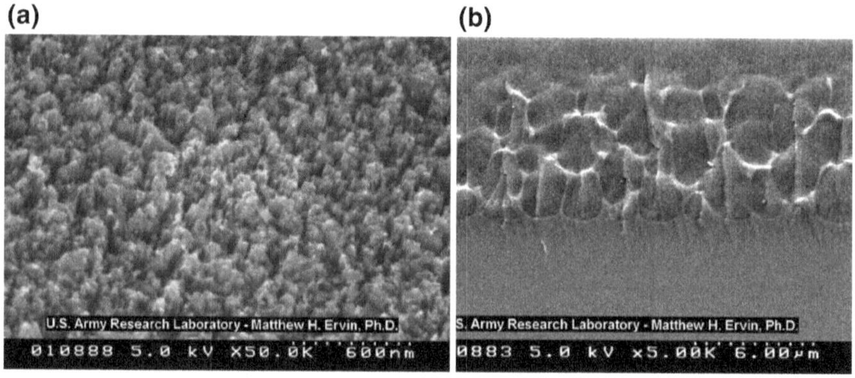

Fig. 4.11 SEM images of nano-PSi. **a** Top view **b** cross section. Reprinted with kind permission from Currano et al. (2009)

oxidizer which enhance the intimacy between Si and O_2. The activation energy released from the reaction of the PSi/O_2 substance can be adjusted before loading of oxidizer by controlling layer of SiO_2. The result is verified by Currano and Churaman (2009).

PSi-based nano-energetic material has advantages in that it has a compatible use in MEMS applications. So nano-energetic materials (nEMS) is widely used for fabrication of MEMS devices using silicon as a base material. Ultimately, it provides good compatibility for MEMS-based application at a low production cost. One more advantage is that semiconductor circuitries are easily made by utilizing PSi-based nano-energetic materials.

c. *"Bottom-Up" Approaches*: To improve the homogeneity and control the contact area of the final material of the fuel/oxidizer, bottom-up approaches have been explored and found suited by the researcher.

Molecular self-assembly is one of the interesting techniques used widely. In this technique, the nanoparticles of fuel are arranged around the oxidizer on its own or by means of outside forces in a restricted way. In some cases, inorganic solution or polymers are used to arrange the molecules of oxidizer and fuel. A method is developed by Kim and Zachariah (2004) to synthesize MIC in which electrostatic forces control the self-assembly of Al and Fe_2O_3 existed in charged particles of aerosol. The results of the reaction of Al/Fe_2O_3 have been compared with a randomly assembled Al/Fe_2O_3 by some researcher. The surface of Fe_2O_3 particles is surrounded by the Al nanoparticles. The contact between Al and Fe_2O_3 is intimate as shown in Fig. 4.12b. In Fig. 4.12a (Kim and Zachariah 2004), it is shown the TEM image of synthesized nano-Al/Fe_2O_3 particles by the random assembly. As per the results reported by researcher, the material produced electrostatically assembled process such as Al/Fe_2O_3; MIC is more powerful than the material produced by random assembly process due to interfacial surface of oxidizer and fuel.

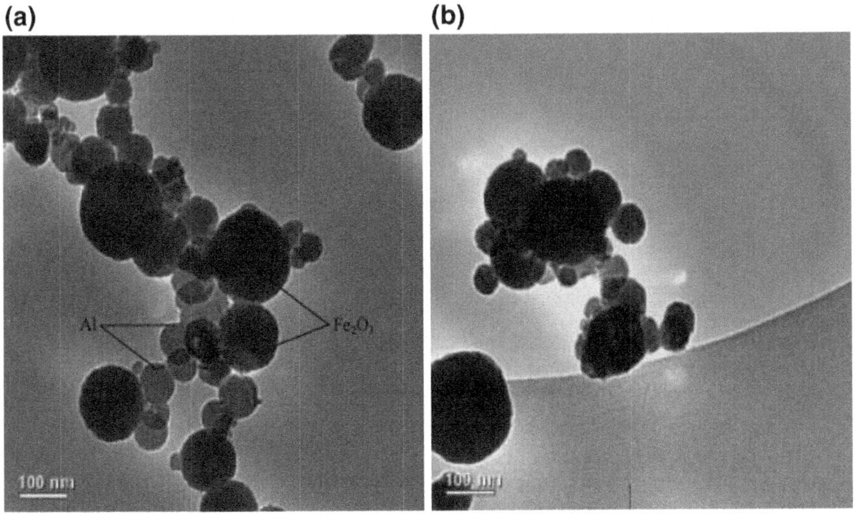

Fig. 4.12 Al/Fe_2O_3 MIC by **a** random assembly and **b** electrostatic assembly. Reprinted with permission from Kim and Zachariah et al. (2004)

4.7 Propellants and Explosives in Defense Application

Propellants and explosives are extensively used in defense for different purposes such as rocket propulsion, warheads, munitions. The following paragraphs present some of the important use of propellants and explosives in defense.

4.7.1 Rocket Propulsion

As discussed, the most generic way to produce nano-energetic material is to mix nanoparticles of fuel (e.g., Al powder) and oxidizer (e.g., SnO_2) with proper proportion. The power density of combustion will be improved at a faster burning rate of materials. This will be possible when the particles are more precisely divided. Let us take the diameter of the nanoparticle as $d(t)$, at $d(t = 0)$, as d_0 we have

$$d(t)^2 = d_0^2 - k_t \tag{4.1}$$

where

k = coefficient of combustion rate, normally taken 1 mm^2/s
t = propulsion time
d_0 = diameter of particle at $t = 0$

When two substances are more finely divided, the ignition delay time is reduced. Hence, nanofication is surely a beneficial work. In a monomolecular energetic material because of ultimate degree of miniaturization, every molecule will have a moiety of fuel and oxidizing elements. The energy density normally varies from 10 to 12 kJ/cm^3 which is about half of the mixture. The energy density can be improved by using polymer-based mono-macromolecular energetic materials. Although such mixtures are very useful, there is a practical difficulty of dispersion of substances. The surface energetic properties of a nano-object mostly nanoparticles play an important role because of the extremely high surface to volume ratio. Material can be characterized into its three components of surface tension as γ^{LW}, $\gamma^{(-)}$, and $\gamma^{(+)}$. These components are related to three different parameters such as $\gamma^{(+)}$ related to electron acceptor potential (+), $\gamma^{(-)}$ related to electron donor potential (−), and γ^{LW} related to Lifshitz–van der Waals (LW) potential. The adhesive potential (so-called interfacial tension) is denoted by $\boldsymbol{\gamma_{12}}$, where subscript 1 is denoted for fuel proportion and subscript 2 is for the oxidizer. Interfacial tension is found out by summation of the surface tension contributed by LW effect and donor–acceptor (da) (Ramsden 2012)

$$\gamma_{12} = \gamma_{12}^{LW} + \gamma_{12}^{(da)} \tag{4.2}$$

where LW interfacial tension can be calculated as

$$\gamma_{12}^{LW} = \left(\sqrt{\gamma_1^{LW}} + \sqrt{\gamma_2^{LW}}\right)^2 \tag{4.3}$$

and similarly donor–acceptor (da) interfacial tension can be calculated as

$$\gamma_{12}^{(da)} = 2\left(\sqrt{\gamma_1^{(-)}\gamma_1^{(+)}} + \sqrt{\gamma_2^{(-)}\gamma_2^{(+)}} - \sqrt{\gamma_1^{(+)}\gamma_2^{(-)}} - \sqrt{\gamma_1^{(-)}\gamma_2^{(+)}}\right) \tag{4.4}$$

The surface tension of single-substance components can also be found out by measuring the angle of contact of droplets of three different liquids such as α-bromonaphthalene, water, and dimethylsulfoxide which has certain viscosities on a material flat surface of fuel or oxidizer.

By Bragg–Williams expression, the interaction energy ν of the powdered fuel–oxidizer mixture is given by

$$\vartheta = \vartheta_{11} + \vartheta_{22} - 2\vartheta_{12} \tag{4.5}$$

where

ϑ_{11} = interaction energy for fuel particle with each other,
ϑ_{22} = interaction energy for oxidizer particle with each other,
ϑ_{12} = interaction energy for fuel particle with oxidizer particle

The cohesive energies ϑ_{11} and ϑ_{22} are given by Dupré law as follows;

$$\vartheta_{11} = -2\gamma_1 A \tag{4.6}$$

$$\vartheta_{12} = -(\gamma_1 + \gamma_2 - \gamma_{12})A \tag{4.7}$$

where

A = is the interfacial area,
ϑ_{11} & ϑ_{12} = adhesive energy for a single-substance surface tension,
γ = surface tension.

If interaction energy (v) < 0, then the mixture strongly dispersed,
If interaction energy (v) > 0, then the mixture is poorly dispersed,
If interaction energy (v) = 0, then the mixture has an entropic drive.

In order to enhance combustion enthalpy of the fuel, the addition of oxidizable nanoparticles like aluminum, boron to the fuel is used. Another possibility is fullerenes (Adams 2006), C60, C70, C7, etc., the diameters of which are of the order of 1 nm.

The smaller-sized particle has higher combustion rate and erosion severity of the walls where fuel flows are also less. This example is an undiminished benefit in nanofication. There are no such absolute requirements in case of nanoparticle fabrication to obtain atomic precision.

As known, aluminum readily forms a thin layer of aluminum oxide (alumina) on its surface; hence, alumina with surface tension parameters should be inserted into equations such as stated in (4).

The Lawrence Livermore National Laboratory in the USA has developed an approach to make solgel porous materials. The gel is created from fuel particle, and pores are created from oxidizer particle. This is an alternative of powder mixing process as discussed earlier. To establish, a long-time stability of nano-energetic materials as a commercial material requires more research on it. All the above is at technology level 2–3. In recent years, Rebinder effect has been neglected for no particularly good reason; it would appear. Such substances would be especially important for influencing the mechanical sensitivity of explosives. This area of technology is only at level 0 or 1.

4.7.2 Warheads

The explosive material used as warheads should detonate effectively. It is a big task to know how nanoscale conversion (nanofication) affects the deflagration–detonation transition. The scope of more research is available in this field. In the present scenario, there is no such perfect theory of releasing energy methods which can be applicable to real 3D materials (Adams 2006).

4.7.2.1 Insensitive Energetic Materials

Explosive sensitivity can be reduced in many ways such as by enhancing the crystal utility, eliminating voids (pores), decreasing defects in molecule and crystal as well, eliminating multiple phases, and also impurities removal from chemical (Walley et al. 2006). The distribution pattern of particles plays an important role in sensitivity value. All the above features carry a potential effect in nano-science to produce energetic material with less sensitivity.

The main idea for nanofication is to enhance the combustion rate subjected to particles; fuel and oxidizer are finely dispersed. Largely increase in surface to volume ratio also has a significant effect on the reaction. This affects not only the reaction rate of fuel and oxidizer, but also the surface becomes slightly passivated.

Energetic materials are covered with a specific ultrathin coating in order to achieve relative insensitivity. Rationally designed materials can be assembled atom by atom precisely as required if large-scale nano-facture at reasonable cost becomes feasible.

4.7.2.2 Munitions Containing Nanoparticles

Though the release of nanoparticles in an explosion can cause potential health issues, nanoparticles are having a great variety of biological effects. There are two main kinds of nanoparticles: (i) nanoparticles made from a material that is chemically toxic and which is released in the human body through dissolution of the particle and (ii) insoluble nanoparticles or nano-fibers that cannot be destroyed by the body's defense systems and which, therefore, remains sites of persistent inflammation, with potential adverse secondary effects such as the growth of neoplasm. For category (ii), the surface properties are often decisive in determining the nature of the interaction of a nano-object with its biological, protein-containing environment; hence, an approach based on interfacial tension (Sect. 7.1) can be used to predict such interaction.

4.7.2.3 Switchable Explosivity

As mentioned (Sect. 7.2.1), creating insensitive explosives is to mitigate ignition sensitivity by placing appropriately inert coatings around the particles of explosive material. In view of the fact that the theory of explosions is still in a fairly rudimentary state, design of an explosive with switchable sensitivity is rationally difficult at present. However, the degree of ignition of the coating can be mitigated by an external environmental parameter.

Switchable sensitivity would be useful to enable an insensitive fuse until activated close to anticipated demand. Nanotechnology is expected to facilitate the achievement of this property through an atomic-detail understanding of the overall process.

4.8 Conclusion

In this article, fundamental aspects of energetic materials are covered along with the description of contemporary reported literature over this unique category of materials. This article also focuses on the design and synthesis of various types of nano-energetic materials. Since propellants and explosives are extensively used in defense for different purposes such as rocket propulsion, warheads, munitions, its importance for defense applications has also been illustrated in the article.

References

Adams C (2006) Inventor; Lockheed Martin Corp, assignee. Explosive/energetic fullerenes. United States patent US 7,025,840, 11 Apr 2006

Albini A (1991) Heterocyclic N-oxides. CRC Press, Boca Raton

Aumann CE, Skofronick GL, Martin JA (1995) Oxidation behavior of aluminum nanopowders. J Vac Sci Technol B: Microelectron Nanometer Struct Process Meas Phenomena 13 (3):1178–1183

Badgujar DM, Talawar MB, Asthana SN, Mahulikar PP (2008) Advances in science and technology of modern energetic materials: an overview. J Hazard Mater 151(2–3):289–305

Baek YW, An YJ (2011) Microbial toxicity of metal oxide nanoparticles (CuO, NiO, ZnO, and Sb_2O_3) to *Escherichia coli*, *Bacillus subtilis*, and *Streptococcus aureus*. Sci Total Environ 409 (8):1603–1608

Bartuch H, Clément D, Kovalev D, Laucht H (2004) Silicon initiator, from the idea to functional tests. In: 7th international symposium and exhibition on sophisticated car occupant safety systems, Karlsruhe

Blobaum KJ, Reiss ME, Plitzko JM, Weihs TP (2003) Deposition and characterization of a self-propagating CuO x/Al thermite reaction in a multilayer foil geometry. J Appl Phys 94 (5):2915–2922

Bockmon BS, Pantoya ML, Son SF, Asay BW, Mang JT (2005) Combustion velocities and propagation mechanisms of metastable interstitial composites. J Appl Phys 98(6):064903

Bottaro JC, Penwell PE, Schmitt RJ (1997) 1, 1, 3, 3-Tetraoxo-1, 2, 3-triazapropene anion, a new oxyanion of nitrogen: the dinitramide anion and its salts. J Am Chem Soc 119(40):9405–9410

Brown ME, Taylor SJ, Tribelhorn MJ (1998) Fuel—oxidant particle contact in binary pyrotechnic reactions. Propellants Explos Pyrotech 23(6):320–327

Chavez DE, Hiskey MA (1998) Synthesis of the bi-heterocyclic parent ring system 1,2,4-triazole [4,3-b][1,2,4,5] tetrazine and some 3, 6-disubstituted derivatives. J Heterocycl Chem 35 (6):1329–1332

Chavez DE, Hiskey MA (1999) 1,2,4,5-tetrazine based energetic materials. J Energ Mater 17 (4):357–377

Chavez D, Hill L, Hiskey M, Kinkead S (2000) Preparation and explosive properties of azo-and azoxy-furazans. J Energ Mater 18(2–3):219–236

Chung SW, Guliants EA, Bunker CE, Hammerstroem DW, Deng Y, Burgers MA, Jelliss PA, Buckner SW (2009) Capping and passivation of aluminum nanoparticles using alkyl-substituted epoxides. Langmuir 25(16):8883–8887

Coburn MD (1968) Picrylamino-substituted heterocycles. II. Furazans. J Heterocycl Chem 5 (1):83–87

Coburn MD, Hiskey MA, Lee KY, Ott DG, Stinecipher MM (1993) Oxidations of 3,6-diamino-1,2,4,5-tetrazine and 3,6-bis (s,s-dimethylsulfilimino)-1,2,4,5-tetrazine. J Heterocycl Chem 30(6):1593–1595

Currano LJ, Churaman WA (2009) Energetic nanoporous silicon devices. J Microelectromech Syst 18(4):799–807

Delpuech A, Cheville J, Michaud C (1981) Molecular electronic structure and initiation of secondary explosives. In: Proceedings of the 7th symposium (international) on detonation Jun 16, pp 65–74

Ding L, Xuebiao L, Zhengzhuo Z, Mingxin Q, Chenglu L (1987) Laser-initiated aluminothermic reaction applied to prepare the mo-si film on silicon substrates. In: MRS Online Proceedings Library Archive, p 101

Dremin AN, Shvedov KK (1964) Estimation of Chapman-Jouget pressure and time of reaction in detonation waves of powerful explosives. J Appl Mech Tech Phys 2:154–159

Durães L, Campos J, Portugal A (2006) Radial combustion propagation in iron (III) oxide/aluminum thermite mixtures. Propellants Explos Pyrotech 31(1):42–49

Eichhorn B, Zachariah MR, Aksay IA, Selloni A, Car R, Dabbs DM, Yetter RA, Son SF, Thynell S, Groven LJ (2012) Smart Functional Nano-energetic Materials. Purdue Univ Lafayette In

Ferguson JD, Buechler KJ, Weimer AW, George SM (2005) SnO2 atomic layer deposition on ZrO_2 and Al nanoparticles: pathway to enhanced thermite materials. Powder Technol 156(2–3):154–163

Fried LE, Manaa MR, Pagoria PF, Simpson RL (2001) Design and synthesis of energetic materials. Annu Rev Mater Res 31(1):291–321

Gao C, Xu ZC, Deng SR, Wan J, Chen Y, Liu R, Huq E, Qu XP (2011) Silicon nanowires by combined nanoimprint and angle deposition for gas sensing applications. Microelectron Eng 88 (8):2100–2104

Gash AE, Tillotson TM, Jr Satcher JH, Poco JF, Hrubesh LW, Simpson RL (2001) Use of epoxides in the sol-gel synthesis of porous iron (III) oxide monoliths from Fe (III) salts. Chem Mater 13(3):999–1007

Granier JJ, Pantoya ML (2004) Laser ignition of nanocomposite thermites. Combust Flame 138 (4):373–383

Hollins RA, Merwin LH, Nissan RA, Wilson WS, Gilardi RD (1995) Aminonitroheterocyclic n-oxides–a new class of insensitive energetic materials. In: MRS online proceedings library archive, p 418

Hollins RA, Merwin LH, Nissan RA, Wilson WS, Gilardi R (1996) Aminonitropyridines and their N-oxides. J Heterocycl Chem 33(3):895–904

Ismail B, Abaab M, Rezig B (2001) Structural and electrical properties of ZnO films prepared by screen printing technique. Thin Solid Films 383(1–2):92–94

Ivanov GV, Tepper F (1997) 'Activated' aluminum as a stored energy source for propellants. Int J Energ Mater Chem Propul 4(1–6)

Kim SH, Zachariah MR (2004) Enhancing the rate of energy release from nano-energetic materials by the electrostatically enhanced assembly. Adv Mater 16(20):1821–1825

Klapötke TM (2017) Chemistry of high-energy materials. Walter de Gruyter GmbH & Co KG, Berlin, 21 Aug

Kondo K, Tanaka S, Habu H, Tokudome SI, Hori K, Saito H, Itoh A, Watanabe M, Esashi M (2004) Vacuum test of a micro-solid propellant rocket array thruster. IEICE Electron Express 1 (8):222–227

Kumar SG, Rao KK (2017) Comparison of modification strategies towards enhanced charge carrier separation and photocatalytic degradation activity of metal oxide semiconductors (TiO_2, WO_3 and ZnO). Appl Surf Sci 1(391):124–148

Laucht H, Bartuch H, Kovalev D (2004) Silicon initiator, from the idea to functional tests. In: Proceedings of 7th international symposium and exhibit. Sophisticated Car Occupant Safety System, pp 12–16

Lee KY, Storm CB, Hiskey MA, Coburn MD (1991) An improved synthesis of 5-amino-3-nitro-1 H-1, 2, 4-triazole (ANTA), a useful intermediate for the preparation of insensitive high explosives. J Energ Mater 9(5):415–428
Li XJ, Xie XH, Li RY (2005) Detonation synthesis for nano-metallic oxide powders. Explos Shock Waves 25(3):271
Li Y, Cott DJ, Mertens S, Peys N, Heyns M, De Gendt S, Groeseneken G, Vereecken PM (2011) Integration and electrical characterization of carbon nanotube via interconnects. Microelectron Eng 88(5):837–843
Martin AN, Pinkerton AA, Gilardi RD, Bottaro JC (1997) Energetic materials: the preparation and structural characterization of three biguanidiniumdinitramides. Acta Crystallogr B Struct Sci 53 (3):504–512
Martirosyan KS, Ramazanova Z, Zyskin M (2013) Nanoscale energetic materials: theoretical and experimental updates. In: Proceedings of the 8th Pacific Rim international congress on advanced materials and processing, Springer, Cham, pp 57–63
Mattox DM (2010) Handbook of physical vapor deposition (PVD) processing. William Andrew, 29 Apr 2010
McCord P, Yau SL, Bard AJ (1992) Chemiluminescence of anodized and etched silicon: evidence for a luminescent siloxene-like layer on porous silicon. Science 257(5066):68–69
Mench MM, Yeh CL, Kuo KK (1998) Propellant burning rate enhancement and thermal behavior of ultra-fine aluminum powders (Alex). In: Energetic materials—production, processing, and characterization, pp 30–31
Millar DI (2011) Energetic materials at extreme conditions. Springer Science & Business Media, 24 Sep
Miziolek A (2002) Nanoenergetics: an emerging technology area of national importance. Amptiac Q 6(1):43–48
Mukae K, Tsuda K, Nagasawa I (1977) Non-ohmic properties of ZnO-rare earth metal oxide-Co_3O_4 ceramics. Jpn J Appl Phys 16(8):1361
Ou Y, Chen B, Li J, Dong S, Jia H (1994) Synthesis of nitro derivatives of triazoles. Chem Inform 25(44)
Pagoria PF, Mitchell AR, Schmidt RD (1998) Synthesis, scale-up and experimental testing of LLM-105. In: Insensitive munitions and energetic materials technology symposium. San Diego, CA
Pagoria PF, Lee GS, Mitchell AR, Schmidt RD (2002) A review of energetic materials synthesis. Thermochim Acta 384(1–2):187–204
Pennarun P, Rossi C, Estève D, Bourrier D (2005) Design, fabrication and characterization of a MEMS safe pyrotechnical igniter integrating arming, disarming and sterilization functions. J Micromech Microeng 16(1):92
Pevzner MS, Kulibabina TN, Povarova NA, Kilina LV (1979) Heterocyclic nitrocompounds. 24. Nitration of 5-amino-1, 2, 4-triazole and 5-acetamino-1, 2, 4-triazole by acetylnitrate and nitronium salts. Khimiya Geterotsiklicheskikh Soedinenii 1(8):1132–1135
Pierson HO (1999) Handbook of chemical vapor deposition: principles, technology and applications. William Andrew, 1 Sept 1999
Prakash A, McCormick AV, Zachariah MR (2005) Synthesis and reactivity of a super-reactive metastable intermolecular composite formulation of Al/KMnO4. Adv Mater 17(7):900–903
Proud WG (2014) Ignition and detonation in energetic materials: An introduction. STO-EN-AVT-214, 3
Ramsden JJ (2012) Nanotechnology for military applications. Collegium 30:99
Ritter H, Licht HH (1995) Synthesis and reactions of dinitrate amino and diaminopyridines. J Heterocycl Chem 32(2):585–590
Rossi C, Esteve D (1997) Pyrotechnic microactuators. In: Proceedings of 11th EUROSENSORS XI, vol 2, pp 771–774, 21 Sep 1997
Rossi C, Estève D (2005) Micropyrotechnics, a new technology for making energetic microsystems: review and prospective. Sens Actuators A 120(2):297–310

Rossi C, Esteve D, Mingues C (1999) Pyrotechnic actuator: a new generation of Si integrated actuator. Sens Actuators A 74(1–3):211–215

Rossi C, Briand D, Dumonteuil M, Camps T, Pham PQ, De Rooij NF (2006) The matrix of 10 × 10 addressed solid propellant micro thrusters: review of the technologies. Sens Actuators A 126(1):241–252

Rossi C, Zhang K, Esteve D, Alphonse P, Tailhades P, Vahlas C (2007) Nano-energetic materials for MEMS: a review. J Microelectromech Syst 16(4):919–931

Rugunanan RA, Brown ME (1993) Combustion of binary and ternary silicon/oxidant pyrotechnic systems, part i: Binary systems with fe203 and sn02 as oxidants. Combust Sci Technol 95(1–6):61–83

Sanders VE, Asay BW, Foley TJ, Tappan BC, Pacheco AN, Son SF (2007) Reaction propagation of four nanoscale energetic composites (Al/MoO_3, Al/WO_3, Al/CuO, and $B_{12}O_3$). J Propul Pow 23(4):707–714

Sathiyanathan K, Lee R, Chesser H, Dubois C, Stowe R, Farinaccio R, Ringuette S (2011) Solid propellant microthruster design for nanosatellite applications. J Propul Power 27(6):1288–1294

Schönhuber G, Enzmann E, Nuiding H (2011) US Patent 8,083,259. US Patent and Trademark Office, Washington, DC

Sheremetev AB, Kharitonova OV, Mantseva EV, Kulagina VO, Shatunova EV, Aleksandrova NS, Melnikova TM, Ivanova EA, Dmitriev DE, Eman V, Yudin IL (1999) Nucleophilic substitution in a furazane series. Reaction with O-nucleophiles. Zhurnal Organicheskoi Khimii 35(10):1555–1566

Simonenko VN, Zarko VE (1999) Comparative studying the combustion behavior of composite propellants containing ultrafine aluminum. In: Energetic materials—modelling of phenomena, experimental characterization, environmental engineering, pp 21–31

Simpson RL, Pagoria PF, Mitchell AR, Coon CL (1994) Synthesis, properties, and performance of the high explosive ANTA. Propellants Explos Pyrotech 19(4):174–179

Singh RP, Verma RD, Meshri DT, Shreeve JN (2006) Energetic nitrogen-rich salts and ionic liquids. Angew Chem Int Ed 45(22):3584–3601

Solodyuk GD, Boldyrev MD, Gidaspov BV, Nikolaev VD (1981) Oxidation of 3, 4 diaminofurazan by some peroxide reagents. Chem Inform 12(36)

Son SF, Asay BW (2001) Reaction propagation physics of A1/Mo03 nanocomposite thermites. Los Alamos National Laboratory (LANL), Los Alamos, NM

Son SF, Yetter R, Yang V (2007) Introduction: nanoscale composite energetic materials. J Propul Power 23(4):643–644

Stewart DS (2005) Miniaturization of explosive technology and microdetonics. In: Mechanics of the 21st Century. Springer, Dordrecht, pp 379–385

Suceska M (2012) Test methods for explosives. Springer Science & Business Media, 6 Dec

Tägtström P, Maårtensson P, Jansson U, Carlsson JO (1999) Atomic layer epitaxy of tungsten oxide films using oxyfluorides as metal precursors. J Electrochem Soc 146(8):3139–3143

Tanaka S, Hosokawa R, Tokudome SI, Hori K, Saito H, Watanabe M, Esashi M (2003) MEMS-based solid propellant rocket array thruster. Trans Jpn Soc Aeronaut Space Sci 46 (151):47–51

Tichapondwa SM, Focke WW, Del Fabbro O, Muller E (2012) Suppressing hydrogen evolution by aqueous silicon power dispersions. Ph.D. dissertation, University of Pretoria

Tillotson TM, Hrubesh LW, Simpson RL, Lee RS, Swansiger RW, Simpson LR (1998) Sol-gel processing of energetic materials. J Non-Cryst Solids 1(225):358–363

Tillotson TM, Gash AE, Simpson RL, Hrubesh LW, SatcherJr JH, Poco JF (2001) Nanostructured energetic materials using sol-gel methodologies. J Non-Cryst Solids 285(1–3):338–345

Troianello T (2001) Precision foil resistors used as electro-pyrotechnic initiators. In: Proceedings of 51st electronic components and technology conference, IEEE, pp 1413–1417

Vasylkiv O, Sakka Y, Skorokhod VV (2006) Nano-blast synthesis of nano-size CeO2–Gd2O3 Powders. J Am Ceram Soc 89(6):1822–1826

Walley SM, Field JE, Greenaway MW (2006) Crystal sensitivities of energetic materials. Mater Sci Technol 22(4):402–413

Wang L, Munir ZA, Maximov YM (1993) Thermite reactions: their utilization in the synthesis and processing of materials. J Mater Sci 28(14):3693–3708

Wartenberg C, Charrue P, Laval F (1995) Conception, synthèse et caractérisation d'un nouvel explosif insensible et énergétique: Le DANTNP. Propellants Explos Pyrotech 20(1):23–26

Yarrington CD, Son SF, Foley TJ (2010) Combustion of silicon/Teflon/Viton and aluminum/Teflon/Viton energetic composites. J Propul Power 26(4):734–743

Youngner D, Thai Lu S, Choueiri E, Neidert J, Black III R, Graham K, Fahey D, Lucus R, Zhu X (2000) MEMS mega-pixel micro-thruster arrays for small satellite stationkeeping

Chapter 5
Nano-aluminium as Catalyst in Thermal Decomposition of Energetic Materials

Amit Joshi, K. K. S. Mer, Shantanu Bhattacharya and Vinay K. Patel

Abstract Current state of the art propellant and explosive technology relies on the inclusion of nano-metallic fuels such as aluminium nanopowder. Aluminium nanoparticles, a common fuel ingredient in high-energy material developments, can improve the energy density, rate of energy release, ignition, and combustion performance. In this chapter, the catalytic effect of aluminium nanoparticles on thermal decomposition of various high-energy materials is critically reviewed and presented. The combustion behaviour and ballistic performance of these energetic materials are found to be highly modified and influenced by the aluminium nanopowder. The nanoscale aluminium reflects excellent catalytic activity in thermal decomposition of energetic materials with significant reduction in decomposition temperature. The reaction mechanism of aluminium nanoparticles enhancing the decomposition of high-energy materials is detailed.

Keywords Aluminium nanoparticles · Catalytic activity · Thermal decomposition Nanothermites · Ammonium perchlorate · RDX · HMX · TBX

5.1 Introduction

Current stage of nanoscale technology is multifarious with its application from medicinal science and electronics technology to manufacturing engineering and fashion, etc. The use of nanomaterials as catalyst involves the ratio of surface area (S_A) to volume (V). With stepping down the size of material, the ratio of surface

A. Joshi · K. K. S. Mer · V. K. Patel (✉)
Department of Mechanical Engineering, Govind Ballabh Pant Institute of Engineering and Technology, Ghurdauri, Pauri Garhwal 246001, Uttarakhand, India
e-mail: vinaykrpatel@gmail.com

S. Bhattacharya
Department of Mechanical Engineering, Indian Institute of Technology, Kanpur 208016, Uttar Pradesh, India

S. Bhattacharya et al. (eds.), *Nano-Energetic Materials*, Energy, Environment and Sustainability, https://doi.org/10.1007/978-981-13-3269-2_5

area to its volume increases. This concept has an important significance, especially in chemical reactions. While lighting fire we need kindling to start fire, as they have greater surface area as compared to volume, for example, by throwing sawdust onto a burning fire, a giant flare occurs because it has more surface area as compared to ordinary wood or kindling. The physical phenomenon of increasing the speed of reaction by introducing a catalyst can be due to any or combination of ways: catalyst can play a role of facilitator and thus by bringing the reagents all-together more efficaciously; or catalyst can play a role in lowering the activation energy of the intermediate or overall reaction; or it can increase the yield of one species. Use of nanomaterial is more efficient than conventional catalyst because of their small size leading extremely high surface to volume ratio and when these materials are designed at nanoscale their properties get different than their conventional macroscopic counterpart, for example, gold in macroscale is yellow while in nanoscale it is red or purple.

Aluminium is considered as a reactive metal for energetic applications in propellants, solid rocket fuel owing to its potentially superior exothermic energy (31 kJ/kg), low toxicity, high handling in safety, low oxygen demand, and ease of availability during oxidation reaction as compared to other metals (Arkhipov and Korotkikh 2012; Zhi et al. 2006). Worldwide researchers are now obliged that by substituting the conventional metals (micrometre size) which were conventionally used with energetic materials by Al nanopowders in energetic mixture formation or rocket technology bring about burning rate of propellants increase by factor of 5–30. The basic reasons for this are higher reactivity rate of nanometre-sized particles, higher energy release rate below the melting point and packed surface contact between mixed reagents leads to the better diffusion of reagents and thus yield higher reactivity. By reducing the size of nano-aluminium to a lower scale (5–50 nm), a considerable left-hand shift of melting point is discovered and also with reduction in size the enthalpy of fusion decreases (Sun and Simon 2007). It can be depicted from Fig. 5.1 that with reduction of the weight-average core radius of aluminium nanoparticles, the melting point decreases. With the introduction of

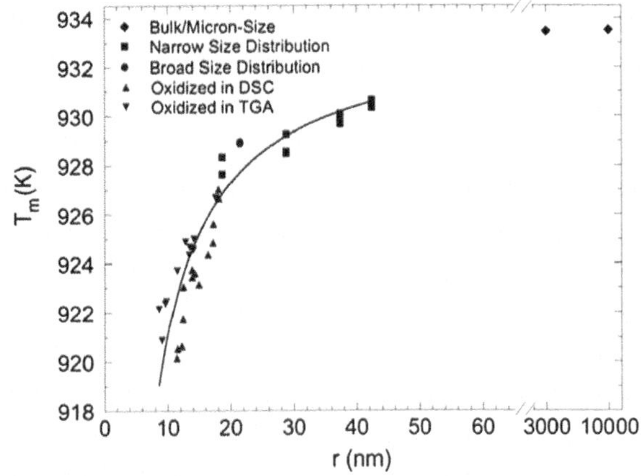

Fig. 5.1 Melting point decrement with respect to weight-average core radius (*r*) *of nanoaluminium with* solid lines fitting to the Gibbs–Thomson equation. Reproduced with permission from Sun and Simon (2007). Copyright (2007) Elsevier Ltd

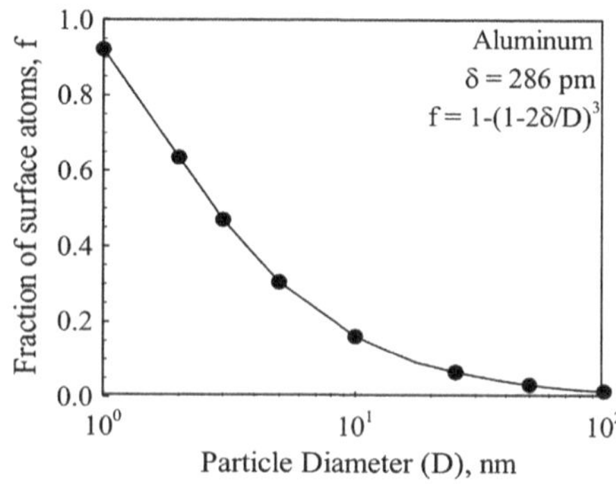

Fig. 5.2 Analytical results of variation of particle size of nano-Al with surface atoms. Reproduced with permission from Sundaram et al. (2017). Copyright (2017) Elsevier Ltd

Table 5.1 Relationship between aluminium particle size with time for completion of burning (Teipel 2006)

Particle size (nm)	Burning time (ms)
10^6	50
10^4	0.5
10^3	0.005
100	0.00005

nano-Al powder in propellant, accelerated speed of burning is obtained because a much shorter time is needed for burning of nano-Al particles leading to emission of a much larger quantity of energy in unit time (Teipel 2006). It is readily accepted that by lowering the size of micrometre-sized aluminium to nanoscale decreases the ignition and burning temperatures (Huang et al. 2007). Significant enhancements in the burning rates are achieved by employing nano-Al instead of micron-sized counterparts in the energetic materials (Yetter et al. 2009). By diminishing the particle-size diameter of a nano-Al, the number of atoms in the periphery of particle goes up to enter the chemical reaction. Figure 5.2 depicts the outcome of nano-Al particle size and the ratio of atoms on the periphery of aluminium layer (Sundaram et al. 2017). From the figure, it can be depicted that by stepping down the particle size of nano-Al, the fraction of surface atoms increases exponentially (Table 5.1).

The combustion of nano-Al particles in detonation has received considerable attention, especially in military-related research. However, detonating nano-Al with explosives has a shortcoming of incomplete combustion of Al particles (Grishkin et al. 1993). So, while aluminizing explosives, combustion efficiency has to be considered, especially at high ratio of nano-Al (greater than 30%) (Trzciński et al. 2007; Jouet et al. 2006).

5.2 Catalytic Activity of Nano-Al on Ammonium Perchlorate

Ammonium perchlorate (AP) is most commonly used as solid propellant because it is inexpensive and is a strong oxidizer that contains a large amount of oxygen that assist in chemical reaction to form stable gaseous products. Much of the attention of researchers in the past has been paid to the study of nanoscale catalytic activity of thermal decomposition of Ammonium perchlorate due to its better catalytic activity (Liu et al. 2004). The main drawback with nanoscale catalyst is their agglomeration due to their small size and high surface area, thus making it difficult to disperse uniformly with other propellants. Thus to improve thermal decomposition, solid propellants are bind primarily with polymeric binder like hydroxyl-terminated poly butadiene (HTPB). Nano-Al particles are deployed in AP to improve the specific impulse of AP (Price and Sigman 2000). Yang et al. studied the effect on specific impulse of AP/HTPB/Al nanocomposite by varying the percentage of HTPB with various content of aluminium loading (Brill et al. 2000). The maximum impulse was for a propellant having 10% HTPB, 18% Al, and 72% AP content. Armstron et al. show that by altering the size of aluminium particles to nanoscale, the burning rate of the AP-based propellants increases considerably from 1 to 100 mm/s (Armstrong et al. 2003). Armstrong et al. performed several comparative experiments at atmospheric as well as at higher pressure on conventional and nanometre-sized aluminium powder and shown in Fig. 5.3.

Liu et al. performed a series of experiments on Al, Cu, and Ni metals having varying size (micrometre-to-nanometre) and based on his study, he reveals that the decomposition temperature of AP decreases drastically with the addition of these nanopowders (Liu et al. 2004a; b). With the addition of nano-Al, nano-Cu, or nano-Ni powders, decomposition temperature of AP decreases by 51.8, 130.2, and 112.9 °C, respectively. Verma et al. prepare an energetic formulation of Al and AP with HTPB as binder by varying the overall size of composite as 15, 70, 200, and 1000 gm (Verma and Ramakrishna 2014). Solid propellant of AP having different sizes i.e., coarse and fine and different packing densities was studied. Observation made by this study is that the burn-rate pressure index increases as the AP size is decreased. In another study by Jayaraman et al., it was observed that with the variation in AP size does not work well with all formulation, but when the nano-Al is embedded in between coarse and fine AP matrix then only better combustion can be obtained (Jayaraman et al. 2007; Jayaraman et al. 2009). Similar findings were made by Dokhan et al., and he suggested that near complete combustion surface nano-Al facilitates in giving heat to the propellant and helps in controlled rate of burning (Dokhan et al. 2002). Addition of nano-Al with AP has drawback of forming alumina (Al_2O_3) during combustion. This alumina by forming agglomerates decelerates reaction gases and flow losses (Boraas 1984). Gaurav et al. established that by mechanically activating aluminium, good burn rates can be achieved (Gaurav and Ramakrishna 2016). In this study, nano-Al is mechanically activated by ball-milling process to reduce the agglomerate mass. By reducing the

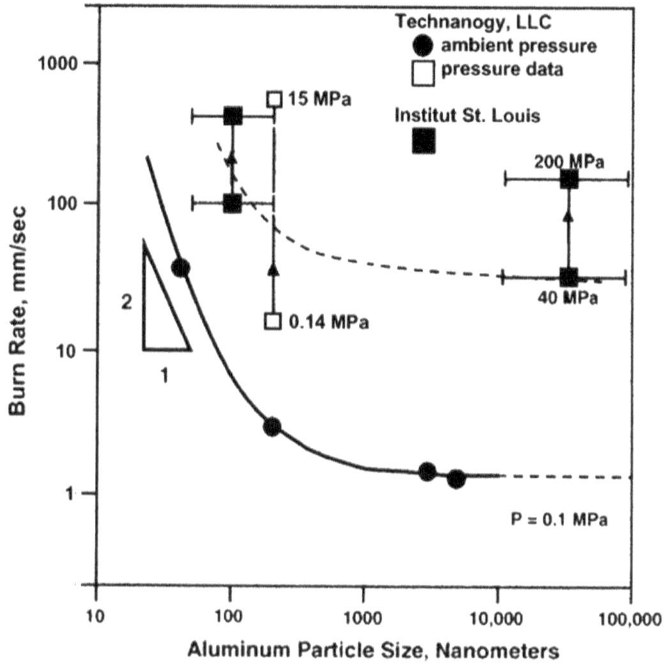

Fig. 5.3 Burning rate with aluminium particle size along with indicated pressure ranges. Reprinted with permission from Armstrong et al. (2003). Copyright (2003) American Chemical Society

agglomeration, burn-rate pressure index reduces and also rate of burning was reported higher as compared to other mixture of nano-Al and Poly tetra fluoro ethylene (PTFE). Pivkina et al. have studied the combustion behaviour of nanoparticles of aluminium and ultra-fine ammonium perchlorate and concluded higher rates of burning as compared to either of the reactants in their micrometre scale (Pivkina et al. 2007). Similar findings were of Romonadova et al. who synthesized the pellets of ammonium perchlorate and nano-sized aluminium mixtures (Romodanova and Pokhil 1970). The gist of above investigation is almost 100% increment of the burning rate of propellants by adding nano-Al. The underlying mechanism for all these investigations is that nano-Al particles that were nearer to the periphery of the propellant show positive increment in the heat feedback to the propellant and burn almost entirely within a small distance. This further causes to accelerate the decomposition reaction of the propellant and thus burning rate.

Nanoparticles are quite sensitive to factors like morphology, composition, size, and passivation methods. Aluminium particles are passivated by an oxide shell of nanometre scale thickness which covers aluminium particle instantaneously as the powder is exposed to air. Many researchers focus on this defect and come to the solution that nano-Al powder to be coated with an organic species like oelic acids (Fernando et al. 2008). Recently, H. Wang et al. have synthesized monodispersed micro particles of nano-Al encapsulated within AP and nitrocellulose (NC) binder (Wang et al. 2017). In his investigation, researchers synthesized Al/AP/NC with different composition of binder and show that with the addition of nitrocellulose (NC) binder, higher flame temperature is reached (500 K higher) than conventional Al/AP and ignition temperature is significantly lower than the melting point of Al. Researchers proposed that the possible reason for this behaviour is due to weakening of protective oxide layer of nano-Al because of gaseous acid released from AP.

5.3 Catalytic Activity of Nano-Al on RDX

Research Department explosive commonly said RDX is an explosive and chemically known as cyclotrimethylenetrinitamine. Efficiency of explosives can be improved with the addition of metal particles to explosives. Aluminium powders are widely used in explosives to enhance their energetic characteristics and blast effects (Antipina et al. 2017). The aluminium particles react with the explosive product and improve explosive heat, blast temperature, and energy release time (Carney et al. 2006). For explosives, a key feature is shock initiation which contains pressure spike and isentropic expansion region (Tarver et al. 2007). For the location of shock, wavefront researchers often measure detonation velocity. With the addition of nano-Al to explosives, different responses on detonation front are exhibited. To validate this detonation characteristic, several experiments are performed like heat of explosion, cylinder test, and flyer plate impact test.

Wang et al. has compared micro-sized Al with nano-Al in explosion test with RDX and showed that heat of explosion for explosives using nano-Al is lesser as compared to micrometre-sized aluminium by adding 20−40 wt% (Wang et al. 2014). In his study, he reveals that by adding 30 wt% micro-Al with 5 wt% nano-Al, heat of explosion of RDX increases by 5.83% and concluded that by adding a little amount of nano-Al can improve rate of oxidation in micro-Al. Huang et al. studied the acceleration ability of aluminium-based explosives having varying diameter of Al (from microsize-to-nanosize) and found that with nano-Al-mixed explosives surface velocity in case of metal plate is higher along with lesser time of reaction as compared to micro-sized aluminium RDX-based explosives (Huang et al. 2006). He concluded that the degree of reaction of aluminium in the reaction zone raises as the particle diameter of Al decreases resulting in rapid release of detonation energy. Conghua et al. studied the thermal decomposition of RDX with nano-Al powder prepared by DC arc plasma method (Hou et al. 2013). By the addition of nano-Al in RDX, peak temperature of decomposition decreases by 4.36 °C along with decrease in energy of decomposition by about 11 kJ/mol.

5.4 Catalytic Activity of Nano-al on HMX

Octahydro-1,3,5,7-tetranitro-1,3,5,7-tetrazocine ($C_4H_8N_8O_8$) is popularly known as HMX. It is an essential ingredient of many propellants and explosives. HMX shows many polymorphs, namely α, β, γ, and δ, out of which β-form shows the highest density and lowest impact sensitivity. HMX explosives show high density, lower toxicity as well as corrosion, and high specific impulse. Due to high-pressure exponent, HMX is not a good candidate for a rocket motor operation. To tailor the HMX properties, addition of aluminium powder can be employed to get higher energy release. Muravyev et al. (2010) studied the influence of particle size of aluminium and HMX with different mixing techniques. With the addition of nano-Al with micrometre-sized HMX, the rate of burning raises 2.5 times along with hike of 4 times combustion completeness. In this work, two mixing techniques were employed conventional (dry) and wet (ultrasonic processing), and it was observed that wet mixing increases 18% burning rate. By adding nano-Al with HMX, metal particles of nano-Al burn along with HMX thus generating higher heat flux into the condensed phase.

5.5 Catalytic Activity of Nano-al on TBX

TBX stands for Thermobaric explosives which are intentionally tailored to enhance the secondary combustion that could further sustain pressure and thermal loading. TBX has received lots of attention because they can be used as a further origin of fatal energy against individuals residing beneath the caves or bunkers (confined spaces). Detonation of TBX has dual action: anaerobic action where air is sucked by primary explosion and then a delayed aerobic burning action of the secondary mixture. TBX are developed by double-base (DB) propellant by incorporating energetic fillers like nitamines (HMX, RDX) (Sun et al. 2015; Wu et al. 2011). Aluminium is added to TBX to improve performance and to enhance the overall energy release, as well as the blast performance. When Aluminium is added to TBX, nano-Al reacts with oxygen and nitrogen of air along with the gases generated by primary decomposition. Following exothermic reactions (5.1–5.5) take place for the secondary reaction pathways (Krier et al. 2011; Conkling and Mocella 2010):

$$2\,\mathrm{Al} + 3/2\mathrm{O}_2 \rightarrow \mathrm{Al}_2\mathrm{O}_3 + 1700\,\mathrm{kJ/mol} \tag{5.1}$$

$$2\mathrm{Al} + 3\mathrm{CO} \rightarrow \mathrm{Al}_2\mathrm{O}_3 + 3\mathrm{C} + 1251\,\mathrm{kJ/mol} \tag{5.2}$$

$$2\mathrm{Al} + 3\mathrm{H}_2\mathrm{O} \rightarrow \mathrm{Al}_2\mathrm{O}_3 + 3\mathrm{H}_2 + 866\,\mathrm{kJ/mol} \tag{5.3}$$

$$2Al + 3CO_2 \rightarrow Al_2O_3 + 3CO + 741\,kJ/mol \tag{5.4}$$

$$2Al + N_2 \rightarrow 2AlN + 346\,kJ/mol \tag{5.5}$$

Nano-Al having large interfacial surface area to initiate the secondary combustion reaction (Conkling and Mocella 2010). Nano-Al readily reacts with air–oxygen and form a compact coating of Al_2O_3 which protects the inner core of aluminium from additional oxidation thus nano-Al can be stored for a prolonged period of time (Conkling and Mocella 2010). Reaction of aluminium in Chapman–Jouguet (CJ) plane is slow due to alumina coating on the surface of Al which impedes its reactivity. Thus by combination of coarse particle and nano-sized particle of aluminium can enhance secondary charge combustion. In TBX detonation, the primary combustion takes place nearly at temperature of 3380 °C, which is fair enough to explode the secondary charge along with release of oxygen and removal of oxide layer. For the development of TBX, optimization of chemical composition so as to get sustained burning and detonation action is an essential factor (Sućeska 1999; Alia and Souli 2006).

Burning rates of flame zone can be enhanced either by raising the combustion temperature or pressure. The combustion pressure and temperature can be regulated by combining double-base propellant with active fuel like aluminium or boron. These active fuels have capacity to catalyse exothermic reactions thus boost burning rate (Bhat et al. 1986). With the combination of oxidizers like AP or AN along with nano-Al in double-base propellant gives high combustion temperature, high heat output, and sturdy burning rate as well as higher specific impulse (Kubota 2002). Sherif et al. in his experiment reported that on mixing nano-Al in double-base propellant of AP/AN were reported to have higher calorific value and better thermal stability (Elbasuney et al. 2017).

5.6 Catalytic Activity of Nano-al Thermite Composite on Ammonium Perchlorate

Nano-Al thermite composite represents a new class of meta-stable intermolecular composites (He et al. 2018) which constitute nanoscale metal fuel (commonly aluminium nanoparticles) and metal oxides like CuO (Patel and Bhattacharya 2013), Bi_2O_3 (Patel et al. 2015a, 2018), and Co_3O_4 (Patel et al. 2015b) as oxidizers. These nanothermites are phenomenally reactive and produce an ultra-fast release of their stored energy for civilian, defence, space, materials processing, etc. applications (Zhou et al. 2014). Many nanoscale metal oxides such as CuO (Patel 2013; Patel and Bhattacharya 2017), ZnO (Patel et al. 2017), and CO_3O_4 (Patel et al. 2017) earlier revealed a high catalytic activity in thermal decomposition of solid propellants like potassium perchlorate (Patel 2013; Patel et al. 2017) and potassium periodate (Patel and Bhattacharya 2017). Li et al. (2015) hybridized one of the nanoscale metal oxide i.e., iron oxide with nanoscale aluminium and boron microparticle and investigated the catalytic activity of nano-Al/micro-B/nano-Fe_2O_3 nanothermite composite on thermal decomposition of ammonium perchlorate. It was observed that adding 7 weight percent of nano-Al/B/Fe_2O_3 composites had decreased the decomposition temperature of AP by 69 °C and enhanced its apparent decomposition heat by 382.3 J/g. They demonstrated the order of catalytic role on individual basis on thermal decomposition of AP and revealed the order of catalytic activity as nano-Fe_2O_3 > nano-Al > micro-B. Thus the nanoscale iron oxide plays a dominant role in thermal decomposition of AP. They found that the catalytic activity of sol–gel-derived nano-Al/micro-B/nano-Fe_2O_3 composites was much higher than the simple mixed one and that was attributed to the nanoporous (high pore volume: 0.5 cm^3/g) structure and large specific surface area (175.0 m^2/g) of sol–gel-derived composites than the simple mixed one (pore volume: 0.16 cm^3/g, specific surface area: 18.2 m^2/g).

5.7 Conclusion

Aluminium owing to its good combustion enthalpy, easy availability, and low toxicity is widely used in propellants, pyrotechnics, and explosives. Nano-Al increases the energy and flame temperature in propellants and thus directly increases specific impulse in propellants. Nano-Al is intentionally mixed with explosives to increase reaction temperatures, air blast, and increase energies in explosives. Owing to high specific area of nano-Al compared to micron-sized material, they are widely used for performance enhancement, the rate of burning as well as combustion efficiency of energetic materials. Contrary to this, in rocket propulsion nano-Al is not used as compared to conventional micron-sized Al because of the formation of agglomeration of Al_2O_3 slag. Nano-aluminium has been emerged as an efficient catalyst in thermal decomposition of various energetic materials such as ammonium perchlorate, RDX, HMX, and TBX. Nano-Aluminium when used as fuel in nanothermite formulation (nano-Al/micro-B/nano-Fe_2O_3) expressed effective catalytic activity in thermal decomposition of ammonium perchlorate.

References

Alia A, Souli M (2006) High explosive simulation using multi-material formulations. Appl Therm Eng 26:1032–1042

Antipina SA, Zmanovskii SV, Gromov AA, Teipel U (2017) Air and water oxidation of aluminum flake particles. Powder Technol 307:184–189

Arkhipov Arkhipov VA, Korotkikh AG (2012) The influence of aluminum powder dispersity on composite solid propellants ignitability by laser radiation. Combust Flame 159:409–415

Armstrong RW, Baschung B, Booth DW, Samirant M (2003) Enhanced propellant combustion with nanoparticles. Nano Lett 3:253–255

Bhat VK, Singh H, Khare RR, Rao KRK (1986) Burning rate studies of energetic double base propellants. Def Sci J 36:71–75

Boraas S (1984) Modeling slag deposition in the space shuttle solid rocket motor. J Spacecr Rockets 21:47–54

Brill TB, Ren WZ, Yang V (eds) (2000) Solid propellant chemistry, combustion, and motor interior ballistics. American Institute of Aeronautics and Astronautics, Reston, Virginia

Carney JR, Miller JS, Gump JC, Pangilinan GI (2006) Time-resolved optical measurements of the post-detonation combustion of aluminized explosives. Rev Sci Instrum 77:063103

Conkling JA, Mocella C (2010) Chemistry of pyrotechnics: basic principles and theory. CRC Press, Boca Raton, FL

Dokhan A, Price EW, Seitzman JM, Sigman RK (2002) The effects of bimodal aluminum with ultrafine aluminum on the burning rates of solid propellants. Proc Combust Inst 29:2939–2946

Elbasuney S, Fahd A, Mostafa HE (2017) Combustion characteristics of extruded double base propellant based on ammonium perchlorate/aluminum binary mixture. Fuel 208:296–304

Fernando KS, Smith MJ, Harruff BA, Lewis WK, Guliants EA, Bunker CE (2008) Sonochemically assisted thermal decomposition of alane N, N-dimethylethylamine with titanium (Iv) isopropoxide in the presence of oleic acid to yield air-stable and size-selective aluminum core—shell nanoparticles. J Phys Chem C 113:500–503

Gaurav M, Ramakrishna PA (2016) Effect of mechanical activation of high specific surface area aluminium with PTFE on composite solid propellant. Combust Flame 166:203–215

Grishkin AM, Dubnov LV, Davidov VY, Levshina YA, Mikhailova TN (1993) Effect of powdered aluminum additives on the detonation parameters of high explosives. Combust Explos Shock Waves 29:239–241

He W, Liu PJ, He GQ, Gozin M, Yan QL (2018) Highly reactive metastable intermixed composites (MICs): preparation and characterization. Adv Mater 3:1706293

Hou C, Geng X, An C, Wang J, Xu W, Li X (2013) Preparation of Al nanoparticles and their influence on the thermal decomposition of RDX. Cent Eur J Energ Mater 10(1):123–133

Huang H, Huang HJ, Huang Y (2006) The influence of aluminum particle size and oxidizer morphology in RDX-based aluminized explosives on their ability to accelerate metals. Explos Shock Waves 26:7

Huang Y, Risha GA, Yang V, Yetter RA (2007) Combustion of bimodal nano/micron-sized aluminum particle dust in air. Proc Combust Inst 31:2001–2009

Jayaraman K, Anand K, Chakravarthy S, Sarathi R (2007) Production and characterization of nano-aluminum and its effect in solid propellant combustion. In: 45th AIAA aerospace sciences meeting and exhibit 1430

Jayaraman K, Anand KV, Bhatt DS, Chakravarthy SR, Sarathi R (2009) Production, characterization, and combustion of nanoaluminum in composite solid propellants. J Propul Power 25:471–481

Jouet RJ, Granholm RH, Sandusky HW, Warren AD (2006) Preparation and shock reactivity analysis of novel perfluoroalkyl-coated aluminum nanocomposites. AIP Conf Proc 845:1527–1530

Krier H, Peuker J, Glumac N (2011) Aluminum combustion in aluminized explosives: aerobic and anaerobic reaction. In 49th AIAA aerospace sciences meeting including the new horizons forum and aerospace exposition 645

Kubota N (ed) (2002) Propellants and explosives: thermochemical aspects of combustion. Wiley-VCH, New York, pp 123–156

Li GP, Shen LH, Xia M, Liu MH, Luo YJ (2015) Preparation of aluminum/boron/iron-oxide nanothermites and catalytic activity for ammonium perchlorate. J Propul Power 31:1635–1641

Liu L, Li F, Tan L, Ming L, Yi Y (2004a) Effects of nanometer Ni, Cu, Al and NiCu powders on the thermal decomposition of ammonium perchlorate. Propellants Explos Pyrotech 29:34–38

Liu L, Li F, Tan L, Li M, Yang Y (2004b) Effects of metal and composite nanopowders on the thermal decomposition of ammonium perchlorate (AP) and the ammonium perchlorate/hydroxyterminated polybutadiene (AP/HTPB) composite solid propellant. Chin J Chem Eng 4:595–598

Muravyev N, Frolov Y, Pivkina A, Monogarov K, Ordzhonikidze O, Bushmarinov I, Korlyukov A (2010) Influence of particle size and mixing technology on combustion of HMX/Al compositions. Propellants Explos Pyrotech 35:226–232

Patel VK (2013) Sonoemulsion synthesis of long CuO nanorods with enhanced catalytic thermal decomposition of potassium perchlorate. J Cluster Sci 24:821–828

Patel VK, Bhattacharya S (2013) High-performance nanothermite composites based on aloe-vera-directed CuO nanorods. ACS Appl Mater Interfaces 5:13364–13374

Patel VK, Bhattacharya S (2017) Solid state green synthesis and catalytic activity of CuO nanorods in thermal decomposition of potassium periodate. Mater Res Exp. 4:095012

Patel VK, Kant R, choudhary A, Painuly M, Bhattacharya S (2018) Performance characterization of Bi_2O_3/Al nanoenergetics blasted micro-forming system. Def Technol. https://doi.org/10.1016/j.dt.2018.07.005

Patel VK, Ganguli A, Kant R, Bhattacharya S (2015a) Micropatterning of nanoenergetic films of Bi_2O_3/Al for pyrotechnics. RSC Adv 5:14967–14973

Patel VK, Saurav JR, Gangopadhyay K, Gangopadhyay S, Bhattacharya S (2015b) Combustion characterization and modeling of novel nanoenergetic composites of Co_3O_4/nAl. RSC Adv 5:21471–21479

Patel VK, Sundriyal P, Bhattacharya S (2017) Aloe vera vs. poly (ethylene) glycol-based synthesis and relative catalytic activity investigations of ZnO nanorods in thermal decomposition of potassium perchlorate. Part Sci Technol 35:361–368

Pivkina AN, Frolov YV, Ivanov DA (2007) Nanosized components of energetic systems: structure, thermal behavior, and combustion. Combust Explos Shock Waves 43:51–55

Price EW, Sigman RK (2000) Combustion of aluminized solid propellants. In: Yang V (ed) Solid propellant chemistry, combustion, and motor interior ballistics, vol 185, pp 663–687

Romodanova LD, Pokhil PK (1970) Action of silica on the burning rates of ammonium perchlorate compositions. Combust Explos Shock Waves 6:258–261

Sućeska M (1999) Evaluation of detonation energy from EXPLO5 computer code results. Propellants Explos Pyrotech 24:280–285

Sun J, Simon SL (2007) The melting behavior of aluminum nanoparticles. Thermochim Acta 463:32–40

Sun C, Xu J, Chen X, Zheng J, Zheng Y, Wang W (2015) Strain rate and temperature dependence of the compressive behavior of a composite modified double-base propellant. Mech Mater 89:35–46

Sundaram D, Yang V, Yetter RA (2017) Metal-based nanoenergetic materials: synthesis, properties, and applications. Pro Energy Combust Sci 61:293–365

Tarver CM, Forbes JW, Urtiew PA (2007) Nonequilibrium Zeldovich-von Neumann-Doring theory and reactive flow modeling of detonation. Russ J Phys Chem B 1:39–45

Teipel U (ed) (2006) Energetic materials: particle processing and characterization. Wiley, London

Trzciński WA, Cudziło S, Szymańczyk L (2007) Studies of detonation characteristics of aluminum enriched RDX compositions. Propellants Explos Pyrotech 32:392–400

Verma S, Ramakrishna PA (2014) Dependence of density and burning rate of composite solid propellant on mixer size. Acta Astronáut 93:130–137

Wang SP, Feng XS, Yao LN, Niu GT, Chao ST, Niu L (2014) The influence of nanometer aluminum on the explosion heat of RDX-based explosive. Initiators Pyrotech 1:21–24

Wang H, Jacob RJ, DeLisio JB, Zachariah MR (2017) Assembly and encapsulation of aluminum NP's within AP/NC matrix and their reactive properties. Combust Flame 180:175–183

Wu XG, Yan QL, Guo X, Qi XF, Li XJ, Wang K-Q (2011) Combustion efficiency and pyrochemical properties of micron-sized metal particles as the components of modified double-base propellant. Acta Astronaut 68:1098–1112

Yetter RA, Risha GA, Son SF (2009) Metal particle combustion and nanotechnology. Proc Combust Inst 32:1819–1838

Zhi J, Shu-Fen L, Feng-Qi Z, Zi-Ru L, Cui-Mei Y, Yang L, Shang-Wen L (2006) Research on the combustion properties of propellants with low content of nano metal powders. Propellants Explos Pyrotech 31:139–147

Zhou X, Torabi M, Lu J, Shen R, Zhang K (2014) Nanostructured energetic composites: synthesis, ignition/combustion modeling, and applications. ACS Appl Mater Interfaces 6:3058–3074

Part II
Fabrication of Nano-energetic Materials

Chapter 6
Nano-energetic Materials on a Chip

Jitendra Kumar Katiyar and Vinay K. Patel

Abstract After integration into an electronic chip, the nano-energetic materials have found a very important role as portable microscale energy systems in many applications such as micro-actuation for micro-fluidics, micro-ignition, micro-propulsion, and onboard micro-power unit. However, there has been an ongoing challenge to fabricate high-reactive and high-energy density nano-energetic materials and integrate it with microelectromechanical system (MEMS) devices. There are various micro-/nanofabrication methodologies to develop and integrate the nano-energetic materials into a chip. This chapter discusses the different approaches and methods of realization of nano-energetic materials on a chip and details about the ignition sensitivity, energy density, and combustion performance. The chapter also particularizes the application of the chip-integrated nano-energetic materials.

Keywords Nano-energetic materials · Nanothermites · Ignition Combustion · Energy

6.1 Introduction

Energetic materials (EMs) are those substances which stored chemical energy. These materials have strained an increasing interest in the different industries and communities such as public industry, combustion industry, and scientific community. These communities are very eye-catching foundations of onboard energy to produce blast, heat, and power. EMs can be classically categorized into three

---

J. K. Katiyar (✉)
Department of Mechanical Engineering, SRM Institute of Science and Technology, Kattankulathur, Chennai 603203, Tamil Nadu, India
e-mail: jitendrakumar.v@ktr.srmuniv.ac.in

V. K. Patel
Department of Mechanical Engineering, Govind Ballabh Pant Institute of Engineering and Technology, Ghurdauri, Pauri Garhwal 246194, Uttarakhand, India

S. Bhattacharya et al. (eds.), *Nano-Energetic Materials*, Energy, Environment and Sustainability, https://doi.org/10.1007/978-981-13-3269-2_6

categories such as propellants, explosives, and pyrotechnics. These categories are the basis of their specific arrangements, enactments, and presentation spaces (Fried et al. 2001; Rossi et al. 2007; Badgujar et al. 2008). Therefore, EMs could have a key impact on micro-/nano-energetic arena of application. These submissions include micro-thruster (Rossi et al. 2006; Youngner et al. 2000; Lewis et al. 2000; Teasdale et al. 2001; Lindsay et al. 2001; Takahashi et al. 2000; Tanaka et al. 2003; Pham et al. 2003; Zhao et al. 2004; Zhang et al. 2005), micro-initiation (Troianello 2001; Di Biaso et al. 2004; Laucht et al. 2004; Hofmann et al. 2006; Barbee et al. 2005), actuation of gases (including injection or moving fluid) (Hinshaw 1995; Rossi and Estève 1997; Rossi et al. 1999; Hong et al. 2003; Norton and Minor 2006), chemical reaction of gases (Li et al. 1988; Pile àgrave 2005; Vasylkiv and Sakka 2005; Vasylkiv et al. 2006), heating system (power) and welding (Stewart 2004), and switching (Pennarun et al. 2006). To achieve the targeted application yield gas, heat, or chemical compound, various researchers have successfully investigated EMs by tailoring in their chemical composition or methods of development. The researchers used the various methods to fabricate the EMs such as chemical treatment, DC reactive magnetron sputtering, electrophoresis. They used Al or Si as a substrate material to develop EMs or structure/cell in EMs. For the development of EMs, researchers used copper oxide, bismuth oxide, iron oxide, molybdenum di- or trioxide, etc. In all developments, a key task is still not really lectured which is compatible nature of the EMs with MEMS systems. Two methods were evolved for handling this issue. Initially, thin film of EMs is grown or deposited at low temperature (<250 °C) but in specific cases EMs film is deposited over the free-standing structures which are in micron range (approximately 1 μm) with controlled stress. These structures are not required at heat treatment process. Moreover, few researchers have already investigated and fabricated semiconductor chip breaker, chip circuit board on silicon surface in which they applied familiar semiconductor dispensation technique such as physical vapor deposition (PVD) with liftoff (Tappan et al. 2003, 2005), display printing (Rossi et al. 2006), or organic reaction (Laucht et al. 2004). In the second method, minimize the heat loss, one of the greatest thought-provoking issues to be explained in MEMS applications. Thus, we believe that apart from traditional Ems, nowadays, nano-energetic material composite is growing for better actuation in ignition, combustion, propulsion, etc. These nano-energetic materials are in the form of shell/structures. The wide application of nano-energetic materials and their different methods of development motivate us for comprehensive study.

Therefore, in this chapter, we focus on nano-energetic composites in the form of structure or film, predominantly progresses over the preceding several years. The chapter has been distributed into six sections. Section 6.1 is the introduction. Section 6.2 comprehensively discusses Al/CuO-based thin energetic film, Al/bismuth oxide (Bi_2O_3)-based thin energetic film, and Al/MoO_x-based thin energetic film. Section 6.3 introduces the Mg/CuO- or MnO_x-based thin energetic films. Section 6.4 comprehensively discusses other materials such as CuPc/MWCNT/$NiCo_2O_4$-based thin energetic film. Section 6.5 presents the development of super-hydrophobic nano-energetic materials (nEMs) for underwater application. Finally, Sect. 6.6 states the conclusions.

6.2 Micro-/Nanofabrication of Thin Energetic Film/Structure

Most common processes used by researchers to fabricate micro-/nanothin energetic films are DC reactive magnetron sputtering process. Apart from sputtering process, researchers also used doctor blading, electrophoresis, COMS procedure, and electrochemical anodizing process. The detailed discussion about the different types of thin micro-/nano-energetic film/structures using above-stated methods is explained in below subsections.

6.2.1 Al/CuO-Based Thin Energetic Film/Structure

There are various methods available for depositing the nano-energetic thin film over substrate. Ideally, silicon wafer is used as a substrate material, but nowadays Al is used as a substrate material for generating energetic film of oxide because of its advantages over silicon. The probable applications of "nano-energetic-on-a-chip" systems are very widespread such as micro-propulsion, vehicles with nanoairbags, nanofluidic, and nanorocket, nanofuze. In this series, Petrantoni et al. (2010) fabricated a multilayered Al/CuO thermite using DC reactive magnetron sputtering method (shown in Fig. 6.1). The process has been designed in such a way that it produces stress less than 50 MPa and each deposited layer produces in the range of 10 nm to 1 μm. Moreover, they have differentiated the micro- and nanostructure by the film thickness, if the film thickness is considered above 500 nm known as microstructure and a nanostructure if the film thickness is considered below 500 nm. From this study, they have also confirmed that microstructured samples decay in two-step reactions: one exothermic reaction of 0.7 kJ/g at 790 K and second exothermic reaction of 1.3 kJ/g among 1036 and 1356 K which confirms a dissimilarity. Therefore, for nanostructured films, only one exothermic reaction (1.2 kJ/g) is required at lower temperatures, about 740 K.

Further, Zhu et al. (2011) fabricated the energy igniter for micro-ignition system using Al and CuO reactive multilayered films (RMFs) with chromium films. They characterized fabricated energy igniter using scanning electron microscopy (SEM), X-ray diffraction (XRD), and differential scanning calorimetry (DSC) analytical methods which were used for determining heat reaction. They observed that the Al/CuO RMFs produce a heat of reaction which is equal to 2760 J/g. In addition, the initial reaction temperature of Al/CuO RMFs drastically declines by 300 °C which is inferior than that of the bulk. Furthermore, Yang et al. (2012) investigated the effect of nanostructures on the exothermic reaction and ignition of Al/CuO_x-based EMs. They have fabricated micro-/nano-EMs using thermal oxidation method in which Cu thin films of 1 μm are deposited onto silicon substrates followed by thermal evaporation method in which Al nanoparticles are embedded over Cu thin film. They characterized produced Al/CuO thin film using SEM, energy-dispersive X-ray analysis (EDX),

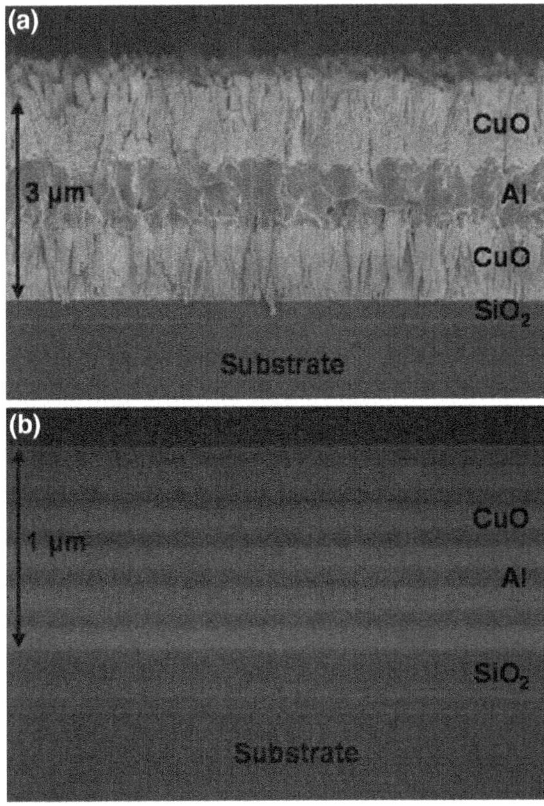

Fig. 6.1 Cross-sectional images of CuO/Al multilayers by SEM. **a** Three layers of CuO/Al/CuO (1 μm each). **b** Ten layers of CuO/Al (100 nm each). Reprinted with kind permission from Petrantoni et al. (2010)

XRD, differential thermal analysis (DTA), and DSC. They observed that an exothermic reaction between CuO nanowires and Al occurred before Al melting. They have mentioned that the CuO nanowires consist of higher surface energy and the deposited nano-Al encourages the solid–solid diffusion process for better thermite reactions and ignition properties. Zhou et al. (2011) studied the influence of Al/CuO reactive multilayer films' additives on exploding foil initiator. The reactive multilayered film was fabricated by using standard microsystem technology and RF magnetron sputtering technology, respectively. A high-speed camera was used for observing the explosion process which has shown an occurrence of a fierce combustion process but simultaneously temperature analysis demonstrated 30% greater temperature of the explosion of Cu/Al/CuO RMFs as compared to Cu film. This finding confirmed that a thermite reaction of Al/CuO RMFs was activated when a large amount of heat is released. Thus, the higher temperature could notionally encourage the development of Cu plasma, which finally would augment the slapper velocity. Furthermore, Zhu et al. (2013) described the ignition characteristics of Al/CuO nano-energetic multilayer films (nEMFs) integrated with semiconductor bridge (SCB). For fabrication of Al/CuO-SCB, they have used the COMS procedure. From COMS, they observed a distinct layered geometry of Al/CuO nEMFs which has been deposited by sputtering. Further, the layered structure has given a reaction heat equal to 2181 J/g. They also conducted firing experiments, from which they have concluded that Al/CuO nEMFs displayed no impact on the electrical properties of SCB. Furthermore, the quick combustion of Al/CuO nEMFs has accomplished to assist SCB producing high-temperature plasma and products by which ignition reliability is enhanced. Furthermore, Bahrami et al. (2014) studied the stoichiometry effect and thickness of various layers of Al/CuO nanolaminates on the rate of burning as well as the release rate. They deposited multilayers by a DC reactive magnetron sputtering method. The range of each deposited layer is 25 nm to 1 μm. They characterized deposited films of Al/CuO using high-resolution transmission electron microscopy (HR-TEM), XRD, and X-ray photoelectron spectroscopy (XPS). From the results, they observed that under stoichiometric conditions, the decrement in bilayer thickness causes a drastic increment in reactivity. The burning rate for 150 nm bilayer thickness has been reported as 80 m/s. Moreover, Shen et al. (2014) investigated the pressure loss and compensation in the combustion process of Al/CuO metastable intermolecular composite (MIC) (nano-energetic) on a micro-heater chip. They proposed a model for pressure change in the combustion process which is ball cell model. This model shows pressure loss properties on rate of reaction and respective energy output efficiency. They concluded that after addition of fine CL-20 particles in Al/CuO MIC, the combustion process reaction has turned into intense deflagration and causes higher maximum pressure and pressurization rate. They optimized the CL-20 particle concentration (50%) on Al/CuO MIC for best combustion process and high pressure. Moreover, the mechanism of these layered structures of Al/CuO nanothermite was addressed by Sui et al. (2017). They also compared Al/CuO nanothermite with Al/NiO nanothermite. They used vacuum filtration method for the deposition of these nanothermite layer structures. After the reaction, they observed that the gap was formed between layers of Al and CuO but layers of Al and NiO remain attached (shown in Fig. 6.2). They also concluded that these structures give rapid heat release and also improve reaction initiation.

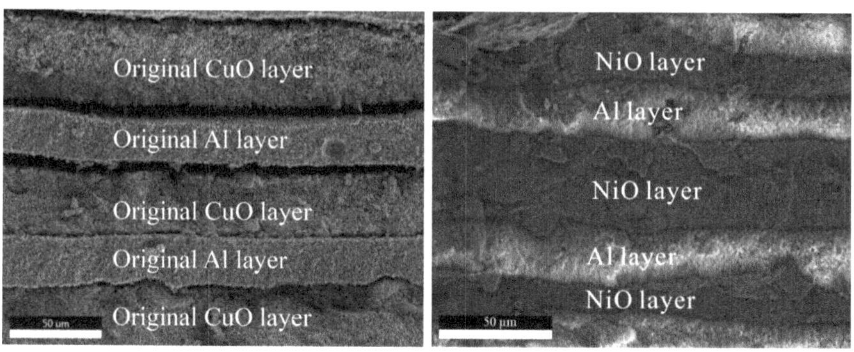

Fig. 6.2 Difference between Al/CuO and Al/NiO multilayered structure after heat reaction. Reprinted with kind permission from Sui et al. (2017)

Apart from the multilayered film of Al/CuO, the tubular structure of highly reactive nEMs is studied by Yin et al. (2017). They prepared the structure by nanotubes and nanorods of diameter 100–200 nm and 5–7 μm typical lengths. They deposited nano-Al by electrophoretic deposition method. They have observed that Al/CuO nanotubes' composites have given suggestively better energy output and combustion performance. This is due to the effective and familiar contact as well as mass transmission between fuel and oxide. Thus, they proposed that nanoarrays with a tubular structure have a superior perspective to enhance the exothermic reaction of nEMs. Further, Yu et al. (2018a) proposed a new method for fabrication of nanostructured Al/CuO thermite film over Cu substrate. They developed nanorods of CuO from Cu substrate using electrochemical anodizing process. Al film is deposited by magnetron sputtering method on developed nanorods. Using this method, a core/shell Al/CuO nanothermite film is formed. They characterized the film using thermal analyses and laser ignition test and observed an outstanding energetic ability. They have also proposed that this technology is compatible with microelectromechanical systems (MEMS). Furthermore, the surface of developed core-/shell-structured Al/CuO nanothermite over Cu foil was functionalized by Xiang et al. (2018) and investigated storage stability for long-term and steady combustion performance. They fabricated dual types of super-hydrophobic Al/CuO core-/shell-structured nanothermites using solution chemistry method followed by magnetron sputtering and surface treatment. Further, they reported that both types of nanothermites demonstrated brilliant thermal performances. They also observed that the functionalized Al/CuO nanothermites show long-term storage stability and improved resistibility to the humid environment. Further, they carried out the combustion analysis of functionalized Al/CuO nanothermites. They reported a stable combustion process with the flame circulation speed around 100 m/s. Finally, Nicollet et al. (2018) integrated the fast circuit breaker (FCB) on Al/CuO nanothermites. These circuit breakers show a vital role in electrical system for protection application. These are shield against overcurrent, external discomposure, and short circuit of a broad range of equipment and systems. The ignition time of the fabricated device is less than 100 μs for few milligram of nanothermite powder. They also confirmed the good procedure and reproducibility of the device (100% of success rate). They obtained ample lower response time with comparison of classical mechanical circuit breakers (CMCB), which are slow. The applications of these types of devices are storage of electricity, manufacturing of aerospace components, human welfare, destruction descend opening, road vehicles, machines which is battery powered, etc.

From the above study, we can understand that the nanothermites' thin film offers a significant energy density, which is very useful when energetic material is integrated with small area. Therefore, Murray et al. (2018) conducted the feasible study of piezoelectric inkjet printing for printing an Al/CuO nanothermite thin film (Schematic diagram is shown in Fig. 6.3). Utilizing the inkjet printer, they deposited the energetic material with varying width and thickness.

Further, they reported that obtainable new technology of depositing energetic material is very useful for micro-propulsion mechanisms and low-scale electronics.

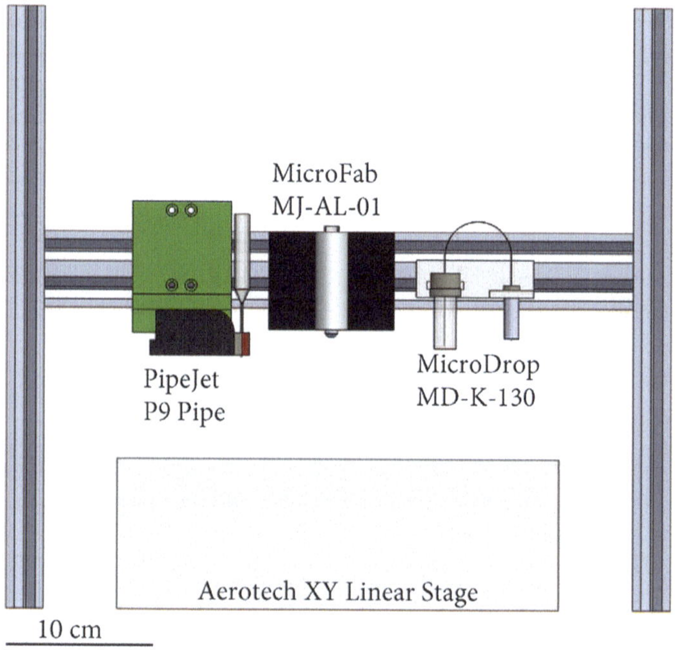

Fig. 6.3 An Aerotech PlanarDL 200-XY, 200 mm travel, 0.5 μm accuracy linear positioning stage mounted beneath a MicroFab MJAL—one piezoelectric nozzle, a MicroDrop MD-K—ten piezoelectric nozzles, and a PipeJet P9 piezoelectrically actuated pipette. Reprinted with kind permission from Murray et al. (2018)

6.2.2 Al/Bismuth Oxide (Bi_2O_3)-Based Thin Energetic Film

From the previous section, we have understood that Al/CuO nanothermites either in the form of films or nanotubes and nanorods are widely used in different applications such as storage of electricity, manufacturing of aerospace components, human welfare, destruction descend opening, road vehicles, machines which is battery powered. But apart from Al/CuO material combination, few researchers used bismuth oxide nanothermites over Al film. In this series, Staley et al. (2011) fabricated a silicon-based bridge wire microchip initiators using Al/Bi_2O_3 nanothermites. They reported improved electrothermal adaptation efficiency of the initiators using a nanoporous silicon bed, and these micro-initiator platforms are profoundly sufficient to be applied toward an extensive spectrum of pyrotechnic MEMS applications. Further, they investigated the electrical behavior of ignition element with or without Bi_2O_3 nanothermites. They found Bi_2O_3/Al-filled micro-initiators showing controllably initiation over a 30–80 microjoule firing energy with a 100% success rate. In case of high power input, the response time less than 2 μs was observed. Furthermore, Patel et al. (2015) demonstrated the micro-patterning of Al/Bi_2O_3 nano-energetic film. In the initial stage, they developed Bi_2O_3 nanosquare tablets with appropriate dimensions. They perform gold-sputtering method to develop nanosquare tablet on silicon substrates followed by chemical bath deposition process (shown in Fig. 6.4). They observed the exothermic reactivity of nano-energetic thin film for different sputter-coated aluminum thicknesses. They achieved this using TG-DSC measurements. They varied the temperature from 50 to 800 °C and performed the test at a constant heating rate of 10 °C min^{-1} under nitrogen atmosphere. They reported that 140-nm thickness nano-energetic film shows highest reaction heat, low commencement temperature, and outstanding pressure–time appearances and their high-resolution micro-patterning on silicon substrates. The demonstrated micro-patterning is very useful for pyrotechnics.

6.2.3 Al/MoO_x-Based Thin Energetic Film

Researcher improves the performance of semiconductor bridge (SB) by the integration of nanothermites. These nanothermites show a metastable intermolecular composite. They are characterized by size of particles and their main constituent, a metal and its oxide, below 100 nm. In this series, Zhu et al. (2014) fabricated a semiconductor bridge with the integration of Al/MoO_x nanothermite film (shown in Fig. 6.5).

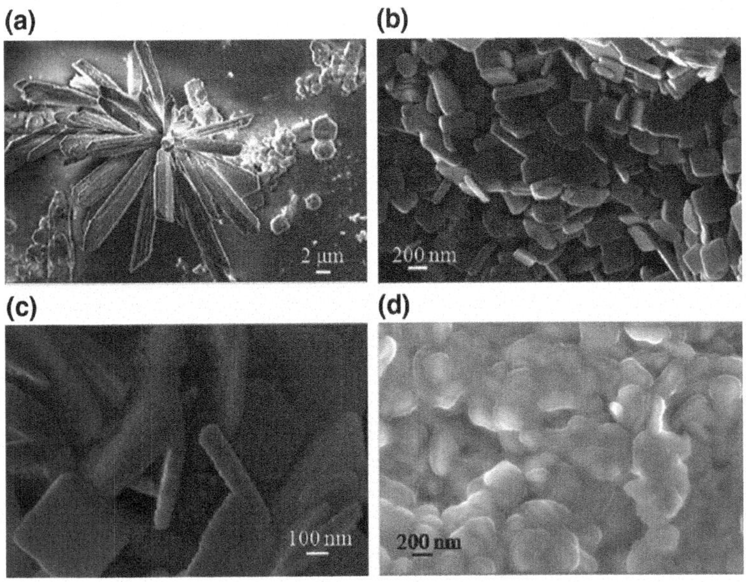

Fig. 6.4 FESEM image of **a** micro-leaf structures of Bi_2O_3 **b** and **c** NSTs of Bi_2O_3 **d** Bi_2O_3 NSTs/nano-Al nEFs, on an Au-coated Si substrate. Reprinted with kind permission from Patel et al. (2015)

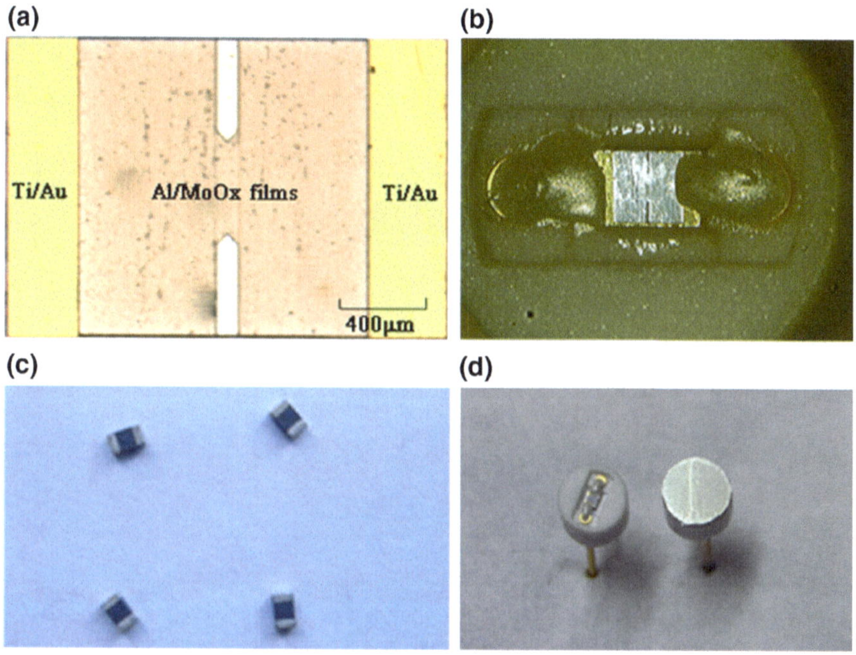

Fig. 6.5 **a** Optical image of the SCB-Al/MoO_x chip (top view). **b** Optical image of the packaged SCB-Al/MoO_x device (top view). **c** Optical image of NTC chips. **d** The SCB-Al/MoO_x device before and after Al/Ni multilayer nanofilms applications. Reprinted with kind permission from Zhu et al. (2014)

These multilayer thin films enhanced the ignition of a conventional pyrotechnics. The fabricated device has shown the capability of interconnection between wire bonds and silver-filled conductive epoxy. They have found that device gives low likely firing energy and insensitivity. Also, no effect on electrical properties of SB has been observed after applying nanothermite multilayer film. Furthermore, Zakiyyan et al. (2018) fabricated the exfoliated 2D molybdenum trioxide (MoO_3) with aluminum powder nano-energetic composite with high interfacial contacts between fuel and oxidizer. From the combustion measurement, they reported highest peak pressure, highest reaction temperature, enhanced pressurization rate, homogeneous distribution of reactants, and linear combustion rate.

6.2.4 *Al/Fe_2O_3-Based Thin Energetic Films*

Apart from the above-stated nanothermites either in the form of multilayered structure or films, very few researchers have tried some other combinations for better ignition and combustion properties such as Zhang et al. (2013) presented 3D-ordered macro-porous-structured Al/Fe_2O_3 nanothermite film for better energy output (shown in Fig. 6.6).

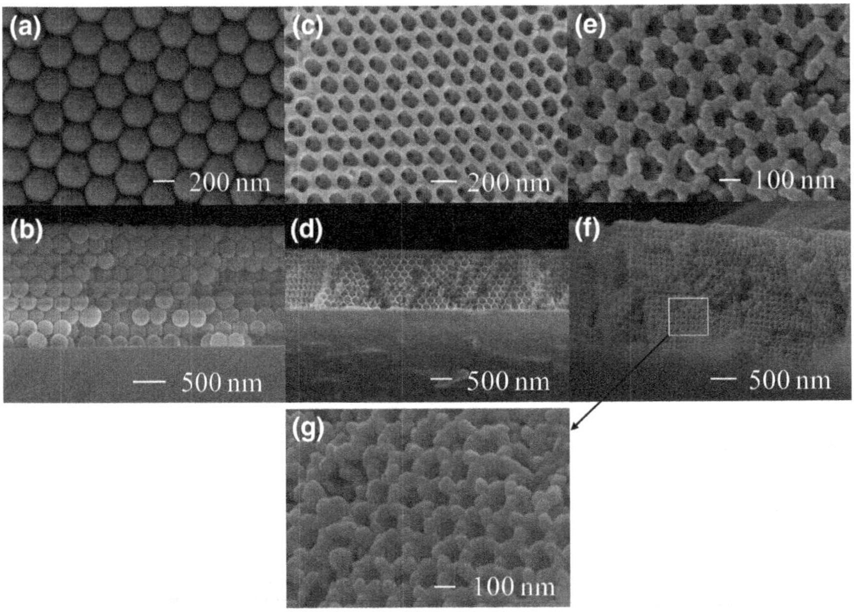

Fig. 6.6 SEM images of **a**, **b** polystyrene spheres template; **c**, **d** 3DOM α-Fe_2O_3 membrane; and **e**, **f** Fe_2O_3/Al membrane after Al deposition; **a**, **c**, **e** surface view and **b**, **d**, **f** cross-sectional view and **g** zoom view from the white square in (**f**). Reprinted with kind permission from Zhang et al. (2013)

They fabricated nanothermite film over polystyrene spheres template and reported that their methodology is fully likeminded with MEMS technology. They have also delivered a competent way to fabricate micrometer thick 3D-ordered nanostructured thermite films with overall spatial consistency.

6.3 Mg/CuO- or MnO_x-Based Thin Energetic Films

From above we have seen that energetic film of oxide was only deposited on Al but when Al-based energetic film comes into contact with the environment then it forms aluminum oxide which degrades the capability of nEMs. Taking this disadvantage of Al-based nEMs into account, Zhou et al. (2013) embedded Mg instead of Al followed byCuO layer and formed the Mg/CuO (core/shell) nano-energetic arrays on silicon (shown in Fig. 6.7). First, they fabricated Mg nanorods using glancing angle deposition and then CuO deposition around Mg nanorods using reactive magnetron sputtering. The various characterization techniques were used for investigation of Mg/CuO core/shell nano-energetic arrays such as SEM, XRD, TEM, and DSC.

They reported that Mg nanorods and CuO mixed uniformly and intimate contact between layers by analysis of heat release curve. Further, they reported that the nano-energetic arrays show a low-onset temperature and high heat of reaction. They

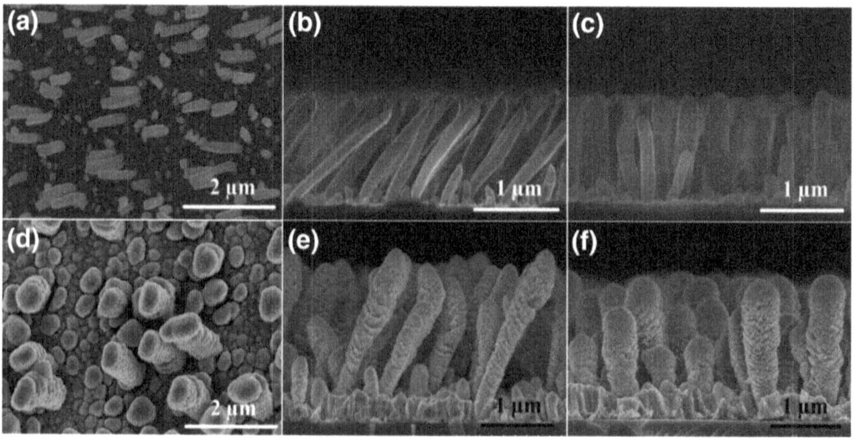

Fig. 6.7 SEM images of (**a–c**) Mg nanorods and (**d–f**) Mg/CuO nano-energetic arrays. **a**, **d** Top view images; **b**, **e** cross-sectional images from the direction parallel to the Mg vapor flux; and **c**, **f** cross-sectional images from the direction vertical to the Mg vapor flux. Reprinted with kind permission from Zhou et al. (2013)

possess long-term storage stability. Furthermore, Meeks et al. (2014) deposited and characterized magnesium (Mg) and manganese dioxide (MnO_2) energetic film using new approach that is called doctor blade casting method. They examined the effect of binder chemistry of polyvinylidene fluoride (PVDF)—methyl pyrrolidone (NMP); VitonÒ fluoroelastomer (Viton A)—acetone; and paraffin–xylene and their concentration on combustion behavior for energetic film. They reported that increase in binder concentration increases calorific output, decreases flame velocity and also delays the energy propagation. They proposed a reason for less conductive nature of binders than metals. Further, they have observed that the Mg/MnO_2/PVDF formulation was utmost effective for fabrication of a thin film coating with better homogeneity of the mixture and smoother surface roughness.

6.4 Other Materials Such as CuPc/MWCNT/$NiCo_2O_4$-Based Thin Energetic Film

Apart from the above-stated oxide-based nanothermite multilayered film or structure, Molodtsova et al. (2014) developed a hybrid organic–inorganic system using Al/copper phthalocyanine (CuPc) thin film. The developed film has been investigated using transmission electron microscopy (TEM) and by photoemission spectroscopy methods. They observed that the charge is transferred from Al to CuPc. Further, they determined the sites of Al lattice. Finally, they reported the establishment of metallic-aluminum-terminated layer on CuPc thin film at high coverage about 64 A ̊. Further, Kim et al. (2017) developed a nano-energetic material on multiwall carbon nanotube (MWCNT)-coated paper chip (shown in Fig. 6.8).

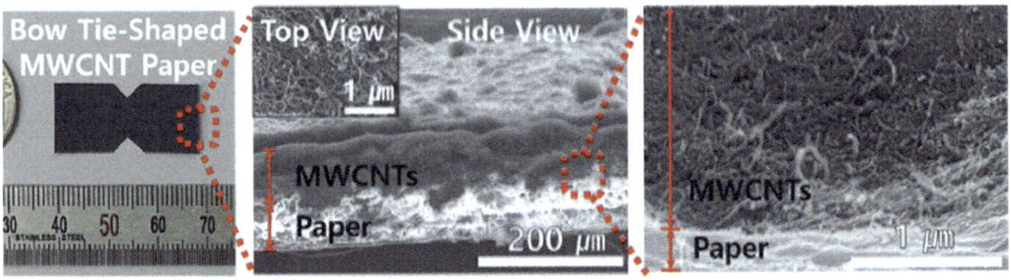

Fig. 6.8 Photograph of the fabricated bow-tie-shaped MWCNT-coated paper electrode, side view SEM image showing the MWCNT thin film coated on the paper substrate (inset is a high-resolution SEM image of the top view of the MWCNT thin film), and high-resolution side view SEM image showing the boundary layer between the MWCNT thin film and the paper substrate. Reprinted with kind permission from Kim et al. (2017)

The electrical resistance of coated paper has been kept constant in all the conditions either bending of the paper or repeated bend and stretch. The ignition and explosion characteristics have been also investigated by them and reported that explosion reactivity of the chip increased by increasing thickness of nano-energetic thin film. At last, they recommended that the developed chip is very compact, flexible, and versatile igniter for various applications in civil and military. In the development of new nEMs, Chen et al. (2018) used $NiCo_2O_4$ as an oxidizer to realize the nanothermite film over Al. They fabricated a uniform structure of $Al/NiCo_2O_4$ core/shell nanowires thermite film using a facile hydrothermal-annealing synthesis method with a controllable magnetron sputtering process. They have reported that the magnitude of total heat release depends upon the quantity of aluminum and that proposed design is extremely compatible with MEMS. Further, it could be applied to understand other nanothermites composed using bimetallic oxide and fuel.

6.5 Super-Hydrophobic Nano-energetic Thin Film

From the above, it can be clearly observed that all the developments of nano-energetic multilayer film or structure is only capable for dry condition. So that further researchers tried to investigate the wettability of developed nano-energetic multilayer film or structure. For underwater application of these nano-energetic materials, the researchers have taken the idea from lotus leaves which have higher contact angle >150° due to which it acts as self-cleaning, anti-wetting, and anti-corrosion. It is observed that nanothermites have very wide application in aqueous surroundings also, including submerged propulsions, torch technologies, metal cuttings, blasting, and generation of power for microscale energy-overwhelming systems. But the above-developed nEMs have only ecological constraints (for species) in the submerged water applications. Therefore, it is immediately essential to carry out penetrating investigations on the research of such

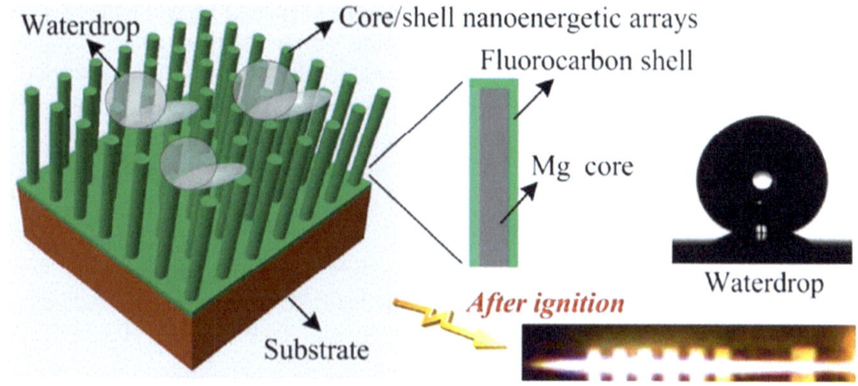

Fig. 6.9 A schematic diagram of the super-hydrophobic Mg/fluorocarbon core/shell nano-energetic arrays. An optical image of a water drop lying on the Mg/fluorocarbon. Reprinted with kind permission from Zhou et al. (2014)

kind of nEMs which can sustain underwater and their super-hydrophobic property. In this series, Zhou et al. (2014) developed an Mg/fluorocarbon core/shell nano-energetic arrays into Si substrate. They deposited Mg nanorods at glancing angle, and after that, Mg nanorods are covered by fluorocarbon using magnetron sputtering deposition process (shown in Fig. 6.9).

They characterized the composition of the Mg/fluorocarbon structures using different techniques which are already used by other researchers and obtained super-hydrophobicity of the Mg/fluorocarbon arrays with a stationary contact angle of 162°, very low reaction temperature, and ultrahigh heat of reaction. Furthermore, Yu et al. (2018b) fabricated an Al/Co_3O_4 core/shell nanowires using a facile hydrothermal-annealing method followed by magnetron sputtering. The developed nanowires have shown super-hydrophobic nature and further tested using laser-induced underwater ignition which exposed that the super-hydrophobic modification successfully leads to the ignition of the nanothermite film. They kept prepared Al/Co_3O_4 core/shell nanowires inside water for 2 days and observed that it maintained ca. 50% of the original chemical energy.

6.6 Conclusions

The different combinations of nEMs multilayer film or structures have been tried by the researchers such as Al/CuO, Al/MoO_x, Al/Bi_2O_3, Al/Fe_2O_3, Mg/CuO, Mg/MnO, Al/CuPc, Al/$NiCo_2O_4$, Mg/fluorocarbon core/shell, Al/Co_3O_4, and MWCNT film on paper. From the study of different types of nEMs films or core/shell, the following conclusions are drawn.

1. Al/CuO nanothermites either in the form of films or nanotubes and nanorods are widely used in different applications such as storage of electricity, manufacturing of aerospace components, human welfare, destruction descend opening, road vehicles, machines which is battery powered.
2. A new method, piezoelectric inkjet printing, is evolved for printing of CuO oxidizer for moderate electronics and/or micro-propulsion mechanisms apart from other techniques such as doctor blading, magnetron sputtering, and electrophoresis.
3. Al/Bi_2O_3 nanothermites show enhanced electrothermal adaptation effectiveness of the initiators by the use of a nanoporous silicon bed, and these micro-initiator platforms are profoundly enough to be functional toward an extensive range of pyrotechnic MEMS applications.
4. Al/Fe_2O_3 nanothermite film over polystyrene spheres template, Mg/CuO nano-energetic arrays, and $Al/NiCo_2O_4$ core/shell nanowires thermite film shows a compatibility with MEMS technology.
5. Mg/fluorocarbon arrays and Al/Co_3O_4 core/shell nanowires exhibit super-hydrophobic property with a static contact angle of 162° and ultrahigh heat reaction. These nano-energetic materials can be used in underwater applications.

References

Badgujar DM, Talawar MB, Asthana SN, Mahulikar PP (2008) Advances in science and technology of modern energetic materials: an overview. J Hazard Mater 151:289–305

Bahrami M, Taton G, Condra V, Salvagnac L, Tenailleau C, Alphonse P, Rossi C (2014) Magnetron sputtered Al-CuO nanolaminates: effect of stoichiometry and layers thickness on energy release and burning rate. Propellants, Explos Pyrotech 39:365–373

Barbee TW, Simpson RL, Gash AE, Satcher JH (2005) Nanolaminate-based ignitors. U.S. Patent WO 2005016850 A2, 24 Feb

Chen Y, Zhang W, Yu C, Ni D, Ma K, Ye J (2018) Controllable synthesis of $NiCo_2O_4$/Al core-shell nanowires thermite film with excellent heat release and short ignition time. Mater Des 155:396–403

Di Biaso H, English BA, Allen MG (2004) Solid-phaseconductivefuels for chemical microactuators. Sens Act A, Phys 111(2/3):260–266

Fried LE, Manaa MR, Pagoria PF, Simpson RL (2001) Design and synthesis of energetic materials. Annu Rev Mater Res 31:291–321

Hinshaw JC (1995) Thermite compositions for use as gas generants. International Patent WO 95/04672

Hofmann A, Laucht H, Kovalev D, Timoshenko VY, Diener J, Kunzner N, Gross E (2006) Explosive composition and its use. U.S. Patent 6 984 274, 10 Jan

Hong CC, Murugesan S, Beaucage G, Choi JW, Ahn CH (2003) A functioning on-chip pressure generator using solid chemical propellant for disposable lab-on-a-chip. Lab Chip 3(4):281–286

Ke X, Zhou X, Gao H, Hao G, Xiao L, Chen T, Liu J, Jiang W (2018) Surface functionalized core/shell structured CuO/Al nanothermite with long-term storage stability and steady combustion performance. Mater Des 140:179–187

Kim KJ, Jung H, Kim JH, Jang NS, Kim JM, Kim SH (2017) Nanoenergetic material-on-multiwalled carbon nanotubes paper chip as compact and flexible igniter. Carbon 114:217–223

Laucht H, Bartuch H, Kovalev D (2004) Silicon initiator, from the idea to functional tests. In: Proceedings of 7th international symposium and exhibition Sophist car occupant safety systems, pp 12–16

Lewis DH, Janson SW, Cohen RB, Antonsson EK (2000) Digital micropropulsion. Sens Act A, Phys 80(2):143–154

Li D, Lu XB, Zhou ZZ, Qiu MX, Lin CG (1988) Laser initiated aluminothermic reaction applied to preparing Mo–Si film on silicon substrates. Mater Res Soc Symp 101:487–490

Lindsay W, Teasdale D, Milanovic V, Pister K, Pello CF (2001) Thrust and electrical power from solid propellant microrockets. In: 14th IEEE International Conference on MEMS, Piscataway, NJ, pp 606–610

Meeks K, Pantoya ML, Apblett C (2014) Deposition and characterization of energetic thin films. Combust Flame 161:1117–1124

Molodtsova OV, Aristova IM, Babenkov SV, Vilkov OV, Aristov VY (2014) Morphology and properties of a hybrid organic-inorganic system: Al nanoparticles embedded into CuPc thin film. J Appl Phys 115:164310

Murray AK, Novotny WA, Fleck TJ, Emre GI, Son SF, Chiua GTC, Rhoads JF (2018) Selectively-deposited energetic materials: a feasibility study of the piezoelectric inkjet printing of nanothermites. Addit Manuf 22:69–74

Nicollet A, Salvagnac L, Baijot V, Estève A, Rossi C (2018) Fast circuit breaker based on integration of Al/CuO nanothermites. Sens Act A: Phys 273:249–255

Norton AA, Minor MA (2006) Pneumatic micro actuator powered by the deflagration of sodium azide. J Microelectromech Syst 15(2):344–354

Patel VK, Ganguli A, Kant R, Bhattacharya S (2015) Micropatterning of nanoenergetic films of Bi_2O_3/Al for pyrotechnics. RSC Adv 5:14967–14973

Pennarun P, Rossi C, Estève D, Bourrier D (2006) Design, fabrication and characterization of a MEMS safe pyrotechnical igniter integrating arming, disarming and sterilization functions. J Micromech Microeng 16(1):92–100

Petrantoni M, Rossi C, Salvagnac L, Conédéra V, Estève A, Tenailleau C, Alphonse P, Chaba YJ (2010) Multilayered Al/CuO thermite formation by reactive magnetron sputtering: Nano versus micro. J Appl Phys 108:084323

Pham PQ, Briand D, Rossi C, De Rooij NF (2003) Downscaling of solid propellant pyrotechnical microsystems. In: 12th international conference on solid-state sensor and act (Transducers), Boston, MA, pp 1423–1426

Pile àgrave (2005) combustible pour l'alimentation d'appareils électroniques, notamment portables. Patent FR2 818 808 (In French)

Rossi C, Estève D (1997) Pyrotechnic micro actuators. In: Proceedings of 11th EUROSENSORS XI, vol. 2. Varsovie, Pologne, pp 771–774

Rossi C, Estève D, Mingués C (1999) Pyrotechnic actuator: a new generation of Si integrated actuator. Sens Act A, Phys 74(1–3):211–215

Rossi C, Briand D, Dumonteuil M, Camps T, Pham PQ, de Rooij NF (2006) Matrix of 10 × 10 addressed solid propellant microthrusters: review of the technologies. Sens Act A, Phys 126 (1):241–252

Rossi C, Zhang K, Esteve D, Alphonse P, Tailhades P, Vahlas C (2007) Nanoenergetic materials for MEMS: a review. J Microelectromech Syst 16:919–931

Shen J, Qiaoa Z, Wang J, Zhang K, Li R, Nie F, Yang G (2014) Pressure loss and compensation in the combustion process of Al–CuO nanoenergetics on a microheater chip. Combust. Flame 161 (11):2975–2981

Staley CS, Morris CJ, Thiruvengadathan R, Apperson SJ, Gangopadhyay K, Gangopadhyay S (2011) Silicon-based bridge wire micro-chip initiators for bismuth oxide–aluminum nanothermit. J Micromech Microeng 21:115015

Stewart DS (2004) Miniaturization of explosive technology and microdetonics. 21st ICTAM, Varsaw, Poland

Sui H, LeSergent L, Wen JZ (2017) Diversity in addressing reaction mechanisms of nano-thermite composites with a layer by layer structure. Adv Eng Mater 20(3):1700822

Takahashi K, Ebisuzaki H, Kajiwara H, Achiwa T, Nagayama K (2000) "Design and testing of mega-bit microthruster arrays," presented Nanotech, Houston, TX, Paper AIAA 2002-5758

Tanaka S, Hosokawa R, Tokudome S, Hori K, Saito H, Watanabe M, Esashi M (2003) MEMS-based solid propellant rocket array thruster with electrical feedthroughs. Trans Jpn Soc Aeronaut Space Sci 46(151):47–51

Tappan AS, Long GT, Renlund AM, Kravitz SH (2003) Microenergetic materials-microscale energetic material processing and testing. In: 41st AIAA aerospace sciences meeting and exhibition, Reno, NV, AIAA-2003-0242

Tappan AS, Long GT, Wroblewski B, Nogan J, Palmer HA, Kravitz SH, Renlund AM (2005) Patterning of regular porosity in PETN microenergetic material thin films. In: 36th international conference on ICT, Karlsruhe, Germany, pp 134–135

Teasdale D, Milanovic V, Chang P, Pister K (2001) Microrockets for smart dust. Smart Mater Struct 10(6):1145–1155

Troianello T (2001) Precision foil resistors used as electro-pyrotechnic initiators. In: Proceedings of 1st electronic components technology conference, Orlando, FL, pp 1413–1417

Vasylkiv O, Sakka Y (2005) Nanoexplosion synthesis of multimetal oxide ceramic nanopowders. Nano Lett 5(12):2598–2604

Vasylkiv O, Sakka Y, Skorokhod VV (2006) Nano-blast synthesis of nano-size CeO_2-Gd_2O_3 powders. J Am Ceram Soc 89(6):1822–1826

Yang Y, Xu D, Zhang K (2012) Effect of nanostructures on the exothermic reaction and ignition of Al/CuO_x based energetic materials. J Mater Sci 47:1296–1305

Yin Y, Li X, Shu Y, Guo X, Zhu Y, Huang X, Bao H, Xu K (2017) Highly-reactive Al/CuO nanoenergetic materials with a tubular structure. Mater Des 117:104–110

Youngner DW, Lu ST, Choueiri E, Neidert JB, Black RE, Graham KJ, Fahey D, Lucus R, Zhu X (2000) MEMS mega-pixel micro-thruster arrays for small satellite stationkeeping. In: 14th annual AIAA/USU conference small satellites, Logan, UT, AIAA Paper SSC00-X-2

Yu C, Zhang W, Hu B, Ni D, Zheng Z, Liu J, Ma K, Ren W (2018a) Core/shell CuO/Al nanorods thermite film based on electrochemical anodization. Nanotechnology 29(36)

Yu C, Zhang W, Yu G, J D, Z Ni, Ye C, Ma K (2018b) The super-hydrophobic thermite film of the Co_3O_4/Al core/shell nanowires for an underwater ignition with a favorable aging-resistance. Chem Eng J 338:99–106

Zakiyyan N, Wang A, Thiruvengadathan R, Staley C, Mathai J, Gangopadhyay K, Maschmann MR, Gangopadhyay S (2018) Combustion of aluminum nanoparticles and exfoliated 2D molybdenum trioxide composites. Combust Flame 187:1–10

Zhang KL, Chou SK, Ang SS, Tang XS (2005) A MEMS-based solid propellant microthruster with Au/Ti igniter. Sens Act A, Phys 122(1):113–123

Zhang W, Baoqing Y, Shen R, Ye J, Thomas JA, Chao Y (2013) Significantly enhanced energy output from 3D ordered macroporous structured Fe_2O_3/Al nanothermite film. ACS Appl Mater Interface 5:239–242

Zhao Y, English BA, Choi Y, Di Biaso H, Yuan G, Allen MG (2004) Polymeric microcombustors for solid-phase conductive fuels. In: Proceedings of 17th IEEE international conference on MEMS, Maastricht, The Netherlands, pp 498–501

Zhou X, Shen R, Ye Y, Zhu P, Hu Y, Wu L (2011) Influence of Al/CuO reactive multilayer films additives on exploding foil initiator. J Appl Phys 110:094505

Zhou X, Xu D, Zhang Q, Lu J, Zhang K (2013) Facile green in situ synthesis of Mg/CuO core/shell nanoenergetic arrays with a superior heat-release property and long-term storage stability. ACS Appl Mater Interface 5:7641–7646

Zhou X, Xu D, Yang G, Zhang Q, Shen J, Lu J, Zhang K (2014) Highly exothermic and superhydrophobic Mg/fluorocarbon core/shell nanoenergetic arrays. ACS Appl Mater Interface 6(13):10497–10505

Zhu P, Shen R, Ye Y, Zhou X, Hu Y (2011) Energetic igniters realized by integrating Al/CuO reactive multilayer films with Cr films. J Appl Phys 110:074513

Zhu P, Shen R, Ye Y, Fu S, Li D (2013) Characterization of Al/CuO nanoenergetic multilayer films integrated with semiconductor bridge for initiator applications. J Appl Phys 113:184505

Zhu P, Jiao J, Shen R, Ye Y, Fu S, Li D (2014) Energetic semiconductor bridge device incorporating Al/MoO_x multilayer nanofilms and negative temperature coefficient thermistor chip. J Appl Phys 115:194502

Chapter 7
Nano-/Micro-engineering for Future Li–Ion Batteries

Prasit Kumar Dutta, Abhinanada Sengupta, Vishwas Goel, P. Preetham, Aakash Ahuja and Sagar Mitra

Abstract Lithium–ion batteries are the key power sources for this technology based world, mainly in the modern mobile world. These batteries have served as the most prominent energy storage devices in case of smartphones, tablets, laptops, electric vehicles and grid-level storage. They offer many advantage in terms of high energy density, moderately high power density, high voltage, low self-discharge, and moderately long cycle life. There are various nano/micro- engineering techniques involved in fabrication of these devices. This chapter discusses in detail the various aspects of Lithium-ion batteries including storage mechanism to installation of various constituents of the battery.

Keywords Lithium-ion battery · Energy storage · Nano/Micro- engineering Energy density

7.1 Introduction

Lithium–ion batteries (LiBs) are one of the key power sources for the modern mobile world. From smartphones, tablets, laptops to electric vehicles and grid-level storage, LiBs are the prime energy storage devices. Their ubiquity in our lives is because of their high energy density, i.e., energy per unit weight or volume, moderately high power density, i.e., power per unit weight or volume, high voltage, low self-discharge, and moderately long cycle life. Table 7.1 compares the performance of lithium–ion batteries with other rechargeable battery systems.

P. K. Dutta · A. Sengupta · V. Goel · P. Preetham · A. Ahuja · S. Mitra (✉)
Department of Energy Science and Engineering, IIT Bombay, Powai, Mumbai 400076, India
e-mail: sagar.mitra@iitb.ac.in

S. Bhattacharya et al. (eds.), *Nano-Energetic Materials*, Energy, Environment and Sustainability, https://doi.org/10.1007/978-981-13-3269-2_7

Table 7.1 Comparison of various battery systems with lithium–ion batteries

Properties	Battery systems			
	NiCd	NiMH	Lead acid	LiB
Gravimetric energy density (Wh/kg)	45–80	60–120	30–50	110–160
Cycle life (to 80% of initial capacity)	1500	500	300	1000
Fast charge time (h)	1	2–4	8–16	2–4
Self-discharge/month (room temperature) (%)	20	30	5	10
Cell voltage (nominal, V)	1.25	1.25	2.0	3.6

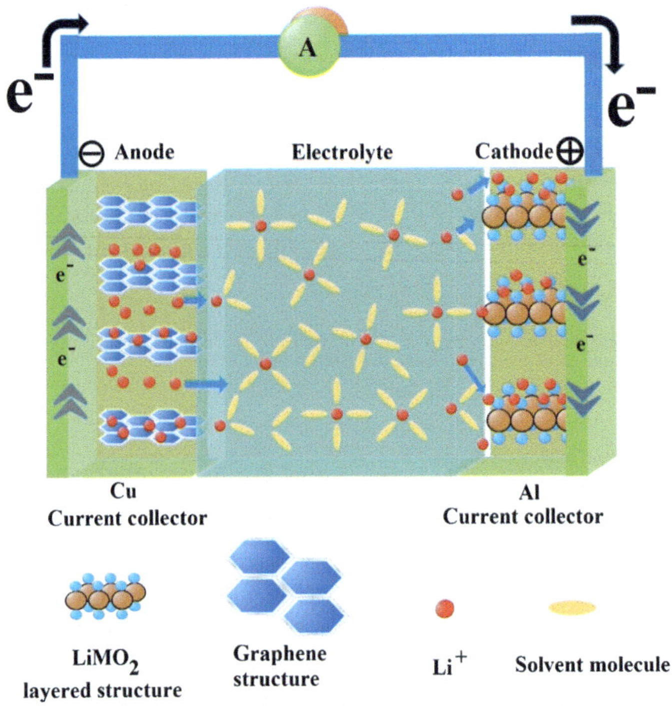

Fig. 7.1 Schematic of a lithium–ion battery (Dunn et al. 2011)

7.1.1 Introduction to the LiB Storage Mechanism

LiBs are rechargeable electrochemical storage devices since they convert electrical energy into chemical energy during charging and vice versa during discharging. In order to understand how a LiB stores energy on charging, one needs to understand the structure of these batteries (Dunn et al. 2011).

Typically, a LiB includes a cathode and an anode cast on the current collector, a separator, and an electrolyte. Figure 7.1 represents a basic layout of a lithium–ion battery.

During charging, electrical energy is used to break the chemical bonds inside the cathode material which generate pairs of lithium–ions and electrons. The ions travel through the electrolyte to the anode side, and the electrons travel via the external circuit to the anode side. At the anode side, both the ions and electrons are consumed in the formation of new bonds with the anode material. The formation of these bonds is what enables energy storage in a LiB. In order to enable rechargeability, the process of bond formation and breakage needs to be reversible. For achieving this reversibility, the active materials inside the electrodes make or break bonds with lithium. Typically, the active material in the cathode is a $LiMO_2$ where M is a or a combination of transition metals like Co, Mn, and Ni. $LiCoO_2$ was the first cathode material invented and commercialized. The active material in the anode is a graphite variant. The following Eqs. (1)–(3) are the reversible electrochemical reactions that occur during charging and discharging:

Cathode half-cell,

$$LiCoO_2 \leftrightharpoons xLi^+ + Li_{1-x}CoO_2 + xe^- \tag{1}$$

Anode half-cell,

$$xLi^+ + C_6 + xe^- \leftrightharpoons Li_xC_6 \tag{2}$$

Full cell reaction,

$$LiCoO_2 + C_6 \leftrightharpoons Li_{1-x}CoO_2 + Li_xC_6 \tag{3}$$

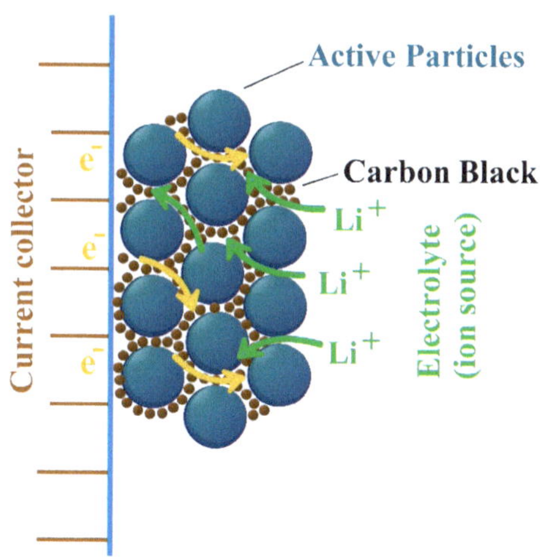

Fig. 7.2 Porous structure of an electrode typically used in LiBs (Gao et al. 2015)

In addition to the active material, the electrodes contain two other major components, viz. a carbon additive and a polymer binder. The role of carbon additive is to enhance the electronic conductivity of the active material. Super P carbon or carbon black, KS series of graphite are few examples of carbon additives typically used in LiBs. Polymer binder is added to tie the particles of carbon additive and active material together and to ensure good adhesion of the tie particles to the current collector. Polyvinylidene fluoride (PVDF) and carboxymethyl cellulose (CMC) are primarily used as a polymer binder in LiBs. All these components are mixed together for fabricating a porous and composite electrode as shown in Fig. 7.2 (Gao et al. 2015).

Traditionally, liquid electrolytes are used in a LiB, which creates a need for them to be concealed in a hermetic pouch package or a metal cylinder. The electrolyte contains a Li salt such as $LiPF_6$, $LiClO_4$ in organic solvents such as ethylene carbonate, dimethyl carbonate, and diethyl carbonate. The high voltage at which LiBs operate makes the aqueous solvents unsuitable for their use; however, research is going on for making aqueous LiBs. The role of the electrolyte is to facilitate movement of lithium–ions inside the battery.

LiBs were firstly commercialized by Sony in the 1990s, since then their performance has been improved hugely especially in terms of the gravimetric and volumetric energy densities. Over the span of 2 decades, from 1994 to 2014, the energy densities have improved by a factor of 3 (Hanson et al. 2017).

These improvements are the result of advancements made in electrode engineering. The term electrode engineering means designing and fabrication of electrodes for achieving particular battery characteristics. The design variables will include the selection of materials used in these electrodes, their physical, chemical, and electrochemical properties; geometric properties of the composite electrode; and the techniques used for fabricating these electrodes. All these variables have a huge impact on the battery characteristics such as battery life, voltage, power and energy densities, safety, and cost. Thus, the design variables are required to be optimized for achieving the desired properties. This chapter explains these variables and their impact on the battery properties in detail along with sharing perspectives on the advancements in the field of electrode engineering and its future outlook.

7.2 Active Materials

Active materials are the source and the sink of lithium–ions in a LiB. They interact with lithium–ions for reversibly forming and breaking bonds. The bulk and surface properties of the active materials play a huge role in determining key battery characteristics such as charging speed, voltage, and cycle life. In this section of the chapter, the relations between the active materials' properties and the battery characteristics are explored.

The bulk properties which primarily influence the battery characteristics are size, morphology, porosity, and composition of the active material particles. The effect of each is individually discussed here.

7.2.1 *Effect of Size of the Active Material Particles*

The transfer of lithium–ions between electrode and electrolyte depends upon the ionic conductivity and the surface area of the active material. Ionic conductivity determines the rate of ion movement inside the particle, and the surface area determines the permeation of the ions between the solid particle and liquid electrolyte. Now since the diffusivity of lithium–ions is constant for a material, its ionic conductivity is a direct function of the diffusion length or the particle size. Thus, reduction in particle size results in higher ionic conductivity and more surface area both of which facilitate faster ion transport and better rate capability of the active material.

Wang et al. (2010) studied the effect of particle size reduction on the rate capability of $LiMnPO_4$ cathode. $LiMnPO_4$ particles with average size ranging from 140 to 270 nm were synthesized and tested at different rates. The smaller particles invariably showed better rate performance as compared to larger particles as shown in Fig. 7.3. Similar results have been reported for various materials. Chen and Cheng (2009) have collated some of such results as summarized in Table 7.2.

However, size reduction comes with its own set of trade-offs. The increase in surface activity leads to an increase in unwanted side reactions. Kobayashi et al. (2009) showed that nanosized $LiFePO_4$ became unstable in the air because of the reaction of Fe^{2+} with humidity and air. Nanosized graphite is also known to exhibit

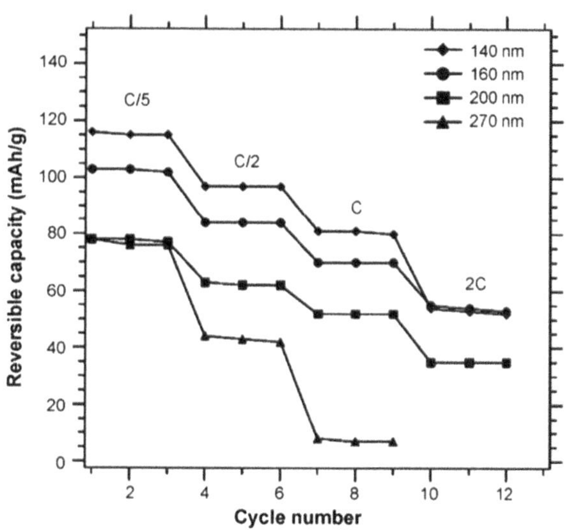

Fig. 7.3 Rate performance of different sizes of $LiMnPO_4$ particles (Wang et al. 2010)

Table 7.2 Summary of the electrode performance of typical nanostructures and their particle counterparts for rechargeable LiBs (Chen and Cheng 2009)

Electrode materials	Capacity in first cycle (mAh g^{-1})	Capacity in 50th cycle (mAh g^{-1})	Current density (mA g^{-1})
$LiNi_{0.8}Co_{0.2}O_2$ nanotube	205	157	10
$LiNi_{0.8}Co_{0.2}O_2$ bulk particle	182	143	
$LiFePO_4$ nanoparticle	162	158	50
$LiFePO_4$ bulk particle	126	119	
Poly(aniline) nanotube	75.7	73.2	20
Poly(aniline) bulk particle	54	50	
Si porous nanosphere	3052	1095	2000
Si bulk particle	1600	52	
Co_3O_4 nanotube	850	530	50
Co_3O_4 nanoparticle	635	321	
SnS_2 nanoparticle	1323	390	50
SnS_2 bulk particle	979	284	

higher first cycle losses and poor cycle life as compared to micro-sized graphite. Moreover, nanoparticles are in general costlier to synthesize and harder to handle which render them less desirable by the industry. Thus, a particle as a variable warrants optimization as per the application.

7.2.2 *Effect of Morphology and Structure of the Active Material Particle*

Morphology and structure of the active material plays a huge role in determination of the battery characteristics such as rate capability, cyclability, and energy density. Crystalline materials are usually anisotropic in their physical properties; thus, morphologies influence physical process such as lithium–ion transport within the particle. Bruce et al. (2008) reported that nanowires of TiO_2-(B) perform much better than its nanoparticles of similar size (the size of the particles is equal to the diameter of nanowires), as shown in Fig. 7.4. The reason for such behavior is attributed to the better conductivity of lithium–ion along the length of nanowires.

The rate capability and cyclability are further improved by introducing porosity into the active materials. The rate capability is enhanced because of the increased access of the particle to the electrolyte, and cyclability is enhanced because of

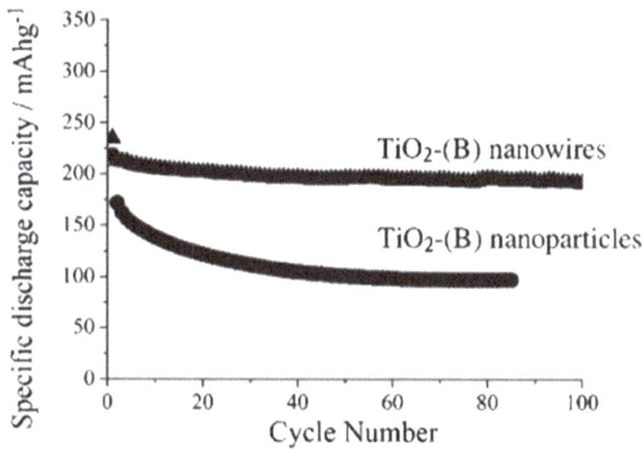

Fig. 7.4 Cycle life of TiO_2-(B) nanowires (Bruce et al. 2008)

availability of voids for accommodating large volume changes during lithiation/de-lithiation processes, especially in conversion-based active materials. Ji et al. synthesized and tested porous MnO_x fibers against Li metal and reported a stable capacity of 600 mAh g^{-1} for more than 50 cycles (2009). Figure 7.5 shows a TEM image of a mesoporous MnO_x fiber and the cycle performance of the synthesized electrode.

Lastly, the morphology of the deposited active material on the current collector also influences its electrochemical performance. Chan et al. demonstrated this phenomenon by depositing an array of silicon nanowires onto the current collector (2008a) as shown in the inset of the figure, using template-free vapor deposition method. The unique design of the deposited material rendered its high cyclability as compared to the bulk deposited material. The enhancement in the cyclability is attributed to the availability of void spaces for accommodating the volume change in silicon during lithiation/de-lithiation. Hence, the structure of both the particles and the electrode as a whole has an effect on the electrochemical performance of the material.

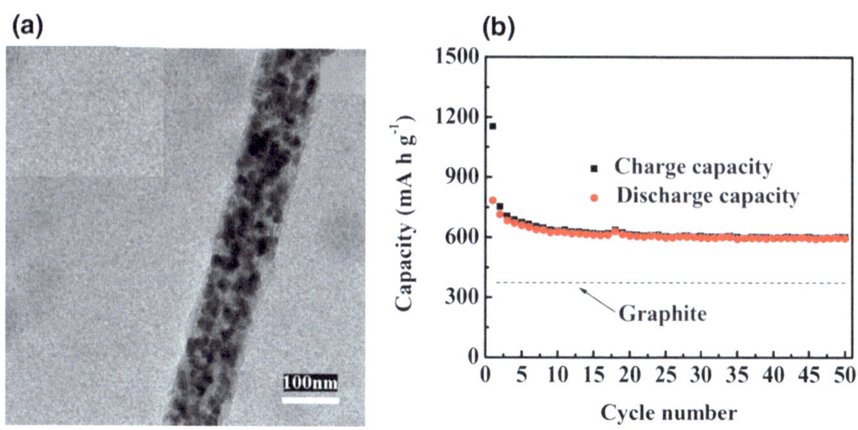

Fig. 7.5 **a** TEM image of MnO_x mesoporous fiber and **b** cycle performance of the synthesized electrode (Ji et al. 2009)

7.2.3 *Effect of Active Material Composition*

The composition of the active material influences the battery characteristics such as safety, charge capacity, and rate capability. This can be altered in two ways, one is to change the inherent composition of the active material, e.g., use of mixed cations in cathode intercalation materials like $LiNi_{1-x-y}Co_xMn_yO_2$, and the other is to use a physical mixture of two active materials, e.g., a mixture of $LiCoO_2$ and $LiFePO_4$.

The change in the inherent composition of the active material leads to alteration of the electrochemical performance of the material. Liu et al. (2015) have collated results from various studies regarding the effect of cation mixing in intercalation-based cathodic active material as shown in Fig. 7.6. According to that figure, increasing Co content increases the rate capability of the material but it also increases the cost; increasing Ni content increases the charge capacity of the material at low cost but it dramatically reduces the cycle life as shown in Fig. 7.6; increasing Mn content enhances safety and reduces the cost of the material but it losses on the capacity. Thus, application-based optimization is required for achieving desired performance out of the battery.

Physical mixing of two or more active materials is another way in which the electrochemical behavior of the electrodes can be influenced. $LiFePO_4$ (LFP) has low electronic conductivity (1.8×10^{-6} S cm^{-1}), slow lithium–ion diffusion (10^{-16} to 10^{-14} m^2 s^{-1}), and low operation voltage (3.3 V). However, when it is physically mixed with $Li_3V_2(PO_4)_3$ (LVP) which has higher ionic and electrical conductivities but low cycle life, the resultant properties of the hybrid electrode improve drastically. Xiang et al. (2010) conducted a detailed study on the effect of the ratio of LFP to LVP on the charge capacities of the hybrid electrode, and the results are shown in Fig. 7.7. According to the study, the ratio of 9:1 exhibited the highest discharge capacity of 168 mAh/g at 1 °C rate and very stable cycling performance at rates as high as 15 °C.

In addition to the bulk properties, the other factors which influence the performance of active material are mainly its surface properties. Typically, the surface of the active materials is modified for achieving either of the following: higher

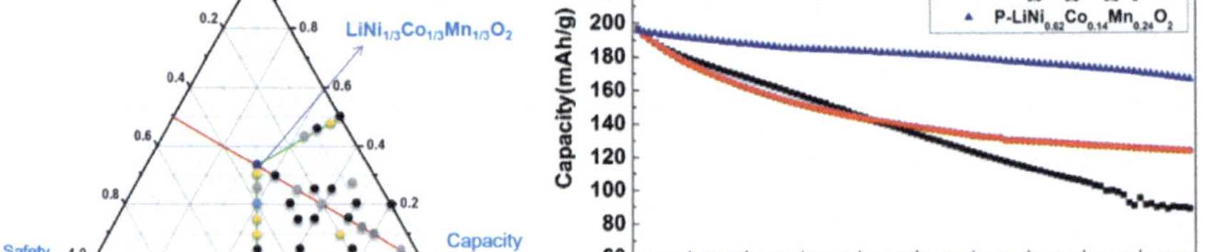

Fig. 7.6 **a** Effect of cation mixing on the performance of intercalation-based cathode material and **b** poor cycle life of cathode material with the relatively higher amount of Ni (Liu et al. 2015)

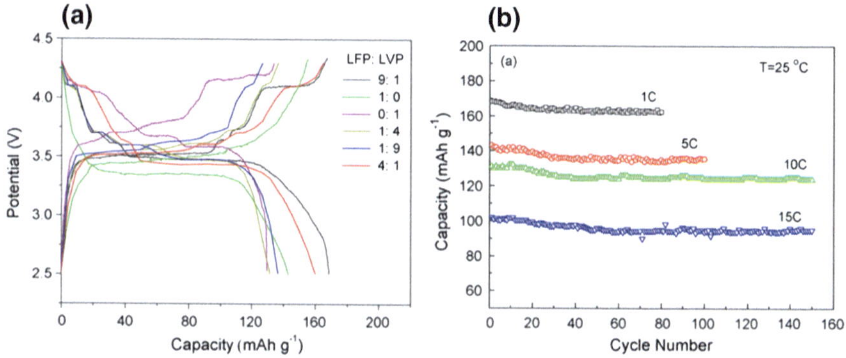

Fig. 7.7 **a** Charge discharge curves of hybrid electrodes containing various ratios of LFP and LVP and **b** rate capability of hybrid electrode containing LFP and LVP in 9:1 ratio (Xiang et al. 2010)

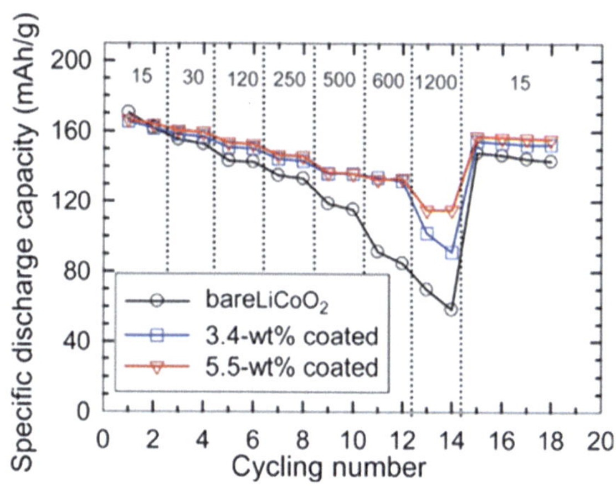

Fig. 7.8 Rate capability of bare and Li_3VO_4-coated $LiCoO_2$ (Pu and Yu 2012)

electronic conductivity, higher ionic conductivity, or suppression of unwanted reactions. The improvement in electronic conductivity is achieved by coating the particles with a good conductor such a carbon black—this approach is discussed in more detail in the section on carbon additives. Similarly, the ionic conductivity is improved by coating the particles with a good conductor of lithium–ions such as Li_3PO_4 (2×10^{-6} S cm^{-1}) and Li_3VO_4 (1.24×10^{-4} S cm^{-1}) (Wang et al. 2015). Pu and Yu (2012) synthesized Li_3VO_4-coated $LiCoO_2$ and demonstrated an improvement in rate capacity as shown in Fig. 7.8. In a similar fashion, the side reactions of active material with by-products of electrolyte decomposition such as HF can be avoided by coating the active material with an inert material such as Al_2O_3, ZrO_2, and B_2O_3 (Wang et al. 2015).

7.3 Binders

To achieve such stable and durable performances described above, it is important to maintain adequate mechanical integrity in between the electro-active particles during the entire operation of the battery. A binder having sufficient viscoelastic properties is a need to provide better integrity during the volume change due to lithium–ion insertion and de-insertion. It is well known that the energy density of a battery is decided by the capability of an active material and the conducting agent plays a major role in improving the rate capability by fueling more electrons. While the binder connects the active material and conducting agent to the current collector, it provides a mechanical integrity thoughtout the battery operation. Hence, the cycle life of the battery is heavily dependent on the binder since the peeling off of the active material from the current collector results in a catastrophic degradation in the capacity.

Due to the critical role, binder plays an important role in the battery performance. A binder should have stable structural integrity during functioning of the battery, less resistance, and a connecting pathway which connects various parts of the electrode (Shi et al. 2017). Conventional binders' mechanical integrity collapses during the functioning of the battery due to the high stress developed during the volumetric expansion and contraction taking place in the electrode thus hampering the performance. Hence, the optimum material for the binder can be the one which can perform both as an efficient binder and as a conductive agent and hence the focus of researchers is on multifunctional binders.

Polyvinylidene difluoride (PVDF) is one of the most prominent binders used in lithium systems due to its properties like favorable electrochemical stability, ability to bind the electrode materials to the current collector efficiently as well as increased uptake electrolyte which favors the transport of lithium–ions into the active material. Even the cost of battery production due to the use of toxic and expensive solvents for PVDF binder has resulted in exploring better binders with favorable properties (Ling et al. 2015). However, the weak van der Waals interaction between the binder and the active material results in electrode losing its integrity during high volumetric expansion, thus hindering its application in the material with large volumetric expansion. Most of these problems are rampant in low-cost materials such as $LiNiMnCoO_2$, $LiMnO_2$, and $LiFePO_4$ which have the capacity to satisfy the needs. Hence, it is important to address this critical issue in the binders of this type to increase its ability to meet the requirements as depicted in Fig. 7.9. Here we look at various approaches carried out recently to increase the binder performance.

7.3.1 *Binders with High Carboxylic Group*

In recent years, carboxymethyl cellulose (CMC) (Shim et al. 2002), alginate (Kovalenko et al. 2011), and polyacrylic acid (Mazouzi et al. 2009) have gained importance due to its higher carboxylic groups which are attached to the polymeric

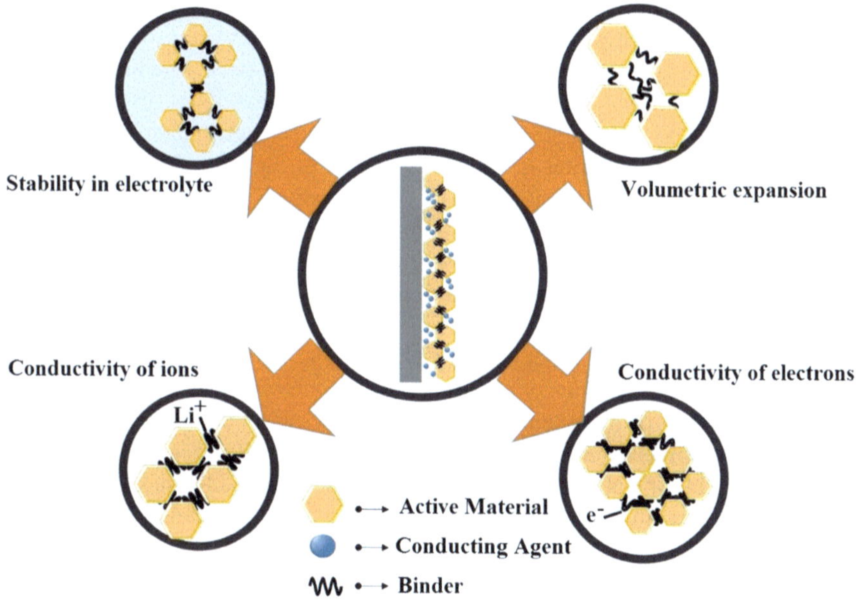

Fig. 7.9 Criteria for good binder

chain. It is found that surface modification due to the development of chemical bonds leading to a stable SEI layer reduces the stress developed due to volumetric expansion, thus helping in increasing the cycle life of the battery (Shi et al. 2017).

Kovalenko et al. (2011) reported alginate binder (sometimes referred as Na-alginate) with a high modulus which has increased the number of carboxylic group leading to increased bonding between silicon and binder, thus controlling its volumetric expansion. Chemically, alginate is a copolymer of 1 → 4 linked b-D-mannuronic acid (M) and a-L-guluronic acid (G), the M and G mono-blocks have been linked to different properties related to physical changes, and this has been optimized for a particular environment by the brown algae. This study also showed that when immersed in an electrolyte, the stiffness of alginate did not change compared to PVDF film which became 50 times softer. Alginate does not show considerable swelling in contact with liquid electrolyte but due to the hopping of lithium–ions through the carboxylic group, it shows good lithium–ion conductivity. The high polarity of the functional group has also resulted in its adhesion to the material with a high volumetric expansion like silicon, thus aiding the formation of a stable SEI layer which controls the electrode integrity. It has also been noted that the surface hydroxyl group on the metal oxide (Fe_2O_3) and the carboxyl group with the alginate binder interacted with weak hydrogen bonding which is increasing the rate performance of lithium–ion battery (Veluri and Mitra 2013).

Li et al. (2008) reported that the electrochemical performance of Fe_2O_3-based electrodes which have the problem of volumetric expansion of the active material showed good long-term cycling using carboxymethyl cellulose binder (CMC).

Studies from IIT Bombay has explored the electrochemical performance of $NH_4V_4O_{10}$ with the various binders such as Na-alginate and carboxymethyl cellulose and concluded that this water-soluble binder was electrochemically active compared to that of polyvinylidene fluoride which is concluded to be inactive (Sarkar et al. 2014). In this study, it is also concluded that though PVDF gives stability, its capacity retention is less as compared to alginate and CMC (Sarkar et al. 2014). It has also been noted that $CoFe_2O_4$ with alginate binder showed a superior capacity of 470 mAh g^{-1} at a high current rate of 20 °C and delivered a capacity of 890 mAh g^{-1} at 0.1 °C which was stable during cycling (Mitra et al. 2014). Here, the FTIR experiments carried on the electrode films show that there is an interaction between the OH groups of $CoFe_2O_4$ and COOH groups of alginate. This interaction was a confirmation of hydrogen bonding.

Furthermore, the properties of carboxylic group rich binder can be increased by the further addition of functional groups on the polymeric chain backbone. Choi et al. (2017) introduced a new binder by adding a small amount of ring-slide polyrotaxane with polyacrylic acid (PAA), thus adding an additional elastic property due to the free movement of PAA polyrotaxane ring components which results in less tension in the polymer network thus increasing the stability of the binder.

7.3.2 Nanoscale Polymer Multifunctional Binder

By reducing the structure of binder to the nanoscale, we gain unusual properties pertaining to electrical conductivity, ionic conductivity, etc., which are derived because of its confined dimensions. Due to its size, it may be easy to mix it during electrode processing as well as increase electrode–electrolyte contact due to its relatively small size. This strategy will also lead to a better ionic and electronic conductivity in the electrode and will also be effective in controlling the volumetric expansion of electrodes (Li et al. 2015).

7.3.3 Conducting Polymer Gels as the Binder

Recently, many innovative strategies have also been explored like the development of polymer gels which have 3D nanostructures synthesized through a dopant molecule cross-linking method which has proved to be an efficient binder for battery electrodes (Pan et al. 2012). These polymer gels connect electrode material as well as provide room for ion transport and electrolyte diffusion, thus effectively connecting the electrode (Wu et al. 2013). In this strategy, polymer gels were designed to have nanostructure network like high surface area, compatibility, and porosity, and this enables them to perform multiple functions like having an excellent conducting pathway for electronic and ionic conductivity, increased uptake of electrolyte, uniform binder coating on the active material, thus suppressing the stress which is generated in the material thus increasing the overall cycle life of the battery. With the increase in the electronic conductivity of the binder material, we can increase the rate capability thus enhancing the overall performance of the battery. There is a possibility that these materials can be 3D printed controlling its microscopic structures, thus tuning its electrical/ionic conductivity and thus mass produced in near future. However, the mechanism of ion transport in these types of materials needs to be critically evaluated.

7.3.4 *Conducting Polymer with Functional Groups*

The possibility of the addition of functional groups on the polymeric backbone may result in unique mechanical properties added to the binder along with increased electrolyte uptake and prevent aggregation without compromising or in some case increasing the electronic conductivity of the binder. To carry out this procedure, it is necessary to study the properties of each functional group in different working conductions, thus understanding its area of application. The possibility of tailoring the molecular structure of binder to add certain qualities which are suited for the requirement is also being pursued by several researchers due to its enormous potential.

Another efficient strategy to develop an efficient multifunctional binder is by hybridizing the structure of conducting polymer with functional groups of required properties, thus inculcating properties such as ionic conductivity, electrical conductivity onto the matrix of conducting polymer. Ling et al. (2015) functionalized sodium alginate with 3,4-propylene dioxythiophene-2,5-dicarboxylic acid thus not only retaining the binding capabilities of sodium alginate but also adding additional properties to the electrode like increasing the ionic conductivity as well as structural integrity of the electrode.

7.4 Carbon Additives

Conductive carbon plays a major role in the engineering of the electrode even if it is used in small amounts since it affects various aspects of batteries such as ionic resistivity, electrical resistivity, and the density of the electrode. These factors directly connected to the performance of the battery are co-related to the morphology, conductivity, and the particle size of the battery which is described in detail in Fig. 7.10 (Spahr et al. 2011). Hence, it is important to conduct studies on this aspect to develop a battery (Wang et al. 2008). It is also noted that the highly conductive carbon black gets into the polymer binder matrix which is used in the electrode and this results in an increase in the conductivity of the polymer binder, thus affecting even the binder in a positive way (Wang et al. 2012). Hence, it is important to develop newly optimized carbon in terms of purity, performance, and

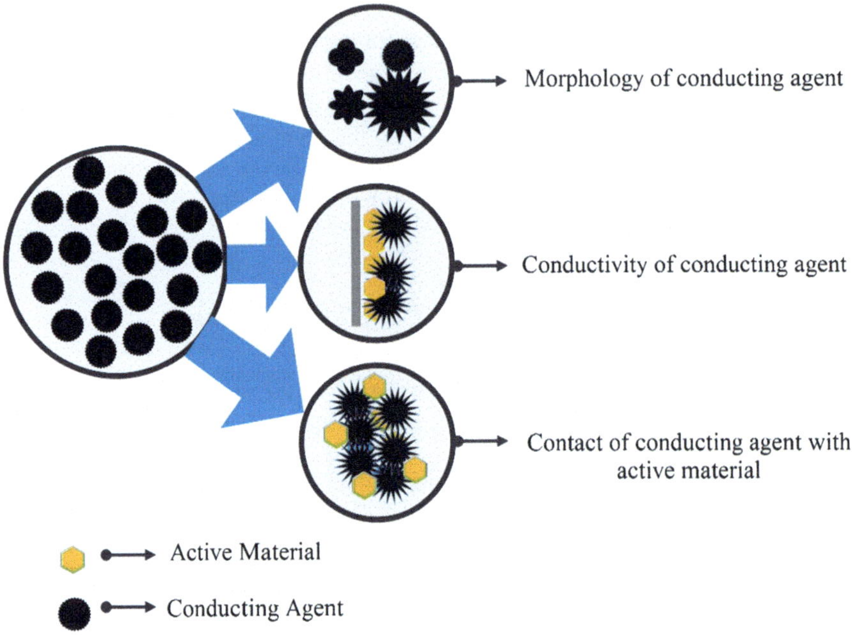

Fig. 7.10 Criteria for efficient carbon additive

processing thus further improving the performance of batteries contributing significantly to ameliorate the energy storage research (Wang et al. 2012).

The conducting agent does not contribute toward energy density in some cases but adds toward the power density of the battery by reducing the internal cell resistance. This is because the additive agents are electron conducting pathways, and hence, it controls the rate of storing or giving power from lithium–ion batteries. But in the case of sodium ion batteries, we have reported that C-65 conductive carbon did add toward storage of Na–ions (Dutta and Mitra 2017). If we increase the conducting agent amount, then the energy density gets reduced; hence, there is a trade-off between energy density and power density (Spahr et al. 2011). The addition of conductive agent in terms of volume fraction to the electrode does not increase the electrode resistivity suddenly since the carbon added may not be connected to form a pathway of electrons in the electrode. But if the volume fraction is kept on increasing, then at a particular volume fraction, the electrode resistivity drops suddenly. Further adding the conducting agent does not reduce the resistivity level significantly because it has reached beyond the percolation threshold of the electrical resistivity curve (Spahr et al. 2011). Hence, it is important to note this critical volume fraction necessary for optimum performance. Besides this critical volumetric ratio, the surface area and morphology also play a very important role in helping to reach the percolation threshold of the electrode material.

7.4.1 *Role of Carbon Nanostructures*

Recent research has shown that by reducing the particle size of material from bulk to nanoscale will result in modification of crystal structure, thus changing the way in which volumetric expansion is generated in the material (Chan et al. 2008b). Furthermore, if electrode materials are combined with nanostructured carbon additives like graphene and carbon nanotube, then an electrically conductive network in the nanoscale can be established which will reduce the lithium–ion diffusion length, increase electron transfer, thus helping in increasing the rate capability of the electrode. This network can also be established to form a free-standing electrode, thus eliminating the need for a current collector.

7.4.2 *Carbon Coating*

Carbon coating onto the active material is another strategy to improve the conductivity of an electrode. This technique is applied by researchers since coating directly may buffer the volumetric expansion of the active material meanwhile increasing the electrical conductivity. Liu et al. (2009) applied carbon coating on ZnO nanorod arrays and used it as active material directly without any extra carbon additive. They found that the coating not increased the contact of ZnO with current collector enhancing the charge transfer but also reduced the strain created during volumetric expansion of the ZnO nanorod, thus protecting electrode integrity. But, increasing the carbon coating may reduce the lithium–ion diffusion into the active material, thus reducing the discharge capacity. This is because the pores of the carbon coating slow down the lithium–ion penetration and hinder its electrochemical utilization (Wang et al. 2012). Thus, a right balance between the thicknesses of the coating needs to be maintained in order to design an efficient electrode for the targeted battery system.

7.5 Electrode Parameters

To achieve such high energy-dense lithium–ion battery, various parameters need to be considered apart from the materials used in electrodes. Parameters such as the thickness of the electrodes, composition involving the active material, binder, and conductive additive, and several process-related parameters pursue an important role in the electrochemical performance of the battery. Hence, a battery developer always objects to optimize such parameters to manufacture an ideal lithium–ion battery.

7.5.1 The Thickness of the Electrode

It can be easily understood that the thickness of the electrode determines the overall thickness of a battery. Hence, a manufacturer always targets to achieve thinner electrodes. Generally, in a commercial cell, the electrodes are either stacked or jelly rolled. So, in order to reduce the thickness, we need to reduce the number of electrodes used that brings to the choice of adjusting the thickness of the electrodes as per requirement. However, results from various studies have shown that electrode thickness can highly influence the battery parameters like the energy density, temperature response, capacity fading rate, overall heat generation, and distribution and proportion of heat sources (Zhao et al. 2015). Just as a coin has two sides, increasing the thickness of the electrodes also has its pros and cons. Although it reduces the battery size and the time required for the complete cell assembly, at the same time, it gives an adverse effect. With increasing electrode thickness, the mass transport limitations of lithium–ions in the electrolyte phase as well as the impedance for electrons in the solid phase of the electrode become dominating. This reduces the capacity of the cell due to overpotentials upon charging and discharging within fixed voltage limits. Simultaneously, the geometric current density in the separator is higher at a given C-rate for thicker electrodes compared to a thinner one which certainly causes an increase in overpotential. The term overpotential can be simplified as the difference between average oxidation potential and average reduction potential (Taberna et al. 2006). Also at the time of charging lithium, plating can happen onto graphite anode (Singh et al. 2015).

7.5.2 Composition of the Materials

As we already know the main components of the electrode are the active material, the conductive additive, and binder. A proper ratio of these materials has a significant effect on limiting the energy density, power density, safety and lifetime of a typical lithium–ion battery (Gallagher et al. 2016). Generally, the active material is taken in bulk and then the rest of it is divided into a binder and a conductive additive. It is generally recommended to have less amount of carbon in order to make the batteries eco-friendlier. Changing the ratio of the composition of the active material, binder and carbon greatly affect the performance. For example, if we reduce the conductive additive in the mixture of the electrode materials, then the conductivity of the electrode will definitely be reduced thereby deteriorating the electrochemical performance of the cell.

7.5.3 The Porosity of the Electrodes

To design a highly efficient and perfectly working battery, it is very important to understand the structural properties of an electrode like porosity and tortuosity which have a significant influence on the electrode's performance. Porosity, also known as void fraction, measures the empty space in a material and represents it in terms of percentage. It is generally measured through the determination of the gas/liquid volume which flows into the electrode material and fills the previously empty void (Wasz 2009; Goriparti et al. 2014; Chung et al. 2014; Röder et al. 2016; Magasinski et al. 2010; Smekens et al. 2016).

In order to ensure the lithium–ion transport between the active materials of the electrode with the help of the electrolyte within the cell, electrode materials need to be porous in nature. On controlling the porosity, the interactions between the electrode and the conductive diluent increases which again increases the electrode conductivity, along with proper lithium–ion intercalation. Introducing nanostructures in the electrode will definitely increase the porosity of the materials, allowing the pores to act as buffers for the volume changes, which in turn increases the performance of the battery. Particle size and shape also affects the porosity of the electrode materials because it has been observed that a reduction in the particle size can reduce the volume change upon intercalation (Wasz 2009; Goriparti et al. 2014; Chung et al. 2014; Röder et al. 2016; Magasinski et al. 2010; Smekens et al. 2016).

When the liquid slurry is cast on the current collector foil, contaminations may occur irrespective of the precautions and inertness maintained, as a result of errors like bubbles and spatters, porosity in the casting occurs, which also affects the uniformity in thickness as well. Porosity is one of the most important factors if we want to achieve optimum utilization of the electrode. It is observed that the voltage drop decreases with the decrease in porosity. Also, from various studies it is seen that porosity more than 51% (for anode) and more than 40% (for cathode) does not enhance the electrochemical performance. This could be due to the decrease in mechanical properties during the cycling which in turn also brings in variation in the energy density. Variation in resistance and electrode–electrolyte interface area is observed while varying the porosity, especially during fast discharge test measurements. Higher porosity can also affect the stability of the electrodes during volume change during lithium interaction–de-intercalation. Basically, higher porosity means more amount of electrolyte is in touch with the electrode, thus more solid electrolyte interface (SEI) formation takes place leading to loss of active material, hence, lowering the energy density. But higher porosity also has an advantage of power capability. So, keeping all the challenges and benefits in mind, we can engineer our electrode according to the requirement of the battery (Singh et al. 2016).

The porosity of an electrode can be estimated using the following formula;

$$\text{Porosity} = \frac{L - \left[W\left(\frac{C1}{D1} + \frac{C2}{D2} + \frac{C3}{D3} + \cdots\right)\right]}{L} \tag{4}$$

where

L is the real thickness of the electrode laminate/coating (without current collector foil),
W is the weight of the laminate per unit area,
C_1, C_2, C_3... are the percentage of the active material, binder, and conductive additive,
D_1, D_2, D_3... are the true density of the active material, binder, and conductive additive.

Theoretically, porosity can be found out using Eq. (4) which would be useful in electrode engineering.

7.5.4 Electrode Manufacturing Processes

There has been an extensive effort for the improvement of lithium–ion battery performance since its evolution in the early 1990s (Nishi 2001). The focus has been mostly on electrochemistry and the materials used, but the manufacturing process of the electrodes plays a significant role. The electrode fabrication processes govern the morphology of the electrodes along with improvement in the techniques which eventually leads to a reduction in cost, better battery capacity, and cyclability. On a research and development scale, various methods have been employed in lithium–ion battery electrode fabrication, which includes chemical vapor deposition, spin coating, spray deposition, laser deposition, molten carbonate method, inkjet printing, screen printing, dry coating of the active material, binder and carbon directly on the current collector (Nishi 2001; Li et al. 2011). As of now, the manufacturing process of the electrodes for lithium–ion batteries can be broadly classified into two categories of slurry deposition and vapor deposition. Among these, generally slurry-based technique is used for commercial scale lithium–ion batteries. Since the electrode fabrication processes influence the morphology (Kraytsberg and Ein-Eli 2016) of the electrode, hence right from the first step of manufacturing all the parameters and processes should be given importance at par if not more while improving the electrode performance.

7.5.5 Slurry Deposition Method

The process of mixing of the electrode materials that is the active materials and the conducting carbon is depicted in Fig. 7.11 where the dry powders are mixed either using a solvent medium or in the bare form.

Once the mixing of powder is over, the slurry preparation port is initiated. The uniformity of a coating is highly affected by the sequence of mixing, solution preparation, mixing devices, and operating conditions, as assessed by its rheological

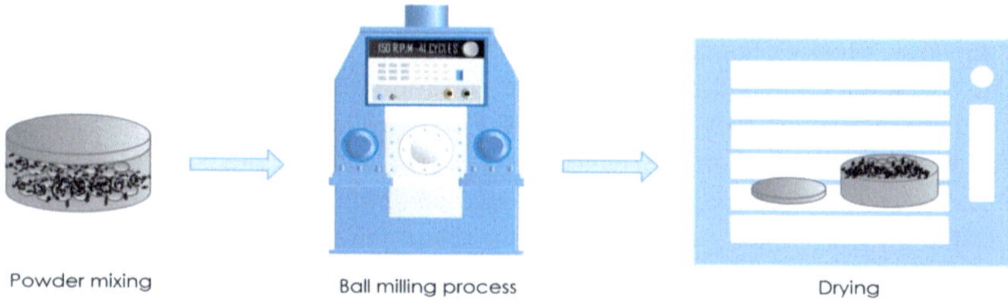

Fig. 7.11 General process of mixing of the electrode materials in laboratory scale

behavior (Liu et al. 2014), hence defining how crucial this stage of manufacturing is. Once the slurry is ready, it is cast on the metal current collector after which the wet electrodes are left for drying. As we know porosity affects the electrode performance, it should be kept in a check which is done by compressing the electrodes through a couple of rollers, commonly known as 'calendaring' (Kraytsberg and Ein-Eli 2016). The next step is to cut the electrodes in their desired dimensions. Generally, in laboratories, it is cut into diodes which are then stacked keeping the separator in between a cathode and an anode, whereas, in industries, a slot-die is used as they wind the electrodes along with the separator in the form of a jelly roll. These stacks or rolls are then packed in a pouch or cylindrical casing, respectively, followed by the welding of the tabs. As soon as the cells are assembled, they are taken to the glove box for electrolyte filling and then finally sealed. The cell is then taken to the testing room for electrochemical formation charge.

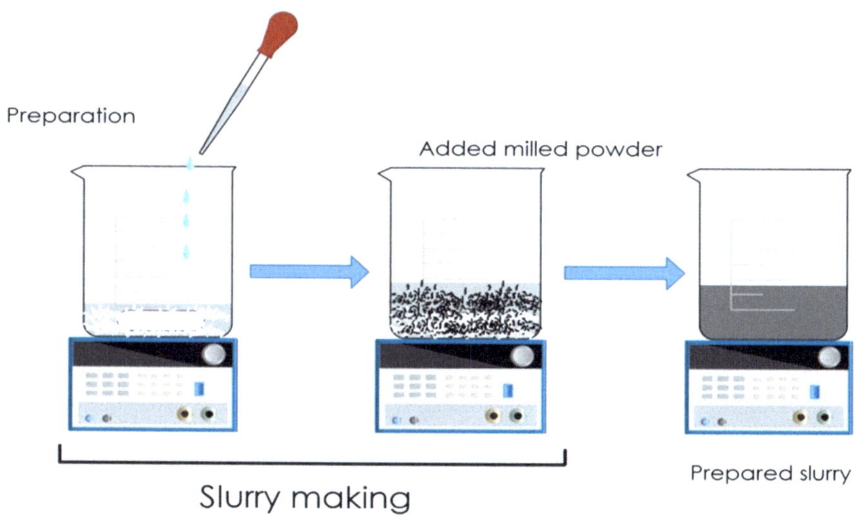

Fig. 7.12 Process of slurry preparation in laboratory scale

The solvents used in the slurry can be mainly classified as either organic solvents or aqueous. Mostly organic solvents are used for research purposes because of its ability to unlock the full potential of the electrode fabrication (Liu et al. 2014). Generally, *N*-methyl 2-pyrrolidine *N*-methyl-2-pyrrolidone (NMP) is used as an organic solvent in lithium–ion batteries. However, the aqueous solvent provides lubricative advantages like cost-effectiveness, environmental benefits, and also its use during high-humidity environment, but it still remains as a challenge when it comes to the potential of the electrodes (Liu et al. 2014). In laboratory scale, the slurry is generally prepared as shown in Fig. 7.12.

7.5.6 Mixing of Materials and Slurry Preparation

Ball milling-based slurry preparation

For mixing the active material and the conductive carbon material, the process of ball milling is used. It is a process of grinding the nanosized materials into fine powders. During the process, the rigid balls usually made of ceramics, flint pebbles, and stainless steel in a concealed container make collisions which in turn creates localized high pressure (Sun et al. 2018). The process of mixing of the active material along with the conductive additive can be achieved by either hand milling in a mortar pestle for laboratory-scale researches or by ball milling them.

Mixing in a ball miller results in an intimate mixture of the reactants in a short time creating new surfaces with defects on the surface of the powders, resulting in an efficient electrochemical performance (Zhang et al. 2012). Among many

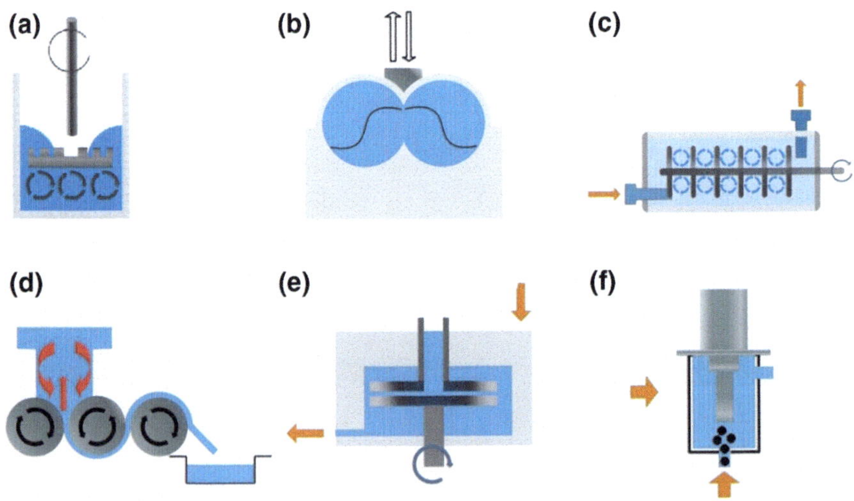

Fig. 7.13 A schematic representation of different mixers available for slurry preparation. **a** Hydrodynamic shear mixer or dissolver, **b** kneaders, **c** stirred ball mills, **d** 3-roller mills, **e** disk mills, and **f** ultrasonic homogenizers (Kraytsberg and Ein-Eli 2016)

different types of ball millers available, the high-energy and planetary ball mill are the most widely used ball millers for lithium–ion batteries. The particle clusters are generally found in two structures, suspended larger agglomerates and smaller aggregates made of primary particles. To obtain a uniform viable lithium–ion battery slurry the dispersing and reduction of the particle size of the active material and the conductive additive is of utmost importance (Kraytsberg and Ein-Eli 2016).

The ball milling in a planetary ball miller can be done in two ways: dry and wet pathways. In case of dry ball milling, the active material and conductive additive are taken in a closed crucible with the balls, whereas in wet mixing the active material and the conductive carbon material are taken in a crucible in which the appropriate amount of organic solvents like isopropyl alcohol, acetone, ethanol, or benzene is added along with balls for proper mixing.

Instead of grinding the materials and then making a slurry out of it, there are various other ways to mix the materials as well (Fig. 7.13).

Hydrodynamic shear mixing-based slurry preparation

The mixing in this type of process takes place by hydrodynamic shear-based mixers like low energy magnetic stirrers, rotary drum mixers, high-energy homogenizers and Rushton turbines, and static stirrers. The slurry preparation is based on hydrodynamic forces, where the forces are controlled by the flow shear rate, cluster cross-sectional area, and the dynamic viscosity of the fluid. It involves a two-step procedure; the cluster break up and the suspended clusters re-association (Kraytsberg and Ein-Eli 2016).

The cluster break-up mechanism consists of three parts—erosion, rupture, and shattering (Moussa et al. 2007). This process solely depends on the particulars of the particles, slurry solvent–particle interaction, and the applied shear force. In the first part, erosion, the larger agglomerates are broken down with the applied shear force. As the name suggests the second stage, rupturing is a high-energy input where the clusters are further broken down and separated into comparable sizes. Finally, shattering is done, which is a special variant of rupturing where the clusters simultaneously separate into smaller fragments (Pieper et al. 2013). This process provides slurries with agglomerates not smaller than ~100 nm, and thus, this type allows complete dispersion of the powder (Bubakova et al. 2013).

For a lithium–ion battery, the morphology plays a significant role. It is to be noted that the rebuild slurry agglomerates are denser than the precursor clusters, thus giving a low-porosity electrode (Kraytsberg and Ein-Eli 2016; Xiao et al. 2015). In such a case, the hydrodynamic mixing is better in respect to the other available mixing techniques.

Ultrasonic mixing-based slurry preparation

Presently, ultrasound has been used for mixing in micro-scale solids and macro-scale liquid flow (Kraytsberg and Ein-Eli 2016). The micro-scale mixing action works basically on the transient acoustic cavitation, where the cavitation takes place at a quite high-intensity ultrasound. The micro-bubbles form and grow through the low-pressure half period of the sound waves; these bubbles keep on growing till the critical bubble size is reached after which it collapses during the high-pressure half-period sound wave, thus creating shock waves for a very short time (Ding and Pacek 2009). For macro-scale liquid flow, the principle is little different as it is based on the concentration of the cavitation bubbles. The concentration of the cavitation bubbles decreases outwards along the axis, and then the bubble diffuses toward the low-concentration bubble areas. This liquid flow is sufficient enough for providing a mixing action inside the slurry using the ultrasound waves (Ding and Pacek 2009).

The main advantage of using an ultrasound mixer is that the slurry is well dispersed in a low solvent content, also the preparation of a concentrated slurry might be even more energy efficient than that of a diluted one (Kitada et al. 2016). In terms of the lithium–ion battery, the higher the slurry concentration, the shorter will be the drying time and the smaller would be the unwanted active material concentration changes (Kraytsberg and Ein-Eli 2016). Reducing the drying time means better adhesion of the active material and the conductive additive as long drying time affects the binder inhomogeneity as well. This kind of mixing is benefitted for lithium–ion battery slurry mixing because uneven distribution of the slurry components, i.e., the active material, conductive additive, and binder, will lead to inhomogeneous pore distribution across the electrode thickness while drying. This inhomogeneity aggravates the shortage of lithium–ions in the active material because in the electrode surface accessibility of lithium–ions become lesser compared to the bulk (Kitada et al. 2016).

The application of high-intensity ultrasound results in degradation in polymeric molecules. So, this kind of slurry mixing might not be good for binders such as methyl cellulose, poly acrylic acid, and polyvinyl alcohol (Kraytsberg and Ein-Eli 2016).

7.5.7 The Casting of the Slurry

Electrode coating or casting on the conduction electrode foil is another important part of lithium–ion battery fabrication which decides the performance of the battery electrodes. Casting can be of two types, tape casting and commercial large-scale casting at present. There are various other casting techniques that are emerging now like spray coating, inkjet printing, and powder coating.

Tape casting often referred to as doctor's blade casting or knife casting is used to cast slurry on the current collector by spreading the liquid slurry or slurry paste with the doctor's blade on the surface, thus obtaining a film with thickness varying from 5 μm to few millimeters. The advantage of using tape casting is the use of it on a flexible substrate. This kind of casting is mostly used for non-aqueous slurries but it is also used for aqueous slurries as well, which is rather gaining more attention these days. The aqueous slurry casting involves some specific techniques like slow drying, flocculation, higher crack sensitivity, and poor wetting (Bensebaa 2013).

Fig. 7.14 Electrode coater at electrochemical energy laboratory, IIT Bombay

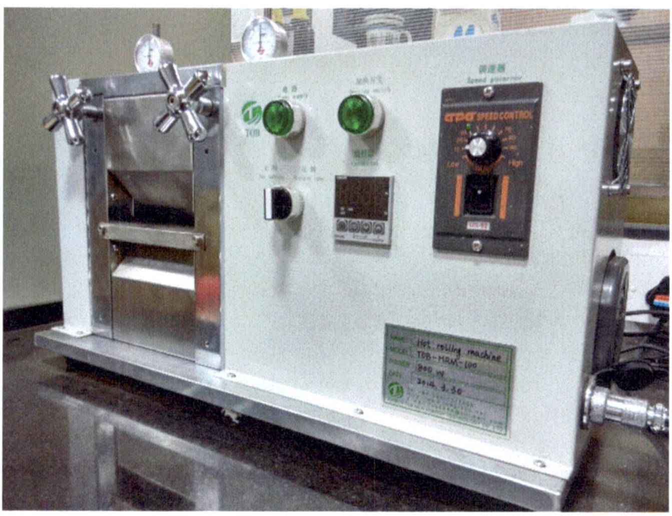

Fig. 7.15 Calendaring machine at electrochemical energy laboratory, IIT Bombay

The other coating processes used in pilot level are slot-die, roll, comma bar coating technique, etc. Generally, for pouch cell fabrication we use small coating machines comprising of these techniques (Fig. 7.14). Slot-die coating is mostly preferred over roll or comma bar coating in industries because of its advantages in terms of high precision, wide coating window, reliability, easily scalable parameters, and its closed feed system (Schmitt et al. 2013, 2014). The limitations for slot-die coating are the bead pressure and the low flow limit even in the non-Newtonian fluid (Schmitt et al. 2013).

7.5.8 Calendaring of the Coated Electrodes

As understandable from the flowchart (Fig. 7.15), calendaring of the electrodes follows after the process of casting. Calendaring is the process of compaction for lithium–ion electrodes which has an impact on the pore structure, particle deformation at the surface, the overall reduction of pore volume, and hence on the electrochemical performance of the battery (Meyer et al. 2017; Haselrieder et al. 2013). Energy density is one of the most important characteristics of energy storage devices for which the coating porosity has a crucial role as it determines the energy content per volume and that is where the compaction step comes in play. There are mainly two types of calendaring machines like uniaxial hydraulic press and roll press among which uniaxial hydraulic presses are generally used for research purposes and roll-press machines are used for small-scale pouch cell fabrication (Fig. 7.15). A continuous compaction is established between the rolls workings in opposite direction. The electrode, a current collector foil coated with the coating on a single side or both sides, gets pulled into the gap (Meyer et al. 2017). The gap size can be adjusted according to the required pore size.

7.5.9 Cutting of Electrodes

Once the electrodes reach its desired porosity, it is cut into different dimensions and sizes depending upon the type of cells to be made. For example, if we are making a pouch cell generally we need an electrode which is generally rectangular in shape as shown in the figure below which can vary in size according to the capacity required and is cut using a semi-automatic die cutter or compact pneumatic die cutter (Fig. 7.16). However, for a Swagelok or coin cell that is used for research purposes, round electrodes are needed to be cut in a suitable small shape.

7.5.10 Cell Assembly

This is the endmost part of cell fabrication where we assemble all the parts into one single battery. For commercial pouch cells or cylindrical cell, the electrodes are stacked or jelly rolled, respectively, with a polymer porous separator and then pouches are made or they are put into the casing of the cells after which the initial sealing is done. The electrolyte is filled inside the glove box, and a final sealing of the battery is done there. The cells are then ready to be electrochemically analyzed.

Fig. 7.16 Electrode cutter machine at electrochemical energy laboratory, IIT Bombay

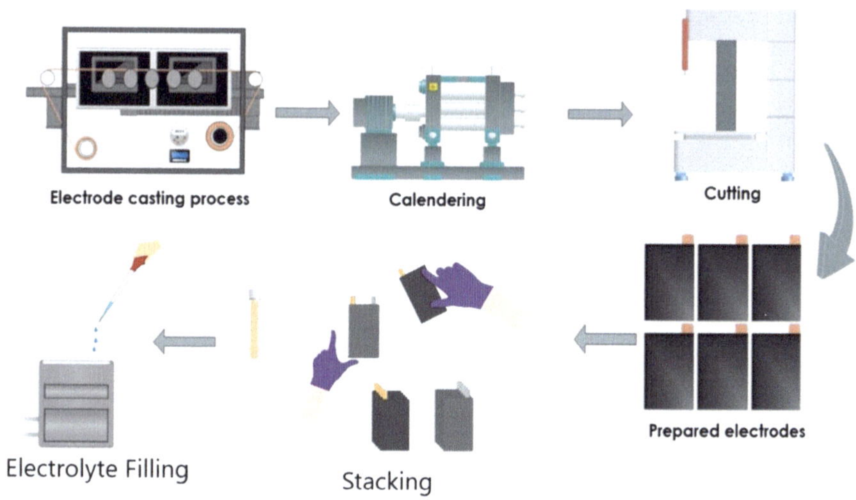

Fig. 7.17 Complete fabrication process of Li–ion pouch cells

For research-oriented purposes, the cells are assembled inside the glove box and then put for testing.

The complete process of slurry deposition technique is represented in the following diagram flowchart (Fig. 7.17).

7.5.11 Chemical Vapor Deposition Method

Chemical vapor deposition (CVD) technique has been there in the research field since 1960 which involves a deposition of a solid thin film on a substrate material by chemical reaction of the vapor phase precursors (Meyer et al. 2017). The reactions include both homogeneous gas phase reactions and a heterogeneous reaction in the vicinity of the substrate by means of heat and high-frequency radiations (Meyer et al. 2017; Haselrieder et al. 2013). After a lot of research and developments, CVD was used in various forms such as Reduced Pressure CVD (RPCVD), Plasma-Enhanced CVD (PECVD), Metal-Organic CVD (MOCVD), Low-Pressure CVD (LPCVD), Plasma-Assisted Metal-organic CVD (PAMOCVD), Ultra-High Vacuum CVD (UHVCVD), Laser-Assisted CVD (LACVD), Very Low-Pressure CVD (VLPCVD), Rapid Thermal Low-Pressure CVD (RTLPCVD), and Electron Cyclotron Resonance CVD (DECRCVD) (Xie 2014).

The process of CVD involves a series of complex reactions and is quite complex in nature. It comprises of the following reactions—thermal decomposition (pyrolysis), reduction, hydrolysis, disproportionation, oxidation, carburization, and nitridation. The process starts with evaporation and transport of reagents, i.e., the precursors in the bulk gas phase and then transport of the gas into the reactor followed by the gas phase reactions of precursors in the reaction zone to produce reactive intermediates and gaseous by-products. Once this is done, the mass transport of reactants to the substrate surface takes place and then adsorption of it on the surface which leads to surface diffusion to growth sites, nucleation and surface chemical reactions leading to film formation. Finally, desorption and mass transport of remaining fragments of the decomposition from the reaction zone occur (Xie 2014).

CVD has various benefits over other processes of depositions such as its ability to produce conformal coatings on complex three-dimensional substrates, including powders because radio transmissions are generally obstructed by this deposition technique. Furthermore, it could be essentially free from nanopores and defects, capable of producing dense and uniform coatings on electrode materials. These advantages extend to the fabrication of composite and hierarchical electrode materials, synthesis of current collectors, manufacture of thin-film batteries, and improvements in the performance of separators. Hence, it can be predicted that the use of chemical vapor deposition method in high energy density and ultra-high energy density battery will grow rapidly in near future (Choy 2003).

7.6 Research and Development (R&D) Trends

Although the lithium–ion battery fabrication technique discussed above is well developed, scientists need to innovate more to meet the requirements of next few decades. From the last decade, the improvement of lithium–ion battery in terms of electrode engineering is going at a very slow pace. This also confirms that this field is almost matured in terms of commercial aspects. Today, sectors like consumer electronics, automobile, and aviation are depended on energy storage, like a lithium–ion battery. These technologies are growing fast by innovating new gadgets, electric vehicle, various small- to large-scale grid storage devices, etc. As these technologies are evolving, the demand for energy storage is increasing as it is an essential integral part of such modern devices. Also, the parameters like energy density, power density, and voltage are in higher need. The traditional lithium–ion battery technology is nearly incapable to meet those requirements. All these factors direct a researcher to avail of scientific advancements accordingly which will certainly involve the use of new materials, modification in the engineering process, etc.

As discussed earlier, a battery electrode contains mainly three components, and hence, few promising trends have been listed below, component-wise, in Table 7.3.

7.6.1 1% Al-Doping on Cathode Side

In most of the commercial cells, four cathodes are used majorly: $LiFePO_4$ (LFP), $LiNi_{1/3}Mn_{1/3}Co_{1/3}O_2$ (NMC), $LiNi_{1/3}Co_{1/3}Al_{1/3}O_2$ (NCA), and $Li_2Mn_2O_4$ (LMO). Each cathode has some advantages and disadvantages which has been discussed in the table. As LFP is a comparatively low-voltage cathode with low specific capacity and higher durability, it is avoided in large-scale applications. Based on the current demand, high-capacity and long-durable cathodes are scavenged which certainly includes a feature of high voltage. Among rest of the three cathodes, LMO is having a tendency of degradation due to the dissolution of Mn–ions in the electrolyte including a negative factor of John Taylor distortion. Hence, in the current lithium–ion and/or lithium polymer batteries, NCA and NMC are used. Till this decade, different composition of NMC and NCA has been tried to achieve better performance. Mostly, R&D sectors are trying to increase the Ni content in both the cathodes. The driving force of such requirement varies based on the energy or

Table 7.3 Probable research trends for the upcoming few decades of lithium–ion battery era

S. No.	Issue at	Details	Modification
1	Cathode	Capacity is limited till 180 mAh g^{-1}	1% Al-doping
2	Anode	Limitation to carbonaceous materials	Second-generation anode
3	Anode	First cycle capacity loss	Dry coating and pre-lithiation
4	Li metal	Dendrite formation	FEC-based electrolyte

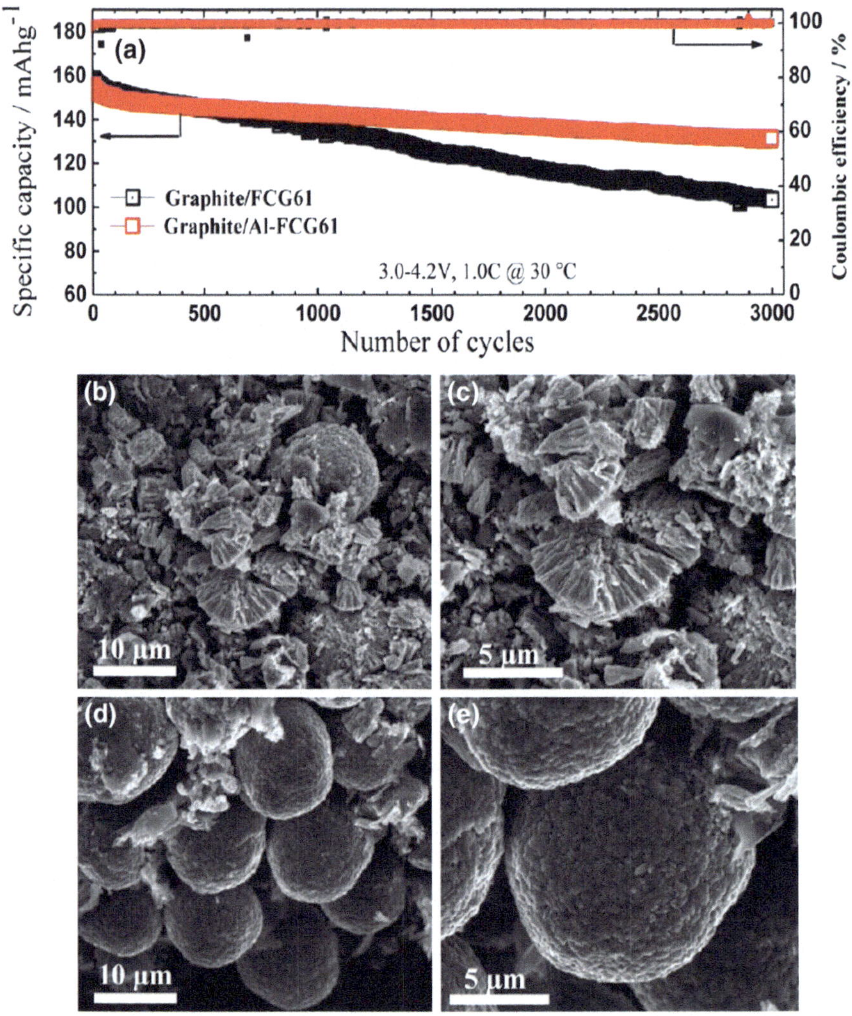

Fig. 7.18 **a** Cycle performance of Al-doped and undoped FCGLi[$Ni_{0.60}Co_{0.12}Mn_{0.27}Al_{0.01}$]$O_2$ cathode at 1.0 °C in a full cell versus graphite in the voltage range of 3.0–4.2 V; SEM images of **b**, **c** the undoped FCG cathode, **d**, **e** the Al-doped FCG cathode after 3000 cycles

power requirement in the device. However, the recent research trend has come out with a new cathode, Al-doped NMCA, merging both NCA and NMC cathodes and optimizing the composition to achieve better energy as well as power requirement featuring higher cycle life. Based on the current R&D research, we feel this Ni-rich material is one of the promising next-generation cathodes which can satisfy a large demand in the next few decades (Etacheri et al. 2011).

While producing such Ni-rich cathodes, NMC was chosen. The composition ratio was taken as Ni:Co:Mn = 61:12:27. This material has been termed as FEG. In this particular case, Ni–ions were partially substituted in FEG by Al–ions to prepare

$LiNi_{0.60}Co_{0.12}Mn_{0.27}Al_{0.1}O_2$. This helps to mitigate degradation even after 3000 cycles which is ideal for EV applications. This research was conducted from Prof. Y-K Soon's laboratory which gauges its viability till ~10 years. Figure 7.18a compares the durability of doped and undoped NMC. Additionally, this cathode can be used to achieve 100% DOD, whereas commercial cathodes like NCA are generally recommended to utilize till 60% DOD.

Surprisingly, the synthesis of this doped cathode is very simple and scalable. Initial NMC matrix can easily be obtained by co-precipitation technique. Finally, LiOH and $Al_2O_3{\cdot}3H_2O$ can be mixed with NMC matrix to achieve NMCA via calcination at 850 °C. This nice material also helps to protect cathode particle from cracking. Figure 7.18b–d compares the morphologies of undoped and doped cathodes after 3000 cycles. If this material can be taken care of by manufacturers, we might realize much durable energy storage devices in upcoming decades (Kim et al. 2016).

7.6.2 Second-Generation Anode

With amelioration of the cathode, a better alternative of the carbon-based system is really required to enhance the demand of volumetric energy density of a battery. Prolonged researches from last few decades have confirmed Si- and Sn-based anodes which can be coupled with an advanced cathode. Additionally, the current LiB anode graphite is bygone which has also proved itself as a lithium dendrite-prone material. Lithium dendrites are well known for catching fire, hence the explosion of the battery which is, unfortunately, growing day by day. Scientifically, this graphite material actually responsible for making a lithium–ion or lithium polymer battery unsafe. Hence, alternate materials need to be scavenged

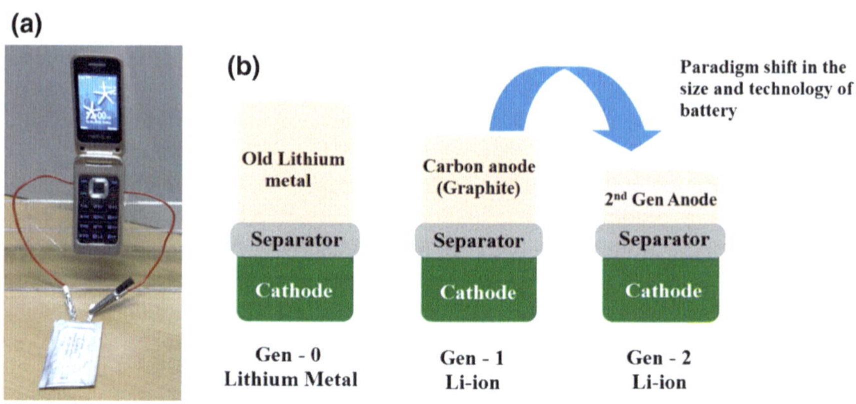

Fig. 7.19 **a** Demonstration of pouch cell developed at IIT Bombay and **b** concept of the energy-dense slimmer battery using a Gen 2 anode material

in order to fix this safety issue as well as to meet the requirement of the next few decades. This means high volumetric energy density with ultra-safety is required in future-generation LiBs. According to the latest reports, Si- and Sn-based anodes are wise choices. These are mostly operated at a safer range, itself declaring almost negligible chances of dendrite formation. Unfortunately, the voltage of the overall cell will be little lower by 0.3–0.5 V. In comparison of capacity, both can provide a minimum of 2–3 times capacity of graphite (372 mAh g^{-1}, theoretical). Hence, less amount of anode material will be needed to manufacture battery which will certainly lead to a slimmer pack. However, a common problem people face is the first cycle capacity loss. This loss can be covered by pre-lithiation technique which is described in the following part.

Researchers at IIT Bombay are highly interested to work on new-generation anode materials. Already, few patents have been filed on this topic which includes the fastest synthesis of Si (Furquan et al. 2018) and SnS (Dutta et al. 2014) anodes including a demonstration of battery prototype used to power a mobile device (Veluri et al. 2015). We believe that these technologies will help to have a paradigm shift providing higher energy density by producing slimmer batteries (Fig. 7.19).

7.6.3 Dry Coating and Pre-lithiation

The next big change is required in terms of electrode fabrication. The NMP used in the slurry making is carcinogenic in nature. Hence, a process without involving such chemical is an imminent mastery as a greener move. Also, it will help to reduce the cost of energy storage devices. Currently, Maxwell Technologies, a San Diego-based ultracapacitor manufacturer, is developing such electrode fabrication technology which can certainly compete with the existing process in terms of energy and power requirements. Their dry electrode production lines operate at high throughput taking a minimal manufacturing footprint. The process does not instigate any volatile waste into the atmosphere. Additionally, the scrap comes out from one manufacturing cycles is collected and reused in successive batches to enhance the efficacy. The overall manufacturing plant is built in a very simple way which also helps in a future expansion. The latest result obtained by Maxwell claims dry-coated electrode having a thickness ranging from ~50 μm to about 1 mm having adequate flexibility, cohesion, adhesion properties, etc., which are superior in comparison to traditional wet coating technology. Also, energy and power density of such electrodes have been improved recently which is quite comparable with the existing technology. In 233rd ECS meeting, they had shared electrochemical performance of dry battery electrode produced at the laboratory level and pilot level in single-layer pouch cell as well as multilayer-stacked pouch cell platform with capacity ranging from about 200 mAh to about 15 mAh (Shin and Duong 2018). A schematic of such a process is described below as Fig. 7.20 (Park et al. 2016). Initially, active material (AM), carbon black (CB), and binder are mixed in blade miller; which is transferred in the chamber to dry spraying. Once

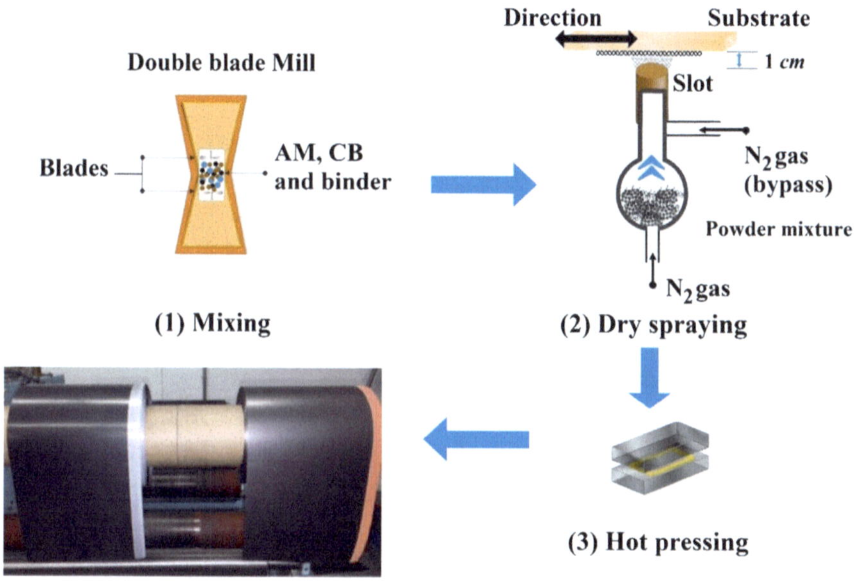

Fig. 7.20 Manufacturing process of dry-coated electrodes (Park et al. 2016)

spraying is done in a nitrogen atmosphere, hot pressing is applied to melt the binder with AM and CB.

In addition to this technique, Maxwell is also working on the pre-lithiation technique which facile the inception of high-capacity anodes. In this technique, anode materials are loaded with 2–3 wt% of lithium physically, before cycling. This helps to mitigate the first cycle capacity loss which comes to 30–60% in the range. After using such techniques, loss becomes minimal and the battery performs much better for adequate cycles.

7.6.4 Modification for Lithium Metal-Based Batteries

While most of the R&D works are going on in the direction of improving power and energy requirements, making the battery more dendrite-free, involving greener technologies, etc., some researches are going on with a very old lithium batteries. One may remember button-type Li cells which are mainly used in motherboards. These batteries involved lithium metal as anode and LMO as a cathode. These are high-voltage lithium cells due to the use of high-voltage cathode. Due to use of lithium metal, these were never looked as a safe device although they produced a higher voltage. Hence, these were used on a very small scale. In this decade, some research is also going on to minimize the hazard of dendrite formation in these systems. Few studies from Prof. Doron Aurbach's groups were presented at 233rd

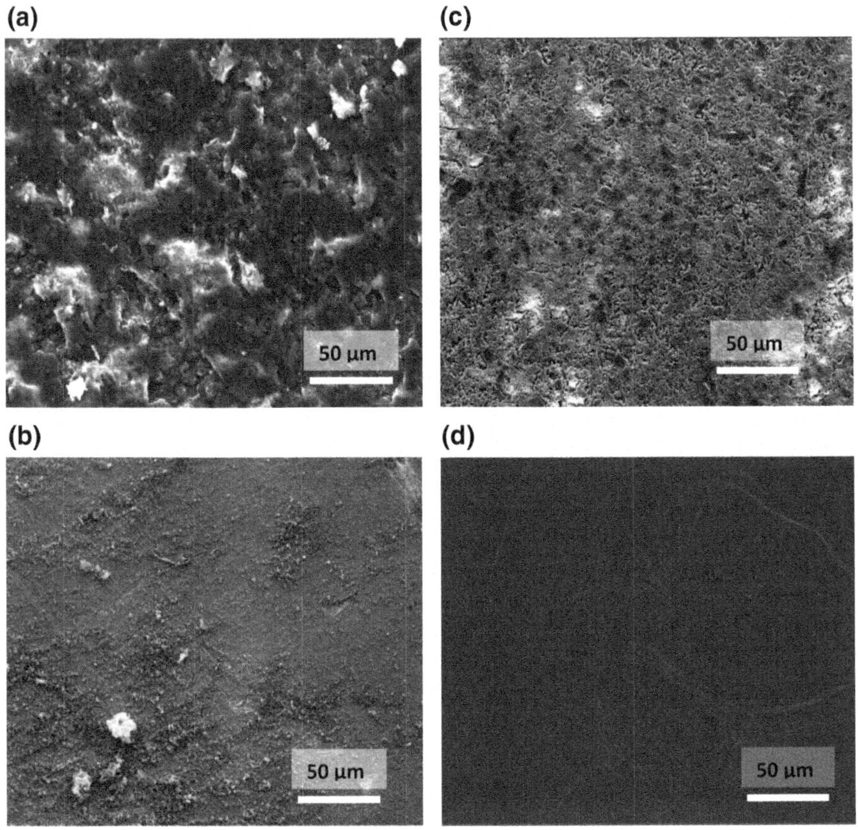

Fig. 7.21 SEM images of cycled lithium electrodes obtained from the **a** central part and **b** periphery of the electrode cycled till 1100 cycles in 1 M $LiPF_6$/FEC/DMC electrolyte solution, **c** electrode cycled for 200 cycles in 1 M $LiPF_6$/EC/DMC electrolyte solution, and **d** pristine Li foil (Markevich et al. 2017)

ECS Meeting, in which FEC was used as the main solvent (Aurbach et al. 2018). The advantage of FEC is a surprising chemistry that does not allow the plate lithium metal in dendritic form. However, the high cost of FEC is a challenge for the future. If this chemistry can be developed at large scale, the energy density of lithium batteries will be enhanced in a few multiples as the theoretical specific capacity of lithium metal is 3860 mAh g^{-1}. In a commercial battery, 1 M $LiPF_6$ is used with EC/DEC or EC/DMC solvent mixture, based on the requirement. However, the EC which is a co-solvent is responsible for SEI formation. SEI is protective layer forms at the first cycle which like a porous skin. A homogenous and stable layer is a rudimentary requirement throughout the battery operation. This determines the lifetime of a battery. Unfortunately, EC is having a tendency to get polymerized over a long time. Hence, after a certain number of cycles, the homogeneous distribution affects. While using lithium metal as anode, striping and plating phenomenon happens to balance the overall redox reaction of a cell. Hence, an inhomogeneous SEI certainly leads to uneven surface morphology. With cycles, these morphologies grow like dendrites and make a battery unsafe. However, FEC gets polymerized after prolonged cycling hence, not creating much change in surface morphologies. As a result, dendrites do not occur even after extensive cycling (Fig. 7.21) (Markevich et al. 2017).

Therefore, it is a hope that some of these techniques will be elevated at the commercial level which will certainly produce advanced and safe energy storage device. However, there are always technical and scientific challenges to elevate a technology from laboratory scale to R&D level to the commercial stage. Hence, all these micro- and macro-engineering need to do in order to make such development.

7.7 Perspective

Electrochemistry is a beautiful gift in the era of modern science, and energy storage like the lithium–ion battery is a successful outcome of that. It is always wished to have a better life while using advanced technologies pursuing human safety and environmental benignity. In such a vision to develop the science and technology of lithium–ion battery, the electrochemical energy storage laboratory at IIT Bombay is progressing to avail indigenously developed smart battery technologies. In a broader way, many promising systems are getting matured simultaneously in the belief that today's science can be the technology of tomorrow.

References

Aurbach D, Markevich E, Salitra G (2018) Fluoroethylene Carbonate-Based Organic Electrolyte Solution for Very Stable Lithium Metal Stripping− Plating at a High Rate and High Areal Capacity. Meeting abstracts, MA2018–01, p 462

Bensebaa F (2013) Nanoparticle Assembling and System Integration. Nanoparticle assembling and system integration. In: Interface science and technology. Elsevier, Amsterdam, pp 185–277

Bruce PG, Scrosati B, Tarascon JM (2008) Nanomaterials for rechargeable lithium batteries. Angew Chemie Int Ed 47:2930–2946

Bubakova P, Pivokonsky M, Filip P (2013) Effect of shear rate on aggregate size and structure in the process of aggregation and at steady state. Powder Technol 235:540–549

Chan CK, Peng H, Liu G, McIlwrath K, Zhang XF, Huggins RA, Cui Y (2008a) High-performance lithium battery anodes using silicon nanowires. Nat Nanotech 3:31

Chan CK, Zhang XF, Cui Y (2008b) High capacity Li ion battery anodes using Ge nanowires. Nano Lett 8:307–309

Chen J, Cheng F (2009) Combination of lightweight elements and nanostructured materials for batteries. Acc Chem Res 42:713–723

Choi S, Kwon T-W, Coskun A, Choi JW (2017) Highly elastic binders integrating polyrotaxanes for silicon microparticle anodes in lithium ion batteries. Science 357:279–283

Choy K (2003)Chemical vapour deposition of coatings. Prog Mater Sci 48:57–170

Chung D-W, Shearing PR, Brandon NP, Harris SJ, García RE (2014) Particle size polydispersity in Li-ion batteries. J Electrochem Soc 161: A422–A430

Ding P, Pacek A (2009) Ultrasonic processing of suspensions of hematite nanopowder stabilized with sodium polyacrylate. AIChE 55:2796–2806

Dunn B, Kamath H, Tarascon J-M (2011) Electrical energy storage for the grid: a battery of choices. Science 334:928–935

Dutta PK, Mitra S (2017) Efficient sodium storage: Experimental study of anode with additive-free ether-based electrolyte system. J Power Soc 349:152–162

Dutta PK, Sen UK, Mitra S (2014) Excellent electrochemical performance of tin monosulphide (SnS) as a sodium-ion battery anode. RSC Adv 4:43155–43159

Etacheri V, Marom R, Elazari R, Salitra G, Aurbach D (2011) Challenges in the development of advanced Li-ion batteries: a review. Energ Environ Sci 4:3243–3262

Furquan M, Vijayalakshmi S, Mitra S (2018) Method of preparing silicon from sand. Google Patents

Gallagher KG, Trask SE, Bauer C, Woehrle T, Lux SF, Tschech M, Lamp P, Polzin BJ, Ha S, Long B (2016) Optimizing areal capacities through understanding the limitations of lithium-ion electrodes. J Electrochem Soc 163:A138–A149

Gao J, Shi S-Q, Li H (2015) Brief overview of electrochemical potential in lithium ion batteries. Chin Phys B 25:018210

Goriparti S, Miele E, De Angelis F, Di Fabrizio E, Zaccaria RP, Capiglia C (2014) Review on recent progress of nanostructured anode materials for Li-ion batteries. J Power Sources 257:421–443

Hanson ED, Mayekar S, Dravid VP (2017) Applying insights from the pharma innovation model to battery commercialization—pros, cons, and pitfalls. MRS Energy Sustain 4

Haselrieder W, Ivanov S, Christen DK, Bockholt H, Kwade A (2013) Impact of the calendering process on the interfacial structure and the related electrochemical performance of secondary lithium-ion batteries. ECS Trans 50:59–70

Ji L, Medford AJ, Zhang X (2009) Porous carbon nanofibers loaded with manganese oxide particles: Formation mechanism and electrochemical performance as energy-storage materials. J Mater Chem 19:5593–5601

Kim UH, Lee EJ, Yoon CS, Myung ST, Sun YK (2016) Compositionally Graded Cathode Material with Long-Term Cycling Stability for Electric Vehicles Application. Adv Energy Mater 6:1601417

Kitada K, Murayama H, Fukuda K, Arai H, Uchimoto Y, Ogumi Z, Matsubara E (2016) Factors determining the packing-limitation of active materials in the composite electrode of lithium-ion batteries. J Power Sources 301:11–17

Kobayashi G, Nishimura S, Park MS, Kanno R, Yashima M, Ida T, Yamada A (2009) Isolation of Solid Solution Phases in Size-Controlled LixFePO4 at Room Temperature. Adv Funct Mater 19:395–403

Kovalenko I, Zdyrko B, Magasinski A, Hertzberg B, Milicev Z, Burtovyy R, Luzinov I, Yushin G (2011) A major constituent of brown algae for use in high-capacity Li-ion batteries. Science 1209150

Kraytsberg A, Ein-Eli Y (2016) Conveying Advanced Li-ion Battery Materials into Practice The Impact of Electrode Slurry Preparation Skills. Adv Energy Mater 6:1600655

Li J, Dahn H, Krause L, Le D-B, Dahn J (2008) Impact of Binder Choice on the Performance of α-Fe2O3 as a Negative Electrode. J Electrochem Soc 155:A812–A816

Li J, Daniel C, Wood D (2011) J Power Sources 196:2452–2460

Li G, Cai W, Liu B, Li Z (2015) Materials processing for lithium-ion batteries. J Power Sources 294:187–192(2011) Science 1209150

Ling M, Qiu J, Li S, Yan C, Kiefel MJ, Liu G, Zhang S (2015) Multifunctional SA-PProDOT binder for lithium ion batteries Nano Lett 15:4440–4447

Liu J, Li Y, Ding R, Jiang J, Hu Y, Ji X, Chi Q, Zhu Z, Huang X (2009) Carbon/ZnO nanorod array electrode with significantly improved lithium storage capability. J Phys Chem C 113:5336–5339

Liu D, Chen L-C, Liu T-J, Fan T, Tsou E-Y, Tiu C (2014) An effective mixing for lithium ion battery slurries. Adv Chem Eng Sci 4:515

Liu W, Oh P, Liu X, Lee MJ, Cho W, Chae S, Kim Y, Cho J (2015) Nickel-rich layered lithium transition-metal oxide for high-energy lithium-ion batteries. Angew Chemie Int Ed 54:4440–4457

Magasinski A, Dixon P, Hertzberg B, Kvit A, Ayala J, Yushin G (2010) High-performance lithium-ion anodes using a hierarchical bottom-up approach. Nat Mater 9:353

Markevich E, Salitra G, Chesneau F, Schmidt M, Aurbach D (2017) Very stable lithium metal stripping–plating at a high rate and high areal capacity in fluoroethylene carbonate-based organic electrolyte solution. ACS Energy Lett 2:1321–1326

Mazouzi D, Lestriez B, Roue L, Guyomard D (2009) Silicon composite electrode with high capacity and long cycle life. Electrochem Solid ST 12:A215–A218

Meyer C, Bockholt H, Haselrieder W, Kwade A (2017) Characterization of the calendering process for compaction of electrodes for lithium-ion batteries. J Mater Process Technol 249:172–178

Mitra S, Veluri PS, Chakraborthy A, Petla RK (2014) Electrochemical properties of spinel cobalt ferrite nanoparticles with sodium alginate as interactive binder. ChemElectroChem 1:1068–1074

Moussa AS, Soos M, Sefcik J, Morbidelli M (2007) Effect of solid volume fraction on aggregation and breakage in colloidal suspensions in batch and continuous stirred tanks. Langmuir 23:1664–1673

Nishi Y (2001) Lithium ion secondary batteries; past 10 years and the future. J Power Sources 100:101–106

Pan L, Yu G, Zhai D, Lee HR, Zhao W, Liu N, Wang H, Tee BC-K, Shi Y, Cui Y (2012) Hierarchical nanostructured conducting polymer hydrogel with high electrochemical activity. Proc Natl Acad Sci 109:9287–9292

Park D-W, Cañas NA, Wagner N, Friedrich KA (2016) Novel solvent-free direct coating process for battery electrodes and their electrochemical performance. J Power Sources 306:758–763

Pieper M, Aman S, Tomas J (2013) Redispersing and stabilizing agglomerates in an annular-gap high shear disperser. Powder Technol 239:381–388

Pu X, Yu C (2012) Enhanced overcharge performance of nano-LiCoO 2 by novel Li 3 VO 4 surface coatings. Nanoscale 4:6743–6747

Röder F, Sonntag S, Schröder D, Krewer U (2016) Simulating the Impact of Particle Size Distribution on the Performance of Graphite Electrodes in Lithium-Ion Batteries. Energy Technol 4:1588–1597

Sarkar S, Veluri P, Mitra S (2014) Morphology controlled synthesis of layered NH4V4O10 and the impact of binder on stable high rate electrochemical performance. Electrochim Acta 132:448–456

Schmitt M, Baunach M, Wengeler L, Peters K, Junges P, Scharfer P, Schabel W (2013) Slot-die processing of lithium-ion battery electrodes—Coating window characterization. Chem Eng Process 68:32–37

Schmitt M, Scharfer P, Schabel W (2014) Slot die coating of lithium-ion battery electrodes: investigations on edge effect issues for stripe and pattern coatings. J Coat Technol Res 11:57–63

Shi Y, Zhou X, Yu G (2017) Material and structural design of novel binder systems for high-energy, high-power lithium-ion batteries. Acc Chem Res 50:2642–2652

Shim J, Kostecki R, Richardson T, Song X, Striebel KA (2002) Electrochemical analysis for cycle performance and capacity fading of a lithium-ion battery cycled at elevated temperature. J Power Sources 112:222–230

Shin J, Duong H (2018) Electrochemical Performance of Dry Battery Electrode. Meeting abstracts, MA2018–01, p 365

Singh M, Kaiser J, Hahn H (2015) Thick electrodes for high energy lithium ion batteries. J Electrochem Soc 162:A1196–A1201

Singh M, Kaiser J, Hahn H (2016) Effect of porosity on the thick electrodes for high energy density lithium ion batteries for stationary applications. Batteries 2:35

Smekens J, Gopalakrishnan R, Steen NV, Omar N, Hegazy O, Hubin A, Van Mierlo J (2016) Influence of electrode density on the performance of Li-ion batteries: Experimental and simulation results. Energies 9:104

Spahr ME, Goers D, Leone A, Stallone S, Grivei E (2011) Development of carbon conductive additives for advanced lithium ion batteries. J Power Sources 196:3404–3413

Sun P, Wang Y, Wang X, Xu Q, Fan Q, Sun Y (2018) Off-stoichiometric Na 3− 3x V 2+ x (PO 4) 3/C nanocomposites as cathode materials for high-performance sodium-ion batteries. RSC Adv 8:20319–20326

Taberna PL, Mitra S, Poizot P, Simon P, Tarascon JM (2006) High rate capabilities Fe3O4-based Cu nano-architectured electrodes for lithium-ion battery applications. Nat Mater 5:567–573

Veluri P, Mitra S (2013) Enhanced high rate performance of α-Fe 2 O 3 nanotubes with alginate binder as a conversion anode. RSC Adv 3:15132–15138

Veluri PS, Shaligram A, Mitra S (2015) Porous α-Fe2O3 nanostructures and their lithium storage properties as full cell configuration against LiFePO4. J Power Sources 293:213–220

Wang H, Umeno T, Mizuma K, Yoshio M (2008) Highly conductive bridges between graphite spheres to improve the cycle performance of a graphite anode in lithium-ion batteries. J Power Sources 175:886–890

Wang Y, Li H, He P, Hosono E, Zhou H (2010) Nano active materials for lithium-ion batteries. Nanoscale 2:1294–1305

Wang G, Zhang L, Zhang J (2012) A review of electrode materials for electrochemical supercapacitors. Chem Soc Rev 41:797–828

Wang KX, Li XH, Chen JS (2015) Adv Mater 27:527–545

Wasz M (2009) 7th international energy conversion engineering conference, p 4503

Wu H, Yu G, Pan L, Liu N, McDowell MT, Bao Z, Cui Y (2013) Stable Li-ion battery anodes by in-situ polymerization of conducting hydrogel to conformally coat silicon nanoparticles. Nat Commun 4:1943

Xiang J, Tu J, Zhang L, Wang X, Zhou Y, Qiao Y, Lu Y (2010) Improved electrochemical performances of 9LiFePO4• Li3V2 (PO4)/C composite prepared by a simple solid-state method. J Power Sources 195:8331–8335

Xiao F, Xu H, Li X-Y, Wang D (2015) Colloids Surf A 468:87–94

Xie J (2014) Syst Eng 6:239–242

Zhang J, Xu S, Li W (2012) High shear mixers: A review of typical applications and studies on power draw, flow pattern, energy dissipation and transfer properties. Chem Eng Process 57:25–41

Zhao R, Liu J, Gu J (2015) The effects of electrode thickness on the electrochemical and thermal characteristics of lithium ion battery. Appl Energy 139:220–229

Chapter 8
Different Approaches to Micro-/Nanofabricate and Pattern Energetic Materials

Amit Joshi, K. K. S. Mer, Shantanu Bhattacharya and Vinay K. Patel

Abstract Nanoscience and nanotechnology have played a tremendous role in stimulating the development of nanoenergetic materials with orders of magnitude increase in the interfacial area and intimacy of contacts of fuel and oxidizer. In the global world, research and development efforts are continuously being made at a very rapid rate to realize high energy density and super-reactive nanoenergetic materials. The reactivity and ignition sensitivity of nanoenergetic materials are significantly enhanced with nanoscaling of reactants (fuel and oxidizers) on the counterpart of microscale reactants. Nano-energetic materials can be easily integrated with the microsystems, which find many promising applications in micro-initiation, micro-ignition, micro-propulsion, micro-power generation, and pressure-mediated gene delivery/transfection. In this chapter, the various fabrication methodologies of nanoenergetic materials have been discussed and detailed with relevance to their energy density, combustion performance (ignition sensitivity, combustion velocity, and pressurization rate), and suitable integration. Further, the formulations and patterning of nanoenergetic materials on silicon substrate are also included as a future potential for micro-electro-mechanical systems energetic devices.

Keywords Nano-energetic materials · Nanothermites · Combustion Pyrotechnics · Aluminum · Microfabrication · Nanofabrication

---

A. Joshi · K. K. S. Mer · V. K. Patel (✉)
Department of Mechanical Engineering, Govind Ballabh Pant Institute of Engineering and Technology, Ghurdauri, Pauri-Garhwal 246001, Uttarakhand, India
e-mail: vinaykrpatel@gmail.com

S. Bhattacharya
Department of Mechanical Engineering, Indian Institute of Technology Kanpur, Kanpur 208016, Uttar Pradesh, India

S. Bhattacharya et al. (eds.), *Nano-Energetic Materials*, Energy, Environment and Sustainability, https://doi.org/10.1007/978-981-13-3269-2_8

8.1 Introduction

On the combustion of energetic materials, a large amount of energy is released instantaneously resulting in high energy release rate and fast combustion. Conventional monomolecular energetic materials like RDX, TNT reaction rate is controlled by chemical kinetics of creation and termination of chemical bonds and thereby depends on energy densities of molecule which are limited (Tillotson et al. 2001). Higher energy density and combustion energy materials like Al and Si are gaining research interest to overcome monomolecular compounds. Composite energetic materials are produced by mixing Al and/or Si energetic materials as a fuel with oxidizers. In composite energetic materials, fuel and oxidizer bind together, and if there is sufficient energy to initiate, the reaction combustion occurs. If after combustion energy (activation energy) obtained is sufficient to cater the reaction, combustion transfers to other surrounding reagents resulting in a continuous combustion and energy transfer. Conventional energetic composites exhibit high gravimetric energy density (31.1 and 32.4 kJ/kg) than TNT and RDX (15 and 9.5 kJ/kg), but lacks in energy release rates compared to the conventional monomolecular material. The mass transport rate is relatively slow and difficult to ignite due to the slow heterogeneous reaction which results in unreacted reactants in large portion. Research in nanomaterials has directed to a bottom-up approach for tailoring of materials from lower (atomic) scale to higher (macroscopic) scale. The key concept for using energetic materials is to increase the surface area and closeness between energetic materials so as to enhance the rate of reaction and to reduce ignition delay. The unique idea of nanoparticles can be understood by the presence of excessive energy which is present on the surface which can be calculated by following Eq. 8.1 (Sundaram et al. 2013):

$$f = 1 - \left(1 - \frac{2\delta}{D_p}\right)^3 \tag{8.1}$$

where f represents fraction of atoms while δ represents the thickness of the surface layer and D_p is the particle diameter of material. As an example, f increases from 0.17 to 16% for 10-nm aluminum particle as compared to 1-μm particle size. The surface atoms have greater energies and lower coordination number as compared to interior region atoms. Thus, properties of nanoparticles depend on size and deviate exponentially as compared to bulk materials. Following these principles, lot of research effort has been done in the field of synthesis and characterization of nanomaterials.

The combustion kinetics as well as energy density of nanomaterials can be significantly controlled by changing the nanotailoring and following various assembling strategies (Zhou et al. 2014). For most applications of energetic materials, it is important to identify methods, crucial challenges, and limitations of the synthesis method used. From the beginning, the focus of researchers was based on nanomaterials based on metals. Various combinations of material are fabricated using various methods in several researches; another field of research focuses on the study of the morphology of nanomaterials which have a vast use in solid propellants and pyrotechnics.

8.2 Synthesis Approaches for Realization of Nano-energetic Materials

8.2.1 Ultrasonic Mixing

Ultrasonic powder mixing is widely used approach for preparation of nEMs, in which the stoichiometry of particles is readily accommodated so as to maximize energy density. Most of the investigation of researchers focuses on mixing of Al, Mg, Zr, B, and Si (Dreizin 2009). This process can be used for synthesizing nanothermites/nanoenergetics of different compositions (Weismiller et al. 2009; Son 2004; Patel et al. 2015). Figure 8.1 displays the ultrasonically mixed nanoenergetic composites of CuO nanorods (by sonoemulsion synthesis method (Patel 2013) with aluminum nanoparticles). Despite easy preparation and distribution of this method, the main disadvantage of this method is nonuniform distribution of reactants and distant interfacial contact; also, there is a necessity of starting with nanoscale particles due to which the combustion performance of reactants considerably diverges and thus the reliability of this method.

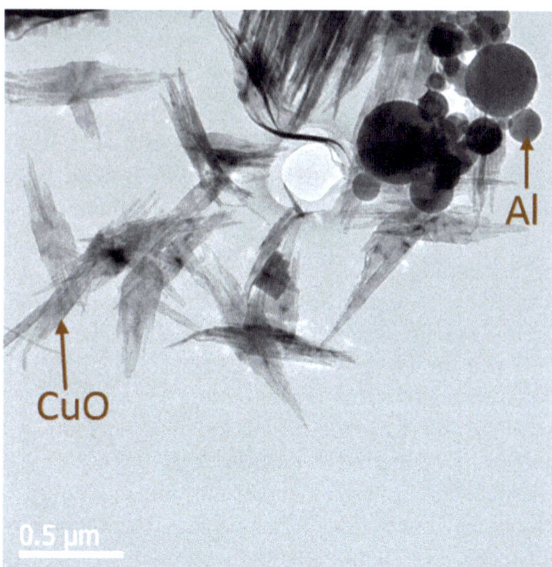

Fig. 8.1 TEM image of ultrasonically mixed CuO nanorods/Al nanoparticles composites

8.2.2 Layered Vapor Deposition

In this method, multilayer nanofoils are generated by alternately depositing different types of reactive materials and are called reactive multilayer nanofoil (RMF). RMFs thus synthesized by this method are mainly based on transition metal oxides (TMO's), e.g., Al/Ni (Gavens et al. 2000) and thermite reactive like Al/CuO (Blobaum et al. 2003a, b). Layered vapor deposition method is efficient in the fabrication of almost all metal, metalloids, and metal oxides by controlling the thickness of layers. In vapor deposition or sputtering, a thin film is chemically grown on to a substrate (mostly silicon) by exposing it to the desired volatile precursors. The process of vapor deposition is a vacuum deposition process. Study on intermetallic reactions focuses mainly on three areas, namely wave propagation, phase transformation, and reaction process. In Al/Ni reactive composites prepared by this method, a varied individual layer (60–300 nm) was deposited by electron beam evaporation process conducted in a 10^{-7} Torr vacuum (Gavens et al. 2000; Ma et al. 1990). Less focus of researchers is on thermite reaction-based RMFs; yet, these reactions offer considerable scope because these reactions give comparatively higher temperature and combustion property as compared to intermetallic reaction (Fischer and Grubelich 1998). Al/CuO was at first investigated by researchers by following the RMFs' route (Blobaum et al. 2003a, b). Magnetron sputtering process has been used to deposit multilayer Al/CuOx nanofoils using two separate targets of Al and CuO. The nanofoils thus developed show heat reaction as high as 3.9 ± 0.9 kJ/g. The technology of layered vapor deposition has certain limitations of cost, scaling, premixing, and under certain cases instability (Lewis et al. 2003).

8.2.3 High Energy Ball Milling (HEBM)

HEBM method can be used for the fast fabrication of materials and efficaciously creates nanoenergetic composites. Arrested reactive milling (ARM) is a wide methodology for this process in which metal and metal oxide micron-sized particles (1–50 μm) are milled together in a ball or planetary shaker mills at room temperature. Hexane or other liquid is mixed with the mill for lubrication and cooling. The milling time is accurately controlled just before the mechanical initiation (triggering) of the exothermic reactions among the reactive constituents, thereby arresting reactions. The nanostructure composites thus synthesized are fully dense. ARM is economical and can be scaled. This process was used with a variety of reactive materials like aluminum–copper oxide, aluminum–molybdenum oxide, aluminum–magnesium, or aluminum–titanium (Umbrajkar et al. 2005, 2006, 2008; Schoenitz et al. 2007). The cryogenic techniques impart uniform mixing of constituents and enhanced reactivity of nanocomposites (Zhang et al. 2010; Badiola et al. 2009). The main drawback of using this route is that sufficient attention is to maintain and a strict safety protocol is to be followed during the milling process.

8.2.4 Solgel Methodology

The solgel technique was first evolved to successfully fabricate nanoenergetic materials made of Fe_2O_3/Al (Tillotson et al. 2001). In solgel methodology, a nanoporous metal precursor like metal alkoxide (X-OR) is hydrolyzed by water, thus forming sol (solid colloidal system in an aqueous solution) and the fuel nanoparticles reside in the nanoscale pores. In the second step, condensation of the sol produces a 3D gel-like solid network (having large viscosity), wherein the open pores are occupied by solvent. This gel is then dehydrated by draining water and or other liquids, thereby transforming it into a microporous structure. Aerogel is formed if the open pores do not collapse during the removal of occupied solvent such as in the case of supercritical drying. Aerogel is more uniform, porous, and lightweight. The main advantage of solgel methodology is the formation of pure and uniform nanocomposites which are synthesized at lower temperature (Clapsaddle et al. 2005). The main drawback of this process is the formation of metal oxide in water, which can be surmounted by preparing oxide precursors (Malynych et al. 2002).

$$X - OR + H_2O \rightarrow X - OH + ROH$$
$$X - OH + RO - X \rightarrow X - O - X + ROH$$
$$X - OH + HO - X \rightarrow X - O - X + H_2O$$

By the solgel methodology, Al/Fe_2O_3 nanoenergetic composites were synthesized having nanoclusters of Fe_2O_3 skelton size about one-order smaller aluminum nanoparticles (Tillotson et al. 2001). Despite solgel method simplicity, the presence of large content organic impurity has remained a challenging task to control. Fe_2O_3/Al formed by solgel route has reported impurities (including water) as high as 25% (Malynych et al. 2002). These impurities retard the combustion reactivity by acting as a heat sink. The Ta/WO_3 synthesized by solgel route was compared with traditional powder mixing route, and the results showed that 30–35% higher heat release was shown by solgel route prepared Ta/WO_3 as compared to the conventional technique (Cervantes et al. 2010).

8.2.5 Formulation by Porous Silicon Embedded with Oxidizers

Porous silicon reacts violently with nitric acid and potassium nitrate, and it was first visible in 1992 (McCord et al. 1992). By controlled anodic etching with hydrofluoric acid, macro-pores and meso-pores are formed on the single crystalline silicon substrates. For pore creation on silicon chip, Becker et al. (2011) first sputtered 170-nm-thick Pt layer on the backside of the silicon for making it as cathode during galvanic etching. He dipped the lithographically patterned specimen into HF-based electrolyte leading to reduction of the oxidizing agent (H_2O_2) at the backside Pt cathode in accordance with the equation shown in Fig. 8.2a and holes are formulated into the silicon chip. Figure 8.2b represents the SEM cross-sectional image of the 20:1 galvanic porous silicon layer having a border of Si_3N_4 mask. Porous silicon has high specific surface area (>900 m^2/gm), and when porous silicon fuel is combined with oxygen source along with heat source, a violent exothermic reaction occurred (Becker et al. 2011). This energetic reaction of porous silicon can be observed with liquid oxygen source (Kovalev et al. 2001), or solid oxygen source (Mikulec et al. 2002), and even with solid sulfur (Clement et al. 2004). The high-oxygen-containing perchlorate (e.g., $NaClO_4$ and NH_4ClO_4) salts are most efficient of highly heated reactions because these salts readily explode with silicon (Clément et al. 2005). The disbursement and contact in between the oxidizers and the silicon are close. By this route, the nanoporous Si/$NaClO_4$ system was developed and showed fastest propagation rate (above 3000 m/s) among solid-phase nanoenergetic system (Becker et al. 2011). Electrochemical etching (Currano and Churaman 2009; Churaman et al. 2010; Wang et al. 2013) and galvanic corrosion etching (Becker et al. 2011; Zhang et al. 2013) methods are two well-known methods that are used to fabricate porous silicon. Porous Si-based nanoenergetic systems have one inherent advantage that they can be used to produce high energy output as well as the flexibility to be easily integrated into micro-electro-mechanical systems (MEMS) to develop micro-devices. Efficiently filling the silicon pores with oxidizers to get higher the filling ratio is the main hurdle of this route, along with longtime storage problem due to hygroscopicity. Besides nanoporous silicon, researchers have successfully synthesized nanoporous copper to fabricate nanoenergetic composites (Zhang et al. 2013).

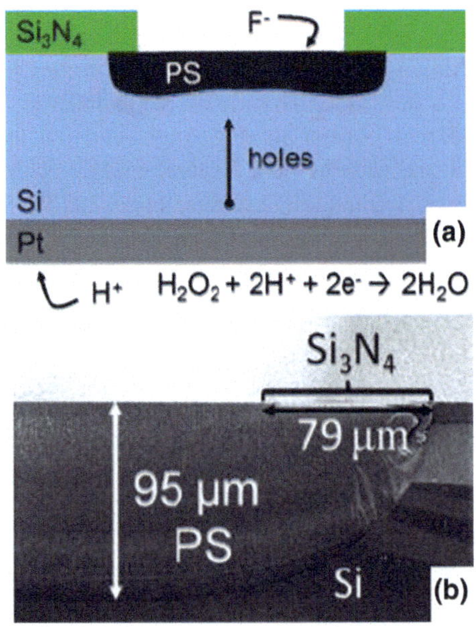

Fig. 8.2 **a** Schematic of galvanic etching mechanism porous silicon formulation and **b** SEM cross-sectional image of the 20:1 galvanic porous silicon layer with silicon nitride mask border. Reproduced with permission from (Becker et al. 2011). Copyright (2011) American Chemical Society

Recently, researchers are successful in fabricating nanoparticles in reduced graphene oxide (RGO) films (Chen et al. 2016). In this ultrafast method, production of nEMs by RGO was done directly by Joule heating (for few milliseconds) and a metal and semiconductor RGO film to elevated temperature (more than 1700 K). At this high temperature, microsized metal is melted which is then cooled rapidly to get nanometal particles embedded in RGO. The agglomerates and segregation of melted particles are done in the crystal lattice plane of RGO which acts as a carrier for nEMs. This process is suitable for any nanoparticles having a melting point lower than 3300 K (melting point of RGO). This method is suitable for aluminum, silicon, tin, etc.

8.2.6 Self-Assembly Methodology

This is an effective method of fabricating nEMs. In this approach, any one or both constituents of nanoenergetic materials comes closer to each other spontaneously or under the influence of some external force and creates intimately mixed nanocomposites. The external forces that are used by researcher in this route are electrostatic force, surface modifiers, and biomolecules.

(i) Electrostatic Self-Assembly

In electrostatic self-assembly approach, the oxidizer and fuel are assembled together under the influence of electrostatic force by creating opposite charges on the fuel and oxidizer. The method results in a significant improvement in the rate of energy release and the heat of reaction on the counterpart of the randomly assembled nanoenergetics (Kim and Zachariah 2004).

(ii) Surface Modifier-Based Self-Assembly

In this approach, the aluminum nanoparticles have been self-assembled with CuO nanorods through the assistance of universal binding agent, i.e., poly (4-vinylpyridine) (P4VP) consisting multiple binding sites. The binding sites derived from pyridyl group of P4VP have strong affinity with both metal and metal oxides (Malynych et al. 2002). The pyridine group forms covalent bond with metals via the donation of electrons and also hydrogen bond with polar groups. In this formation mechanism, first CuO nanorods are coated with P4VP polymer. The excess P4VP polymer is removed by repeated sonication process in 2-propanol solution and centrifugation process. The dried P4VP-coated CuO nanorods are mixed with aluminum nanoparticles in 2-propanol solution by ultrasonication process and finally dried to achieve the self-assembled nanoenergetic composites (Plantier et al. 2005). The self-assembled nanoenergetic composites have developed a higher average combustion velocity of 2300 $\pm$ 100 m/s than that of 1650 m/s achieved with ultrasonically mixed composites (Shende et al. 2008).

(iii) Bio-Inspired Self-Assembly

Recently, deoxyribonucleic acid (DNA)-based self-assembly has been evolved to synthesize aluminum and copper oxide nanoenergetic materials (Séerac et al. 2012; Zhang et al. 2013; He et al. 2018). Two strategies were adopted in binding of oligonucleotides onto CuO and aluminum nanoparticles. Firstly, the oligonucleotides were modified by thiol groups thereby directly grafting of oligonucleotides on the CuO nanoparticles as thiols have strong affinity with CuO. Secondly, Neutravidin was attached to the alumina shell of aluminum, and over this, the biotin-modified oligonucleotides were grafted. Finally, the assembling of as-grafted CuO/Al was completed by DNA hybridization process. The DNA-assembled Al/CuO composites have shown exquisite energetic performance delivering a lower onset temperature (410 °C) and very high exothermic heat of reaction 1.8 kJ/g than there physically mixed counterparts. A biogenic route using the aloe vera leaf extract was utilized to fabricate super-reactive CuO nanorod oxidizers, which were reported to be formulated due to involvements of amino groups, hydroxyl groups, and carboxyl groups and also termed as green synthesis of nanorods of metal oxides (Patel and Bhattacharya 2013).

8.2.7 Core/Shell Patterning of Energetic Materials on Silicon Substrate

Core/shell is a recent and emergent route for research in energetic materials. In this method, a more intimate contact between fuel and oxidizer can be formed. In this methodology, a core mostly made of metal oxide circumvented with metal (shell) is synthesized over a substrate (mostly Si substrate). This methodology is same like RMF, but the structure thus formed by this route consists of core embedded by a shell of different material. Zhang et al. devised a method to prepare core/shell structure of Al/CuO MIC on silicon substrate. The structure thus formed offers several merits as compared to other routes of synthesis, like intimate interfacial contact and customized dimensioning (Zhang et al. 2007). Core/shell nanostructured energetic arrays incorporate better ignition and combustion property because of intimate contact and uniform distribution of reactants (Zhou et al. 2014; Zhang et al. 2013). Moreover, the prepared core/shell can straightforwardly be integrated with MEMS to realize current functional micro-devices (Zhang et al. 2008). Yu et al. recently synthesized core/shell structure by coating nano-Al in the surface of NiO (nanowires and nanorods) (Yu et al. 2016). Xian et al. recently developed non-cracking core/shell MIC composite of Al/CuO on silicon substrate (Zhou et al. 2017). This method is suitable for facile preparation of oxidizing and reducing reagents closely at relatively low temperature.

Recently solvent/non-solvent synthesis approach was utilized in fabrication of core shell morphology in $KClO_4$ @ Al/CuO nanoenergetic materials (Yang et al. 2017). In this study, Yang et al. have done a series of batch experiments to

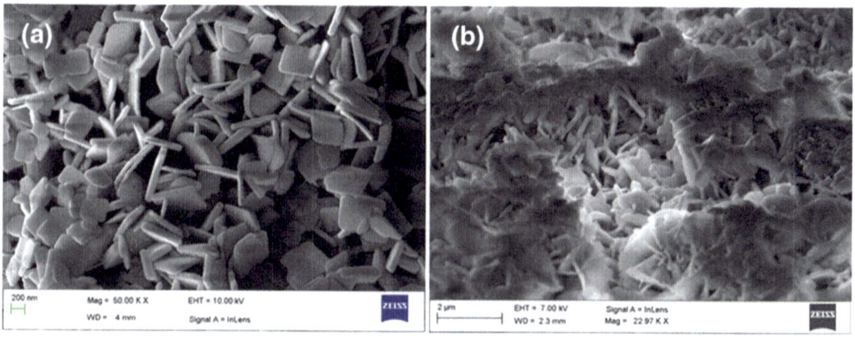

Fig. 8.3 **a** Bi_2O_3 nanosquare tablets and **b** Bi_2O_3/Al nanoenergetic film on Si substrate

manufacture nanosized $KClO_4$ with the help of solvent/non-solvent chemistry method. Nanopowders of aluminum and copper oxide are mixed uniformly by ball milling followed by coating of nanosized $KClO_4$ to get core/shell structure. In these approaches, researchers increase the super-saturation degree of solvent/non-solvent by raising the concentration and temperature difference between solvent/non-solvent, due to which the particle size of $KClO_4$ decreases from micron level to nanolevel. Researchers reported that the nanocomposite thus formed reacts violently having much greater burning velocity and higher rate of energy release because of increased surface area of reactants and shorter mass transfer distance.

Patel et al. (2015) micropatterned nanoenergetic film of Bi_2O_3/Al on silicon substrate. The group developed Bi_2O_3 nanosquare tablets (NSTs) on Au-sputtered Si substrates using a chemical bath deposition process (shown in Fig. 8.3a). They observed the exothermic reactivity of the nanoenergetic film of Bi_2O_3/Al shown in Fig. 8.3b) with varying thickness of 60, 100, and 140 nm, respectively, with the help of TG-DSC measurements, and demonstrated that the nanoenergetic film developed with 140-nm aluminum thickness exhibited increased rate of reaction, low initiation temperature, and excellent pressure–time characteristics and their high-resolution micro-patterning on Si substrates.

8.3 Conclusion

Nanostructured energetic materials are promising means to store energy. The major advantage of nanoenergetic materials over their conventional counterparts is that they offer high reactivity and rapid chemical transformation and thus gives better heat of reaction, combustion efficiency, and impulse. Thus, there is a lot of scope in deploying nanoparticles in energy conversion and energy systems. In the present study, a detailed study on the advances made in the areas of different approaches to micro-/nanofabricate and pattern energetic materials are examined and discussed comprehensively. The special attention must be exercised while preparing a highly reactive and sensitive nanoenergetic composition such as in arrested reactive milling. Thus, with a suitable and optimized selection of different fabrication approaches, the combustion performance, ignition sensitivity, and storage/release energy can be greatly improved.

References

Badiola C, Schoenitz M, Zhu X, Dreizin EL (2009) Nanocomposite thermite powders prepared by cryomilling. J Alloys Compd 488:386–391

Becker CR, Apperson S, Morris CJ, Gangopadhyay S, Currano LJ, Churaman WA, Stoldt CR (2011) Galvanic porous silicon composites for high-velocity nanoenergetics. Nano Lett 11:803–807

Blobaum KJ, Reiss ME, Plitzko JM, Weihs TP (2003a) Deposition and characterization of a self-propagating CuOx = Al thermite reaction in a multilayer foil geometry. J Appl Phys 94 (5):2915–2922

Blobaum KJ, Wagner AJ, Plitzko JM, Heerden DV, Fairbrother DH, Weihs TP (2003b) Investigating the reaction path and growth kinetics in CuOx = Al multilayer foils. J Appl Phys 94(5):2923–2928

Cervantes OG, Kuntz JD, Gash AE, Munir ZA (2010) Heat of combustion of tantalum–tungsten oxide thermite composites. J Combust Flame 157:2326–2332

Chen Y, Egan GC, Wan J, Zhu S, Jacob RJ, Zhou W, Dia J, Wang Y, Danner VA, Yao Y, Fu K, Wang Y, Bao W, Li T, Zachariah MR, Hu L (2016) Ultra-fast self-assembly and stabilization of reactive nanoparticles in reduced graphene oxide films. Nat Commun 7(12332):1–9

Churaman W, Currano L, Becker C (2010) Initiation and reaction tuning of nanoporous energetic silicon. J Phys Chem Solids 71:69–74

Clapsaddle BJ, Zhao L, Prentice D, Pantoya ML, Satcher JHJ, Shea KJ (2005) Formulation and performance of novel energetic nanocomposites and gas generators prepared by sol gel methods. In: 36th international annual conference of ICT, Karlsruhe, Germany

Clement D, Diener J, Kovalev D (2004) Explosive porous silicon: from laboratory accident to industrial application. In: Proceedings of 35th international conference of ICT, pp 5-1–5-11

Clément D, Diener J, Gross E, Künzner1N, Timoshenko VY, Kovalev D (2005) Highly explosive nanosilicon-based composite materials. Phys Status Solid (A) 202:1357–1364

Currano LJ, Churaman WA (2009) Energetic nanoporous silicon devices. J Microelectromech Syst 18:799–807

Dreizin EL (2009) Metal based reactive nanomaterials. J Prog Energy Combust Sci 35:141–167

Fischer SH, Grubelich MC (1998) Theoretical energy release of thermites, intermetallics, and combustible metals. In: Proceedings of the 24th international pyrotechnics seminar, Monterey, CA, 27–31 July

Gavens AJ, Heerden DV, Mann A, Reiss ME, Weihs TP (2000) Effect of intermixing on self-propagating exothermic reactions in Al/Ni nanolaminate foils. J Appl Phys 87:1255

He W, Liu PJ, He GQ, Gozin M, Yan QL (2018) Highly reactive metastable intermixed composites (MICs): preparation and characterization. Adv Mater 3:1706293

Kim SH, Zachariah MR (2004) Enhancing the rate of energy release from nanoenergetic materials by electrostatically enhanced assembly. J Adv Mater 16:1821–1825

Kovalev D, Timoshenko VY, Kunzner N, Gross E, Koch F (2001) Strong explosive interaction of hydrogenated porous silicon with oxygen at cryogenic temperatures. Phys Rev Lett 87:068301–068304

Lewis AC, Josell D, Weihs TP (2003) Stability in thin film multilayers and microlaminates: the role of free energy, structure, and orientation at interfaces and grain boundaries. Scripta Mater 48:1079–1085

Ma E, Thompson CV, Clevenger LA, Tu KN (1990) Self-propagating explosive reactions in Al/Ni multilayer thin films. Appl Phys Lett 57:1262–1264

Malynych S, Luzinov I, Chumanov G (2002) Poly(vinyl pyridine) as a universal modifier for immobilization of nanoparticles. J Phys Chem (B) 106:1280

McCord P, Yau S-L, Bard AJ (1992) Chemiluminescence of anodized and etched silicon: evidence for a luminescent siloxene-like layer on porous silicon. Science 257:68–69

Mikulec FV, Kirtland JD, Sailor MJ (2002) Explosive nanocrystalline porous silicon and its use in atomic emission spectroscopy. J Adv Mater 14:38–41

Patel VK (2013) Sonoemulsion synthesis of long CuO nanorods with enhanced catalytic thermal decomposition of potassium perchlorate. J Cluster Sci 24:821–828

Patel VK, Bhattacharya S (2013) High-performance nanothermite composites based on aloe-vera-directed CuO nanorods. ACS Appl Mater Interfaces 5:13364–13374

Patel VK, Saurav JR, Gangopadhyay K, Gangopadhyay S, Bhattacharya S (2015a) Combustion characterization and modeling of novel nanoenergetic composites of Co_3O_4/nAl. RSC Adv 5:21471–21479

Patel VK, Ganguli A, Kant R, Bhattacharya S (2015b) Micropatterning of nanoenergetic films of Bi_2O_3/Al for pyrotechnics. RSC Adv 5:14967–14973

Plantier KB, Pantoya ML, Gash AE (2005) Combustion wave speeds of nanocomposite Al/Fe_2O_3: the effects of Fe_2O_3 particle synthesis technique. J Combust Flame 140:299–309

Schoenitz M, Umbrajkar SM, Dreizi EL (2007) Kinetic analysis of thermite reactions in Al-MoO_3 nanocomposites. J Prop Power 23:683–687

Séerac F, Alphonse P, Estèe A, Bancaud A, Rossi C (2012) High-energy Al/CuO nanocomposites obtained by DNA-directed assembly. J Adv Funct Mater 22:323–329

Shende R, Subramanian S, Hasan S, Apperson S, Thiruvengadathan R, Gangopadhyay K, Gangopadhyay S, Redner P, Kapoor D, Nicolich S, Balas W (2008) Nanoenergetic composites of CuO nanorods, nanowires, and Al-nanoparticles. Propellants, Explos, Pyrotech 33:122–130

Son SF (2004) Performance and characterization of nanoenergetic materials at Los Alamos. MRS Proc 500

Sundaram DS, Puri P, Yang V (2013) Pyrophoricity of nascent and passivated aluminum particles at nano-scales. J Combust Flame 160:1870–1875

Tillotson TM, Gash AE, Simpson RL, Hrubesh LW, Satcher JH Jr, Poco JF (2001a) Nanostructured energetic materials using solgel methodologies. J Non-Cryst Solids 285:338–345

Tillotson TM, Gash AE, Simpson RL, Hrubesh JHS, Poco JF (2001b) Nanostructured energetic materials using sol-gel methodologies. J Non Cryst Solids 285:338–345

Umbrajkar SM, Schoenitz M, Jones SR, Dreizin EL (2005) Effect of temperature on synthesis and properties of aluminum-magnesium mechanical alloys. J Alloys Compd 40:70–77

Umbrajkar SM, Schoenitz M, Dreizin EL (2006) Exothermic reactions in Al–CuO nanocomposites. Thermochim Acta 451:34–43

Umbrajkar SM, Seshadri S, Schoenitz M, Hoffmann VK, Dreizin EL (2008) Aluminum-rich Al-MoO3 nanocomposite powders prepared by arrested reactive milling. J Propul Power 24:192–198

Wang S, Shen R, Yang C, Ye Y, Hu Y, Li C (2013) Fabrication, characterization, and application in nanoenergetic materials of uncracked nano porous silicon thick films. J Appl Surf Sci 265:4–9

Weismiller MR, Malchi JY, Yetter RA, Foley TJ (2009) Dependence of flame propagation on pressure and pressurizing gas for an Al/Cuo nanoscale thermite. Proc Combust Inst 32:1895–1903

Yang F, Kang X, Luo J, Yi Z, Tang Y (2017) Preparation of core-shell structure $KClO_4$@Al/CuO nanoenergetic material and enhancement of thermal behavior. Sci Rep 3730:1–9

Yu C, Zhang W, Shen R, Xu X, Cheng J, Ye J, Qin Z, Chao Y (2016) 3D ordered macroporous NiO/Al nanothermite film with significantly improved higher heat output, lower ignition temperature and less gas production. J Mater Des 110:304–310

Zhang K, Rossi C, Ardila RGA, Tenailleau C, Alphonse P (2007) Development of a nano-Al/CuO based energetic material on silicon substrate. Appl Phys Lett 91:113117

Zhang K, Rossi C, Petrantoni M, Mauran N (2008) A nano initiator realized by integrating Al/CuO-based nanoenergetic materials with a Au/Pt/Cr microheater. J Microelectromech Syst 17:832–836

Zhang S, Schoenitz M, Dreizin EL (2010) Mechanically alloyed Al–I composite materials. J Phys Chem Solids 71:1213–1220

Zhang F, Wang YL, Fu DX, Li LM, Yin GF (2013a) In-situ preparation of a porous copper based nano-energetic composite and its electrical ignition properties. J Propellants Explos Pyrotech 38:41–47

Zhang Y, Lu F, Yager KG, VanderLelie D, Gang O (2013b) A general strategy for the DNA-mediated self-assembly of functional nanoparticles into heterogeneous system. Nat Nanotechnol 8:865–872

Zhang W, Yin B, Shen R, Ye J, Thomas JA, Chao Y (2013c) Significantly enhanced energy output from 3D ordered macroporous structured Fe_2O_3/Al nanothermite film. J ACS Appl Mater Interfaces 5:239–242

Zhou X, Torabi M, Lu J, Shen R, Zhang K (2014a) Nanostructured energetic composites: synthesis, ignition/combustion modeling and applications. J ACS Appl Mater Interfaces 6:3058–3074

Zhou X, Xu D, Yang G, Zhang Q, Shen J, Lu J, Zhang K (2014b) Highly exothermic and superhydrophobic Mg/fluorocarbon core/shell nanoenergetic arrays. ACS Appl Mater Interfaces 6:10497–10505

Zhou X, Wang Y, Cheng Z, Ke X, Jiang W (2017) Facile preparation and energetic characterization of core-shell Al/CuO metastable intermolecular composite thin film on silicon substrate. J Chem Eng 328:585–590

Part III
Tuning and Characterization of Nano-energetic Materials

Chapter 9
Tuning the Reactivity of Nano-energetic Gas Generators Based on Bismuth and Iodine Oxidizers

Mkhitar A. Hobosyan and Karen S. Martirosyan

Abstract There is a growing interest in novel energetic materials called nano-energetic gas generators (NGGs) which are potential alternatives to traditional energetic materials including pyrotechnics, propellants, primers, and solid rocket fuels. NGGs are formulations that utilize metal powders as a fuel and oxides or hydroxides as oxidizers that can rapidly release a large amount of heat and gaseous products to generate shock waves. The heat and pressure discharge, impact sensitivity, long-term stability, and other critical properties depend on the particle size and shape, as well as assembling procedure and intermixing degree between the components. The extremely high energy density and the ability to tune the dynamic properties of the energetic system makes NGGs ideal candidates to dilute or replace traditional energetic materials for emerging applications. In terms of energy density, performance, and controllability of dynamic properties, the energetic materials based on bismuth and iodine compounds are exceptional among the NGGs. The thermodynamic calculations and experimental study confirm that NGGs based on iodine and bismuth compounds mixed with aluminum nanoparticles are the most powerful formulations to date and can be used potentially in microthrusters technology with high thrust-to-weight ratio with controlled combustion and exhaust velocity for space applications. The resulting nano-thermites generated the significant value of pressure discharge up to 14.8 kPa m^3/g. They can also be integrated with carbon nanotubes to form laminar composite yarns with high power actuation of up to 4700 W/kg or be used in other emerging applications such as biocidal agents to effectively destroy harmful bacteria in seconds, with 22 mg/m^2 minimal content over the infected area.

Keywords Nano-energetic Gas generators · Bismuth · Iodine · Oxidizer
Reactivity tuning

M. A. Hobosyan · K. S. Martirosyan (✉)
Department of Physics and Astronomy, University of Texas at Rio Grande Valley,
Brownsville, TX 78520, USA
e-mail: karen.martirosyan@utrgv.edu

S. Bhattacharya et al. (eds.), *Nano-Energetic Materials*, Energy, Environment
and Sustainability, https://doi.org/10.1007/978-981-13-3269-2_9

9.1 Introduction

Nano-energetic gas generators (NGGs) are thermite-based nano-structured formulations, where the gas generation ability of the system is most important characteristic and is measured by the ability of the system to generate pressure discharge during very short time (order of microseconds). The main difference of nano-structured thermites from so-called traditional thermites is the reduced particle size, where at least one component of the thermite should be in nano-sized particle domain of less than 100 nm (Dlott 2006; Martirosyan 2011; Sullivan and Zachariah 2010). Thermites are well-known pyrotechnic composite mixtures of metal powder and metal or nonmetal oxide, which can produce an exothermic reaction also known as aluminothermic reaction. The traditional thermites have been known since the nineteenth century (Arnáiz et al. 1998).

The class of compositions in thermites can be very diverse, where the metal powders act as a fuel, and the oxidizers are various oxides such as traditional metal oxides including Fe_2O_3, Co_3O_4, MoO_3 (Patel et al. 2015b; Cheng et al. 2010; Umbrajkar et al. 2006). Other highly powerful oxidizers include nano-energetic gas generators (Martirosyan 2011; Martirosyan et al. 2009b), constituents such as I_2O_5 (Martirosyan et al. 2009a; Clark and Pantoya 2010; Hobosyan et al. 2012; Hobosyan and Martirosyan 2017) and Bi_2O_3 (Martirosyan et al. 2009c; Patel et al. 2015a; Puszynski et al. 2007; Wang et al. 2011). The metal fuel in thermites is usually aluminum powder, although occasionally magnesium is also utilized as a fuel metal (Zhou et al. 2014). The reason behind the popularity of Al as a fuel is that it is one of the cheapest metals to produce, and it is highly energetic and reactive and exhibits relatively low melting temperature (660 °C). This implies that the metal can melt and react with oxidizers in liquid form, tremendously accelerating the reaction rate in comparison with solid–solid reactions. Although aluminum is highly reactive, the aluminum oxide layer that naturally forms on aluminum particles protects the aluminum core and makes it safer to use even 20–100 nm particle size.

The thermites at nanoscale have very high chemical reaction rate in comparison with thermites at microscale. When reducing the particle size to nanometer domain, the reaction propagation velocity can increase by up to 2–3 orders of magnitude for some systems (Umbrajkar et al. 2006; Kim and Zachariah 2004; Perry et al. 2004; Dreizin 2009), making them comparable or even better than the traditional energetic materials such as pentaerythritol tetranitrate (PETN), 2,4,6-trinitrotoluene (TNT), or 1,3,5-trinitroperhydro-1,3,5-triazine (RDX). The reason behind is that the nano-thermite systems have higher volumetric energy density compared to traditional energetic materials (Martirosyan 2012). However, this large energy capacity can be only effectively utilized, if the thermites are at nanoscale. When reducing the thermite reactants particle size to nano-sized domain, the contact between reagents' particles becomes much more intimate than in micron-sized particles, and moreover, the size reduction decreases the diffusion and transport limitations (Johnson et al. 2007; Wang et al. 2007; Comet et al. 2010). Thus, the successful energy utilization in nano-thermite mixtures gives them great potential to become the next-generation energetic materials both in military (propellants, primers, pyrotechnics) and civilian applications (energy, mining, etc.) (Miziolek 2002; Puszynski 2009; Rogachev and Mukasyan 2010). The NGGs can create shock waves with velocities up to 2500 m/s (Martirosyan et al. 2009c; Martirosyan 2012), which is opening perspectives for more applications such as medicine, biological sciences, material processing, manufacturing, and microelectronic industries. We should note that the energy capacity has a critical role in determining the pressure discharge capability of energetic materials as demonstrated by Martirosyan et al. in (Martirosyan 2012). Bismuth and iodine oxides mixtures with aluminum have one of the highest energetic capacities per volume, among common nano-thermite systems. Both bismuth oxide and bismuth hydroxide have shown excellent performance in NGG formulations, Al-Bi_2O_3 (Martirosyan et al. 2009c; Wang et al. 2011) and Al-$Bi(OH)_3$ (Hobosyan et al. 2016). The corresponding maximum pressure × volume (PV) per mass (PV/m) value for Al-Bi_2O_3 nano-structured formulation was 8.6 kPa m^3/g (Martirosyan et al. 2009c; Wang et al. 2011), while for the Al-$Bi(OH)_3$ formulation the corresponding value was 5.6 kPa m^3/g. Although the energetic capacity per volume or per mass for Al-$Bi(OH)_3$ system is lower than that for Al-Bi_2O_3 formulation, it can create twice more gaseous products per mass (Hobosyan et al. 2016). This could be advantageous for hydroxide-based systems in comparison with oxide-based thermites, especially in applications requiring large amounts of gaseous products to be released upon charge ignition, such as microthrusters and micropropulsion platforms for space and terrestrial applications (Puchades et al. 2017). The iodine pentoxide-based system has shown even higher PV/m value of 14.8 kPa m^3/g in the nano-structured formulation Al-I_2O_5 and can be used not only as propellant in microthrusters (Martirosyan et al. 2012; Puchades et al. 2014), but also as a very efficient biocidal agent due to highly active atomic iodine released during nano-thermite reaction (Hobosyan et al. 2012).

9.2 Thermodynamic Considerations

For the estimation and control of the thermodynamic properties of the energetic systems and reaction products, knowledge of the adiabatic reaction temperature and equilibrium concentration of solid, liquid, and gas phases is needed. The thermodynamic estimation of the equilibrium composition of multicomponent multiphase systems requires minimization of the thermodynamic free energy (G) subject to mass and energy balances (Greiner et al. 1995). "Thermo" software was used for thermodynamic calculations (Shiryaev 1995), which has thermochemical database for approximately 3000 compounds. In addition, we used the thermochemical software HSC Chemistry 7, which can predict equilibrium compositions and can calculate adiabatic temperature, if input and output species and their amounts are defined. HSC Chemistry 7 has a database of over 25,000 compounds, based on which it can also calculate the theoretical heat balances, by taking molecular amounts of components using the reaction equations at system equilibrium.

The reaction adiabatic temperature and composition of the equilibrium products can be estimated by minimizing the thermodynamic potential. For a system with N(g) gas and N(s) solid number of components, at constant pressure, the concentrations of equilibrium phases can be expressed as:

$$F(\{n_k\},\{n_s\}) = \sum_{k=1}^{N^{(g)}} n_k\left(\ln\frac{p_k}{p} + G_k\right) + \sum_{l=1}^{N^{(s)}} n_l G_l \quad (9.1)$$

where p_k is the partial pressure of the kth gas-phase component, while n_l and G_l are the number of moles and molar Gibbs free energy of components. The adiabatic combustion temperature, T_c^{ad}, is determined by total energy balance:

$$\sum_{i=1}^{N_0} H_i(T_0) = \sum_{k=1}^{N_{(g)}} n_k H_k\left(T_c^{ad}\right) + \sum_{l=1}^{N_{(s)}} n_l H_l\left(T_c^{ad}\right) \quad (9.2)$$

where the enthalpy of each component is

$$H_i(T) = \Delta H_{f,i}^0 + \int_{T_0}^{T} c_{p,i}\mathrm{d}T + \sum \Delta H_{s,i} \quad (9.3)$$

and $\Delta H_{f,i}^0$ is the heat of formation at 1 atm and reference temperature T_0, $c_{p,i}$ is the heat capacity, and $\Delta H_{s,i}$ is the heat of sth phase transition for components (Varma et al. 1998). The combination of both the above-mentioned software programs made it possible to estimate the equilibrium composition, adiabatic temperature, gas generation, and standard enthalpy of formation for each examined system.

The reason of exceptional performance for bismuth- and iodine-based NGGs can be that the boiling temperature of oxide or hydroxide metal (iodine or bismuth) is significantly lower than the combustion adiabatic temperature, which contributes to gaseous products during combustion and improves the pressure generation ability (Martirosyan 2011; Hobosyan et al. 2012; Yolchinyan et al. 2018). The adiabatic combustion temperature of common nano-thermite formulations at stoichiometric ratios along with oxidizer metal boiling temperature and normalized gas generation is presented in Table 9.1. It can be seen that boiling temperature of the final product for systems based on iodine and bismuth oxidizers is significantly lower than adiabatic combustion temperature. They generate significant amounts of gaseous products, with the highest number for iodine pentoxide-based NGGs. The other formulations have the oxidizer metal boiling point higher than the combustion adiabatic temperature, and the gas generation is due to partial decomposition of reaction product aluminum oxide into gaseous aluminum oxides such as Al_2O(g). Therefore, lower boiling point of oxidizer metal appears to be one of the most important factors for high-pressure discharge value.

Table 9.1 Boiling point of metal, maximum combustion adiabatic temperature, and normalized gas generation per initial thermite mass, for thermite formulations

Combustion system	Boiling point of metal (K) (Zhang et al. 2011)	Maximum combustion adiabatic temperature (K)	Normalized gas generation (L/g)
10 Al + 3 I_2O_5	(I) 457	3827	3.1
2 Al + $Bi(OH)_3$	(Bi) 1833	2965	2.1
2 Al + Bi_2O_3	(Bi) 1833	3279	1.1
8 Al + 3 Co_3O_4	(Co) 3200	3174	0.6
2 Al + MoO_3	(Mo) 4921	3808	0.4
2 Al + Fe_2O_3	(Fe) 3134	3130	0.3

The NGGs perform better when both fuel and oxidizer are at nano-structured scale. The fuel Al can be purchased commercially at 100 nm average size (Sigma Aldrich). This size is not very pyrophoric and is covered by approximately 4–5 nm aluminum oxide layer, which allows to safely mix with oxidizers.

The thermite reactions can be accelerated if the oxide layer is removed from Al nanoparticles. At high heating rates, which are characteristic to thermite types of reactions, the aluminum oxide can be removed with polytetrafluoroethylene (PTFE). It is interesting to note that the PTFE can react with the outer Al_2O_3 layer on aluminum particles, which is an exothermic process and can increase the oxidation reaction rate, and ultimately can improve the energy and gas discharge in PTFE containing thermites (Hobosyan et al. 2015). The PTFE-activated thermites were successfully utilized for lunar regolith consolidation purposes (Hobosyan and Martirosyan 2014, 2016). We should emphasize that the activation of thermite reactions with PTFE is useful when the main system has lower combustion adiabatic temperature than the formulation Al-PTFE, such as the system Al_2O_3-PTFE. Moreover, it should be noted that some components which can react with PTFE may change the stoichiometric balance of Al-PTFE reaction and reduce the combustion temperature. Table 9.2 summarizes the results for the calculated adiabatic temperature for each system with or without PTFE. As can be seen, the addition of Al-PTFE system greatly increases the combustion adiabatic temperature for Al_2O_3-PTFE system from 1425 to 3075 K. However, the addition of Al-PTFE formulation in systems containing bismuth oxide, bismuth hydroxide, and iodine pentoxide has negative contribution and reduces the combustion adiabatic temperature by several hundred degrees. The possible reason is due to the partial reaction between oxidizers (Bi_2O_3, $Bi(OH)_3$, and I_2O_5) with PTFE, which shifts the stoichiometric balance of Al-PTFE formulation, and reduces the combustion temperature. Thus, PTFE can be used to activate regolith thermite reactions due to the fact that regolith contains a significant amount of Al_2O_3. Thus, the (Al_2O_3-PTFE)-(Al-PTFE) system has higher energetic resources, as discussed above, and the activation effect of Al-PTFE formulation is significant. However, PTFE should not be used in case of iodine- and bismuth-based oxidizers due to negative contribution in combustion temperature, as presented in Table 9.2.

Table 9.2 Adiabatic combustion temperature for thermites based on PTFE, iodine, and bismuth compounds

Combustion system	Adiabatic temperature (K)
PTFE-Al	3587
PTFE-Al_2O_3	1425
(PTFE-Al_2O_3)-(PTFE-Al)	3075 (large increase)
Bi_2O_3-Al	3284
(Bi_2O_3-Al)-(PTFE-Al)	2552 (decrease)
$Bi(OH)_3$-Al	2970
($Bi(OH)_3$-Al)-(PTFE-Al)	2346 (decrease)
I_2O_5-Al	3830
(I_2O_5-Al)-(PTFE-Al)	3423 (decrease)

9.3 Preparation of Iodine- and Bismuth-Based Oxidizers at Nanoscale

As mentioned above, the best performance for NGGs can be expected when not only fuel, but also oxidizers are at nanoscale. In addition to the particle size, the particle shape may also play an important role when estimating the contact area between reagents, which is one of the most important factors for the reaction rate. The shape of Al fuel nanoparticles is spherical, while the oxidizers can be prepared by various shapes and sizes, which can have a significant effect on NGG performance during pressure discharge characterizations (Yolchinyan et al. 2018). Although Bi_2O_3, Bi $(OH)_3$, and I_2O_5 are commercially available as micrometer-sized powders, the nano-sized powder preparation can be challenging, especially for iodine pentoxide. The high energy ball milling allows to receive iodine pentoxide nano-rods (Hobosyan and Martirosyan 2017) and bismuth hydroxide sub-micrometer and nano-sized particles (Hobosyan et al. 2016), while bismuth oxide can be received at nanoscale using solution combustion method (Martirosyan et al. 2009c). Bismuth oxide and hydroxide with various shapes and sizes can be received using micro-fluidic synthesis approach (Yolchinyan et al. 2018).

For iodine pentoxide nanoparticle preparation, the commercial iodine pentoxide microsized particles with 98% purity (Sigma Aldrich) can be used to produce I_2O_5 nano-rods structures. The particles' compositions are very sensitive from environmental variation, and thus, sample storage, handling, and preparation should be performed under nitrogen environment in glove box, and iodine pentoxide nanoparticles should be stored and used under reduced humidity environment to protect formation of hydrated iodine pentoxide due to water molecules absorption by I_2O_5. For mechanical milling of microsized particles (Fig. 9.1a), a High Energy Shimmy Ball Mill (HSF-3, MTI Co) was used. The ball-to-powder mass ratio in the milling container was 4:1. The particle size of iodine pentoxide was regulated by time controlling of high energy ball milling process via tuning the applied energy dose which is transferred to the milling media. The details of the estimation of energy dose transferred to particles can be calculated using the method described in

(Hobosyan and Martirosyan 2017). The total energy transferred per unit mass of powder, or the specific energy dose, D_E, [J/kg], is given by $D_E = \frac{NtE-(Q+q)}{m_p}$, where N is collision frequency [Hz]; t is the processing time [s]; m_p is the mass of the powder batch [kg]; Q is the thermal heat loss to milling container (stainless steel), calculated as $Q = c_{ps}M\Delta T$, where c_{ps} is the heat capacity of stainless steel, M is the combined mass of metallic container and balls, and ΔT is the change in temperature. q is the thermal heat loss to powder batch calculated as $q = c_{pp}m_p\Delta T$, where c_{pp} is the heat capacity of powder being treated.

The I_2O_5 particles during mechanical treatment are gradually transforming into nano-rods (Fig. 9.1b). In this case, 15 min of mechanical treatment converted the iodine pentoxide micrometer-sized particles into nano-rods, shown in schematics in Fig. 9.1c.

For the bismuth hydroxide, the mechanical treatment method is working similarly (Hobosyan et al. 2016). The initial micrometer-sized particles of commercially available (Acros Organics, 99% purity) bismuth hydroxide (Fig. 9.2a) are converted into sub-micrometer and nano-sized particles (Fig. 9.2b).

For bismuth oxide particle preparation, microfluidic synthesis approach (Fig. 9.3a) was utilized to produce particles with various shapes and estimate the dependence of pressure discharge of NGGs on oxidizer particle shape and size. Bi_2O_3 structures which resemble flowers, brushwood-like structures, and bow ties were successfully synthesized using polyethylene glycol (PEG) with various molecular length as a reaction media (Yolchinyan et al. 2018). The surfactant PEG molecular length was very important during formation of particle morphology, where PEG with 200 molecular weight (MW) results in the production of 1–2-μm flower-like structures, which have self-assembled layers of petals with thickness of 20–30 nm (Fig. 9.3b). Utilization of PEG with 8000 MW results in bowtie- and brushwood-like structures, which are highly crystalline with up to 60 μm size (Fig. 9.3c, d). The nano-thermite mixtures prepared with flower-shaped particles as oxidizers showed higher pressure discharge values than the mixtures prepared with bowtie- and brushwood-like particles.

In all cases, NGGs Al-I_2O_5, Al-Bi_2O_3, and Al-$Bi(OH)_3$ prepared with nano-sized oxidizer particles generated much higher value of pressure discharge than the mixtures prepared with commercial micrometer-sized oxidizer. Table 9.3 summarizes the maximum pressure discharge value for each composition.

Thus, the NGGs based on iodine and bismuth oxidizers exhibited significant pressure discharge properties, which can be finely tuned not only by the size, but also by the shape of oxidizer particles. These tunable pressure generation abilities, as well as the generation of gaseous iodine in the case of Al-I_2O_5 formulation, were successfully utilized in the following emerging applications.

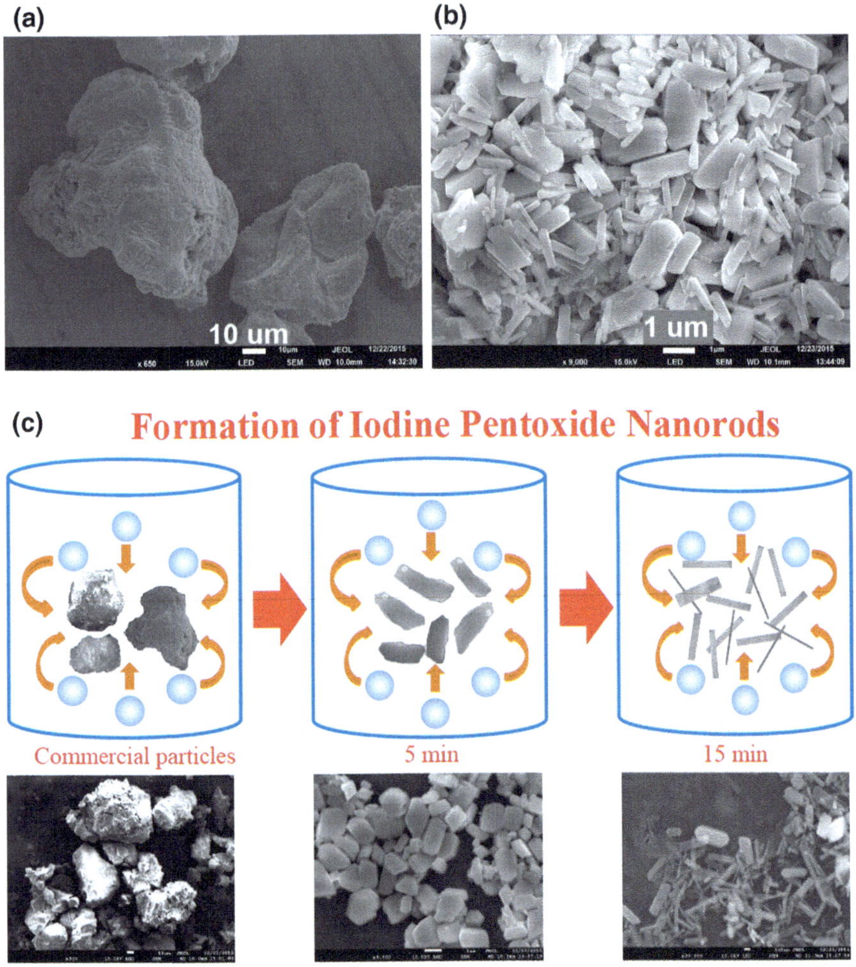

Fig. 9.1 **a** Scanning electron microscopy (SEM) images of typical commercial iodine pentoxide particles; **b** particles after 10 min of mechanical treatment; **c** schematic representation of mechanical treatment, showing the particle morphology after each step

9.4 Emerging Applications of NGGs Based on Iodine and Bismuth Oxidizers

As was outlined above, the rapid generation of copious amounts of gaseous products, which provides extremely high pressure in microseconds, places the NGGs based on iodine and bismuth components in a unique position. Aside from traditional applications as thermites in pyrotechnics, welding, metal cutting, elementary metal production, etc., they can be used in micropropulsion platforms as

Fig. 9.2 Commercially available bismuth hydroxide particles before (**a**) and after (**b**) mechanical treatment in high energy ball mill

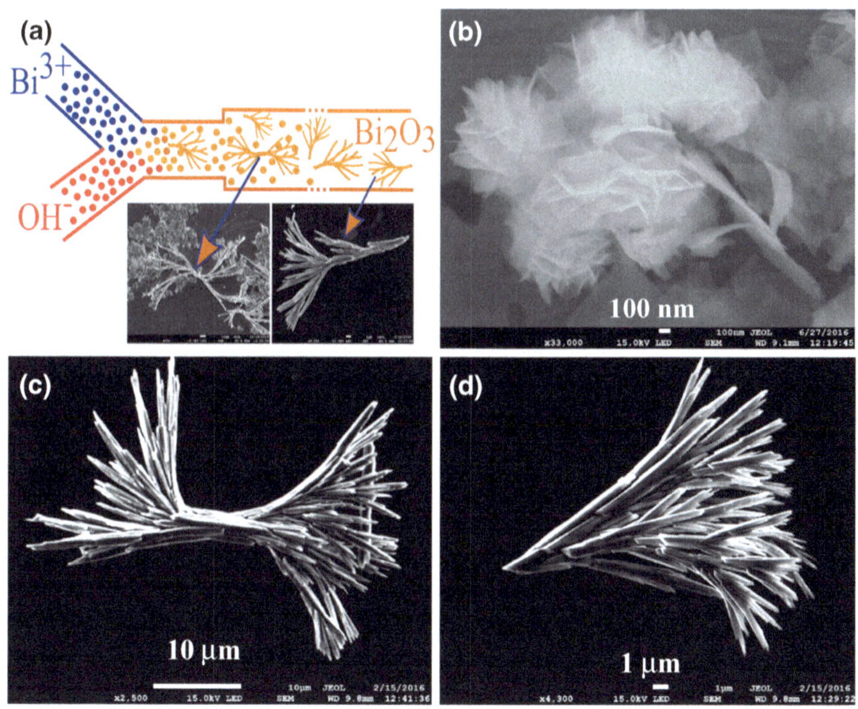

Fig. 9.3 **a** Schematics of microfluidic synthesis, and SEM of Bi_2O_3 product utilizing, **b** PEG-200 surfactant, resulting in flower-like particles, and **c**, **d** PEG-8000 surfactant, showing bowtie-like and brushwood-like structures

Table 9.3 Maximum pressure generation for NGGs prepared with iodine and bismuth oxidizers with commercially available micrometer-sized and nano-sized particles

Combustion system	Maximum pressure for commercial oxidizer, in 0.345 L volume, 0.2 g charge (MPa/g)	Maximum pressure for nano-sized oxidizer, in 0.345 L volume, 0.2 g charge (MPa)	Normalized pressure for nano-sized oxidizer, PV/m (kPa m^3/g)
Al-I_2O_5	4.4	8.68	14.8
Al-$Bi(OH)_3$	3.33	4.34	5.6
Al-Bi_2O_3	1.2	2.86	4.9

new types of propellants, as well as in hybrid actuators and biocidal applications in the case of Al-I_2O_5 system. Each of these new applications is detailed below.

9.4.1 *Nano-energetic Micropropulsion Systems*

The micropropulsion systems utilize microelectromechanical system (MEMS) technology, which are integrated with high energy density nano-energetic materials as solid propellants. These microthrusters can be used as single units or in arrays, in nano- and microsatellites, for applications in payload delivery, navigation, guidance, stabilization, various maneuvering tasks, etc. (Puchades et al. 2014; Rossi et al. 2007; Chaalane et al. 2015). Due to the complex structure and operation of microthrusters with liquid propellants (Ding et al. 2015; Xu et al. 2017), the solid propellants are preferred, which have advantages such as simplicity, adequate functionality, safety. The impulse control should be in mNs resolution (Sathiyanathan et al. 2011). The burning time of propellant in microchamber is a few milliseconds, as the charge mass is in order of micrograms to milligrams. Since the burning time of propellant in microthruster is very short, the polymeric materials can be preferred as structural materials to print 3D-optimized plastic microthrusters, which are lightweight and robust (Puchades et al. 2017). The thermal insulation in polymers is exceptionally good due to low thermal conduction; therefore, nozzle and polymer chamber are not deformed and not melted. The same polymer microthrusters were tested multiple times, and the impulse variations were within 10%.

For the microthruster operation, the total impulse I_{total} and specific impulse I_{sp} are the key characteristics, and they are given as

$$I_{\text{total}} = \int_0^{t_c} F\mathrm{d}t \tag{9.4}$$

$$I_{\text{sp}} = \frac{I}{mg} \tag{9.5}$$

where t_c is combustion time; m is the propellant mass; and F is the thrust.

The specific impulse is proportional to combustion temperature (T_c) of gas products in the chamber and inversely related to the average molecular mass of the gaseous products during combustion W_g (Fut'ko et al. 2011).

$$I_{\mathrm{sp}} \sim \sqrt{\frac{T_c}{W_g}} \tag{9.6}$$

Thus, in order to receive high specific impulse, the combustion temperature should be large, and the average molecular weight of combustion gaseous products should be low. In microthrusters, the heat losses are significant, as the propellant mass is very small, which results in reduced specific impulse values compared to massive rocket engines.

It should be noted that the traditional energetic materials such as nitrocellulose are not suitable to be used in microthrusters due to insufficient thrust generation at extremely small charge mass, which is a requirement for microthrusters. Most solid fuels in a large rocket engine are not suited for microthrusters due to ignition inconsistencies, encapsulation inadequacy, safety, etc. For example, the well-known hydroxyl-terminated polybutadiene (HTPB)–ammonium perchlorate (AP) is not sensitive sufficiently and cannot be used in microthrusters alone. A separate ignitor is necessary, which complicates design and operation (Liu et al. 2015). The examined hybrid composite solid rocket fuels include either HTPB/AP compositions (Sathiyanathan et al. 2011; Liu et al. 2015; Rossi et al. 2005) or traditional energetic materials such as nitrocellulose (NC) combined with thermite types of formulations such as Al-Bi_2O_3-NC (Staley et al. 2013) and Al-CuO-10 wt% NC (Ru et al. 2017). The addition of NC improves the thrust generation ability of thermites until some small (2.5–10 wt%) amount, after which the increase of NC has a negative effect. We note that the thermites can be also combined with HTPB or other binding polymers, which should generate desired thrust and tunability by ranging the concentration of thermites in polymers. The addition of thermite in HTPB should increase sensitivity and provide adequate thrust abilities, without separate ignitors.

As discussed above, the nano-energetic compositions based on iodine and bismuth compounds are very energetic and release large amounts of gaseous products. The reaction Al-Bi_2O_3 has larger heat release than Al-$Bi(OH)_3$, but the later produces almost twice the amounts of gaseous products than Al-Bi_2O_3 (Table 9.1). When comparing the theoretical specific impulse estimates by Eq. (9.6), the reaction Al-Bi_2O_3 yields $\sqrt{\frac{T_c}{W_g}} = 3.96$, while this number for Al-$Bi(OH)_3$ is $\sqrt{\frac{T_c}{W_g}} = 5.65$, which is even higher than the number for reaction Al-I_2O_5, calculated to be $\sqrt{\frac{T_c}{W_g}} = 5.5$. Thus, the bismuth hydroxide NGG formulation has the best propellant energetics. We have tested all three systems, and Al-$Bi(OH)_3$ provided the best thrust generation abilities and specific impulse (Puchades et al. 2014, 2017; Martirosyan et al. 2012).

In order to test the thrust generation abilities of Al-Bi$(OH)_3$ system, single microthrusters and thruster arrays were designed and printed, using thermoplastic polymer Acrylonitrile Butadiene Styrene. Microthrusters were printed by a 3D printer "Printrbot Simple Metal" (Fig. 9.4a) with various nozzle geometries (outer-to-throat diameter ratios). The microthruster arrays can be printed with precise dimensions as presented in Fig. 9.4b. The evaluation of thrust generation for microthrusters, which were integrated with NGGs, is performed using the Phidget force sensors with bridgeboard as presented in Fig. 9.4c. The force sensor was calibrated before each measurement to ensure the measurement accuracy. To eliminate the effect of high-temperature igniter on sensing and imaging, the ignition of microthrusters was initiated using a blue laser with power output of 1.4 W (Puchades et al. 2017).

In microthrusters, the nozzle inner diameter was 0.7 mm and outer diameter was 3 mm. The wall thickness was 1 mm. The microthrusters in array occupy 5 × 5 mm area. The microthruster chamber was cylindrical with 0.7 mm diameter and 2 mm height.

Figure 9.5a shows the measured thrust data, and Fig. 9.5b–h demonstrates the ignition images extracted from high-speed video recording. For the 2 mg mass of loaded Al-Bi$(OH)_3$, nano-thermite generates a peak thrust of 0.31 N. The area

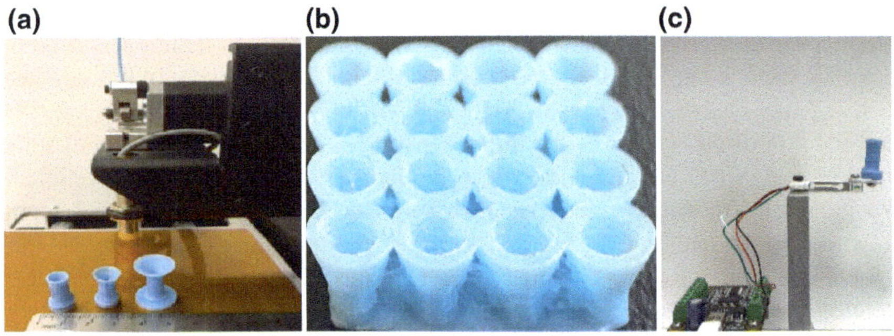

Fig. 9.4 **a** Printrbot simple metal 3D printer, printed microthruster nozzles with three different ratios of outer diameter to inner chamber diameter, **b** side view of a 4 × 4 thruster array, **c** measurement setup with Phidget force sensors

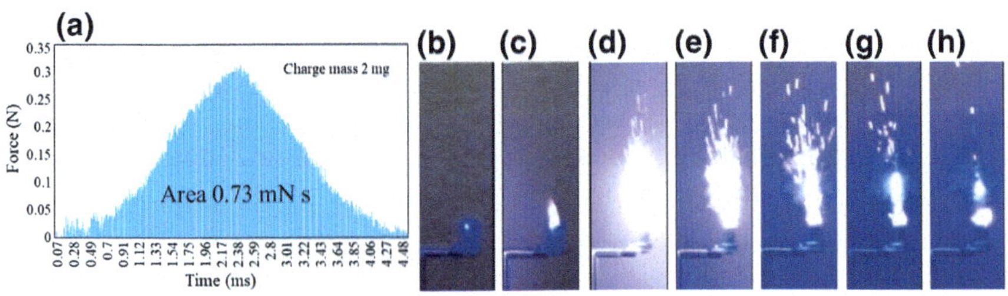

Fig. 9.5 **a** Thrust by a single microthruster with 2-mg Al-Bi$(OH)_3$ nano-thermite and **b–h** consecutive frame snapshots of charge ignition extracted from high speed video recording at 980 fps

under the curve represents the total impulse, summing up to I_{total} = 0.7 mNs with specific impulse I_{sp}= 37 s.

The total impulse can be tuned by changing the mass of thermite and geometry of microthruster chamber. The nano-thermite mass can be reduced to 0.1 mg or increased up to 10 mg for single microthruster to ensure required thrust, I_{total} and I_{sp} for specific flight phases and modes. In case of Al-$Cu(IO_3)_2$ nano-thermite, for 0.1 mg mass the total impulse was 0.2 mNs, which increased up to 20 mNs for the charge mass of 10 mg (Hobosyan et al. 2018). The specific impulse was 216 s for this formulation. The studied single microthruster and thruster arrays can be scaled to nano-, micro-, and mini-satellites and aerial systems.

9.4.2 High Power Output Actuators Based on MWCNT/NGGs Composite Yarns

The NGGs can release high-pressure gases at microsecond timeframe, and the resulting rapid expansion of gaseous products can potentially generate a large work, and a high-force actuation, if the NGGs are incorporated within the yarns made from multiwalled carbon nanotube (MWCNT) sheets (Hobosyan et al. 2017). The MWCNTs possess excellent mechanical and physical properties, which are especially desired for applications where remarkable strength and high stiffness (reaching up to 300 MPa) of single-ply yarns can be utilized (Zhang et al. 2004). The MWCNT sheets extracted from forest can be scrolled to form yarns, which can incorporate various functional materials, where the composite yarns can act as actuators (Baughman et al. 1999; Aliev et al. 2009). This strategy was used in this work, to create nano-composites made from twisted MWCNT yarns that incorporate NGGs based on iodine and bismuth oxidizers. From the side of a MWCNT forest, well-aligned MWCNT sheets were produced by dry spinning using a razor blade. The width of the sheet is proportional to the width of the synthesized MWCNT forest (Fig. 9.6). For a typical yarn actuator, 2-cm-wide MWCNT sheets were placed between two rods separated by a distance of 7 cm. Fifteen MWCNT sheets were stacked on top of each other to increase the sheet strength during NGG coating. The weight of the stacked MWCNT sheets was $\sim$0.8 mg.

MWCNT sheets were coated with NGG materials by utilizing automated dispensing robot EFD-325TT equipped with a Paasche VL-SET airbrush system. The coating solution was prepared by suspending 25 mg/mL of NGG particles in isopropanol and sonicating for 30 min. The robot was programmed to coat the sheets in 30 s time intervals to allow the evaporation of isopropanol and control the MWCNT/NGG weight ratio (Fig. 9.6). The laminar yarns were prepared by twisting the MWCNT/NGG sheets with a low-speed motor under a constant speed of 30 rpm. To form MWCNT/NGG yarns, the coated sheets were twisted into Archimedean type yarns, as described in Lima et al. (2011), with 800–1000 turns/m

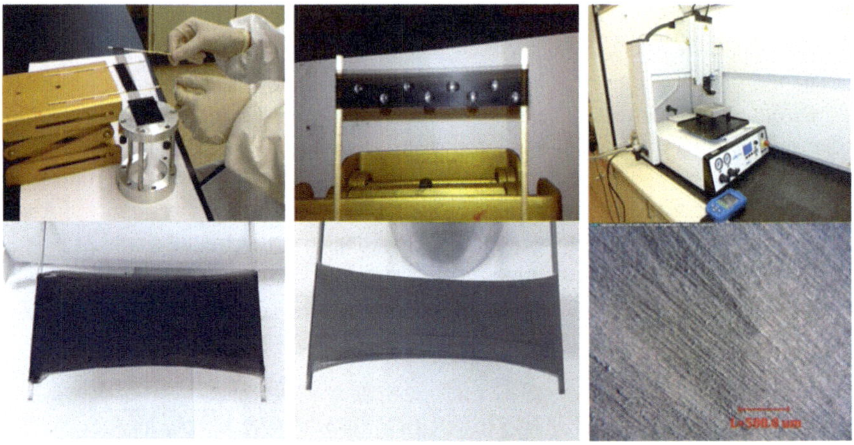

Fig. 9.6 Preparation of MWCNT sheets, and coating with NGG using robot EFD-325TT equipped with a Paasche VL-SET airbrush system

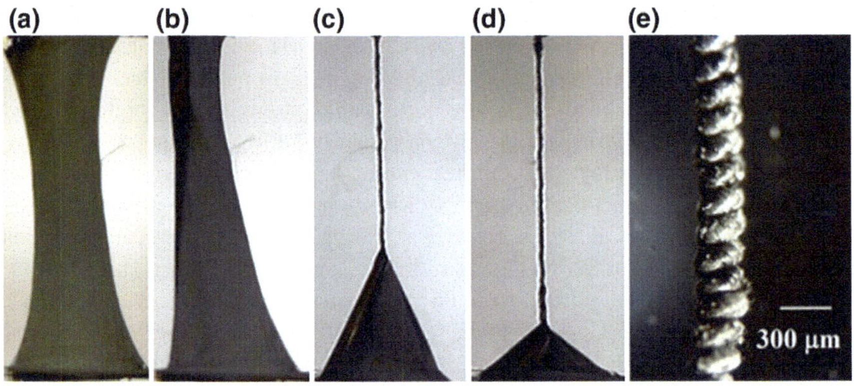

Fig. 9.7 **a–d** Evolution of spin rotation of the NGG-coated MWCNT sheets to form a yarn; **e** micrograph of the coiled yarn

per final yarn length (Fig. 9.7). The yarns with coiled structure were produced by addition of further 2.5×10^4 turns/m (Haines et al. 2014).

The force-stroke measurements were performed using Phidget force sensor calibrated for forces up to 1 N. The frequency of 50 Hz was used to monitor the force change over time. In this case, the yarn length was not allowed to change. An immobile base was used to attach one end of the yarn, while the other end was attached to the force sensor. Ignition of the yarns was initiated by creating 15 V potential difference across the yarn, which caused Joule heating and ignition of the yarn. In another set of experiments, the length of yarn was allowed to change, and the actuation force was measured with pre-calibrated cantilever.

Upon ignition of the MWCNT/NGG twisted and coiled yarns, gaseous iodine rapidly escapes the yarn through spaces between MWCNT sheets, which increases the diameter of the yarn. This process is demonstrated using snapshots from high-speed camera recording and SEM images in Fig. 9.8a, b. During the time-frame of 8–34 ms, the diameter of the yarn increased by about ten times, which caused a force stroke ($\sim$0.45 N) along the yarn length, Fig. 9.8a, shown with blue arrows. The final diameter of yarn increased reaching 520 μm after cooling, which is a 90% increase in comparison with 275 μm diameter value of yarn before the ignition.

The actuation stroke of the MWCNT/NGG twisted and coiled yarns was measured by attaching one end of the yarn to an immobile base, while the other end was attached to the force sensor. MWCNT/NGG twisted yarns which had mass ratio 1:2 generated a 0.45 N stroke force. However, the heavily coiled yarns produced $\sim$0.2 N force, Fig. 9.8c. The twofold reduction of stroke force for coiled yarns compared to twisted yarns can be due to the tight MWCNT structure that does not allow the release of the iodine gas that leads to an increase in diameter (coiled yarns show less than 20% diameter increase, as opposed to twisted yarns with over 90% diameter increase after actuation). For both coiled and twisted yarns, the force impulse duration was between 0.05 and 0.15 s. After combustion, the coiled and twisted yarns have a remnant stress force of 0.08 and 0.18 N, respectively, Fig. 9.8c. Thus, after actuation, both coiled and twisted yarns produce stroke forces and generate a remnant stress along the yarn. When the NGGs contain an excess amount of oxidizer, the yarns produce a higher stroke force of $\sim$0.6 N; however, the excess oxidizer damages $\sim$90% of the MWCNT/NGG yarn integrity after ignition and actuation. In this case, no remnant stress is observed.

In order to evaluate the power of actuation for the MWCNT/NGG twisted and coiled yarns, we used a setup of lifting load of 2 g mass. While in the force-stroke measurement the two ends of the yarn were fixed, in this setup during the actuation the bottom end of the yarn can rotate around the yarn axis. The measured specific power for coiled yarns was higher due to the kinetic energy of torsional untwisting, which produced rotation of mass during uplift. During the yarn actuation, the specific power up to 4700 W/kg was reached. This value is 94 times higher than the power of a typical mammalian muscle (50 W/kg; Madden et al. 2004). The actuation of MWCNT/NGGs composite yarns works in ambient air, vacuum, and inert environment (Hobosyan et al. 2017).

The DSC–TGA analysis of the yarns after combustion/actuation at the heating rate of 20°/min under 100 mL/min air flow was performed to quantitatively estimate the percentage of MWCNT burned (Fig. 9.9). Until 600 °C, only 4.7 wt% mass reduction was observed. This mass can be contributed to iodine, some amount of which still remain in the yarn after actuation. The yarn starts to burn at around 600 °C with 49 wt% mass reduction and 5832 J/g energy release. Thus, in the yarn after actuation about half of the mass are MWCNT.

SEM images and EDS element mapping of the coiled and twisted MWCNT/NGG yarns show the morphology and distribution of the remaining aluminum and

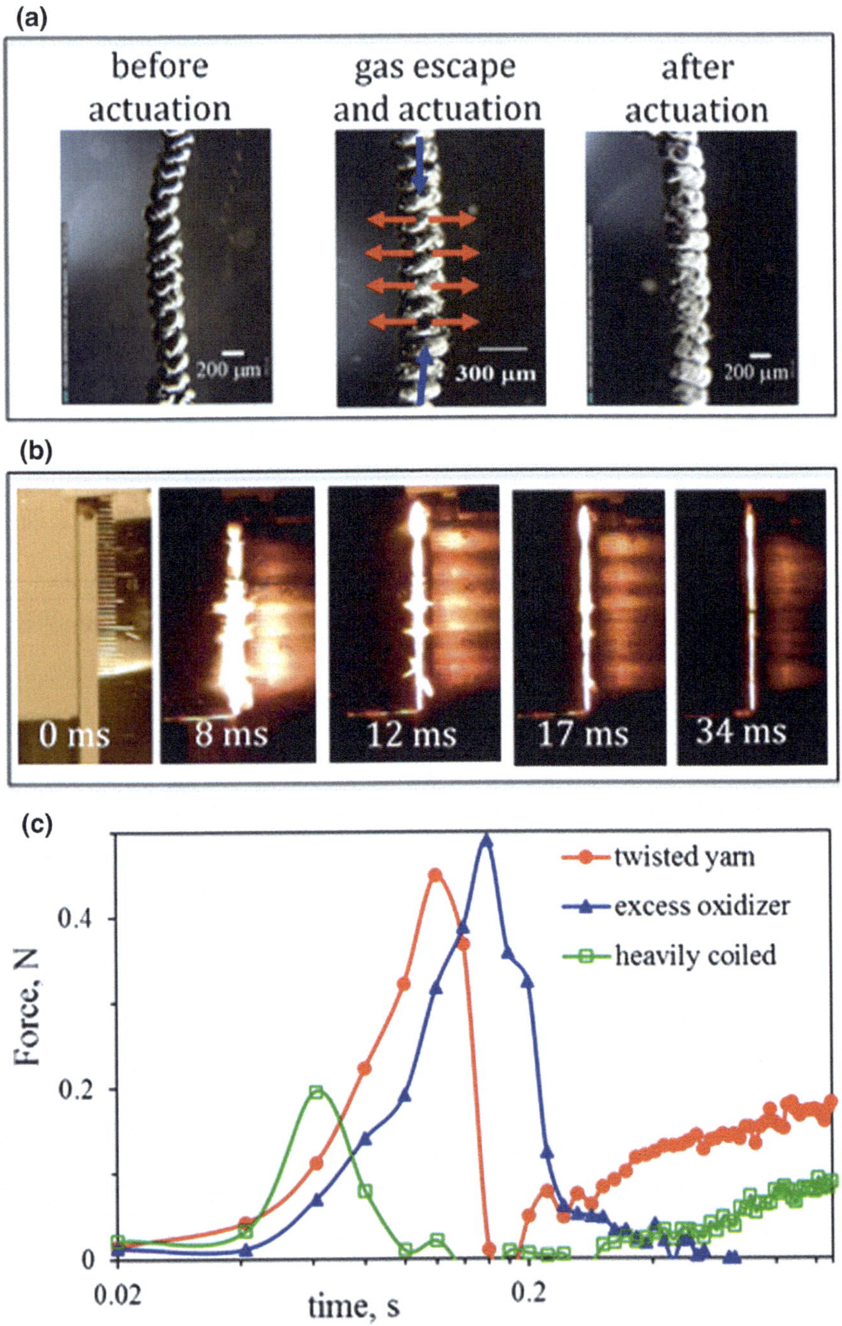

Fig. 9.8 Yarn actuation: **a** the actuator yarn before and after actuation, and the scheme of actuation (blue arrows) due to gas escape (red arrows), **b** the snapshots extracted from 240 fps video recording (IR filtered), demonstrating the change of yarn diameter during gas generation, **c** the force measurement over time for various yarns: simple twisted yarn, CNT/NGGs mass ratio 1:2, twisted yarn made with thermite containing 67 wt% excess I_2O_5 oxidizer, and heavily coiled yarn, CNT/NGGs mass ratio 1:2

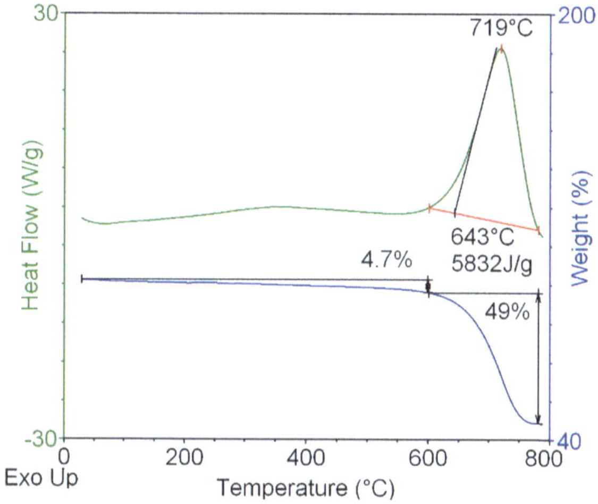

Fig. 9.9 DSC–TGA analysis of yarn after actuation

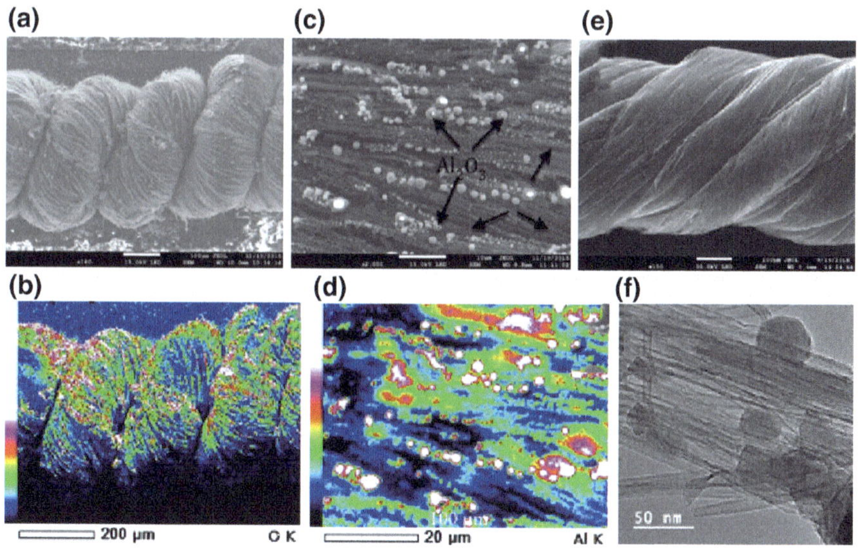

Fig. 9.10 **a** SEM of a coiled yarn after actuation; **b** EDX mapping of oxygen atoms for area shown in (**a**); **c** SEM images at higher magnification for EDX distribution analysis; **d** distribution of Al atoms for area shown in (**c**); **e** SEM of 65 wt% NGG containing twisted yarn after actuation; **f** individual aluminum oxide nanoparticles around MWCNT

oxygen left after combustion along the yarns (Fig. 9.10a–e). TEM image of the individual aluminum oxide nanoparticles reveals a size of ~50 nm and no interconnection with individual carbon nanotubes, Fig. 9.10f.

9.4.3 NGG Based on Iodine Pentoxide Oxidizer for Biocidal Agent Defeat

Traditionally, the iodine-based substances have been effective, simple, and cost effective for disinfection for hundreds of years. Iodine has been used in France during World War I to disinfect water; US Army used Globalin (tetraglycine hydroperiodide) tablets during World War II. The disinfection based on iodine has been used by NASA in space flights (Atwater et al. 1996). Since the early 1950s, extensive research has been conducted exploring the efficiency of disinfection for various forms of iodine (Clark and Pantoya 2010; Sullivan et al. 2010; McDonnell and Russell 2001). Iodine is rapidly bactericidal, fungicidal, tuberculocidal, virucidal, and sporicidal (Gottardi 1991).

For testing the biocidal properties of nano-thermites, plexiglass reaction chamber with a volume of 49.26 L and a wall thickness of 1.2 cm with a weight of 3 kg was used. The samples were placed at various distances and orientation demonstrated in Fig. 9.11a. The charge was placed at 8 cm height and was electrically ignited using a variable autotransformer from Staco Energy Products. An *Escherichia coli* (*E. coli*) bacteria strain HB101 K-12 was used, which is not pathogenic like the strain O157 H7, which has been implicated in some occasions in food poisoning. HB101 K-12 strain has been genetically modified so it can only be grown in an enriched medium. We exposed them to nano-thermite disinfection after 1 h of incubation of bacteria, placing the agar plates coated with bacteria at various orientations in the chamber shown in Fig. 9.11a. The agar plates are placed in an incubator for 24 h at 37 °C after the biocidal treatment. If the treatment is successful, there will not be any visible growth on the agar plate. The *E. coli* colonies

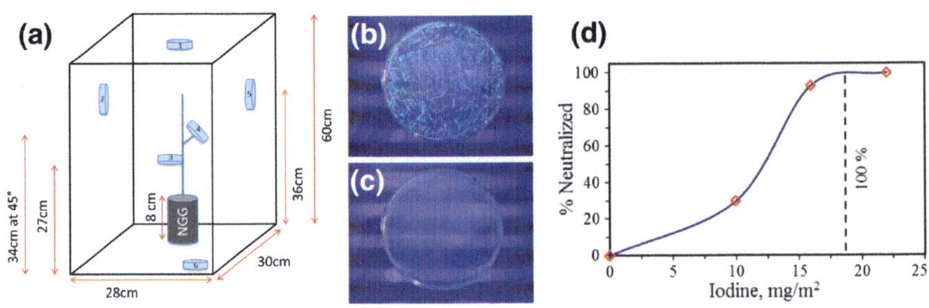

Fig. 9.11 **a** Experimental setup for biocidal performance; **b** control sample; **c** sample exposed to nano-thermite disinfection; **d** statistical curve of killing abilities of active iodine generated by nano-thermite reaction

are counted, and the number of visible colonies is compared with two control agar plates (Fig. 9.11b) that were made but were not exposed to the biocidal gas and were allowed to grow. Thus, the biocidal effect of I_2O_5-Al NGG mixture weight, exposure time, as well as sample distance and orientation from charge, was investigated. According to the calculation, 0.1 g mixture generates 0.53 g atomic iodine (Hobosyan et al. 2012). All samples were completely disinfected, regardless of orientation or distance from the charge (Fig. 9.11c, compared to Fig. 9.11b). Calculations show that minimal effective concentration of iodine over the infected area is 22 mg/m^2 (Fig. 9.11d).

9.5 Summary

This chapter reported the recent research findings of nano-energetic gas generator systems containing bismuth and iodine oxidizers and tuning the reactivity of the systems by altering the oxidizer particle size, as well as shape. The iodine pentoxide, bismuth hydroxide, and bismuth trioxide surpass other traditionally known oxidizers in thermite compositions in terms of providing superior pressure discharge values up to 14.8 kPa m^3/g. These NGG formulations can be used in new emerging technological applications such as microthrusters, providing milli-Newton-second thrust resolution, as well as linear composite actuators with high power output up to 4700 W/kg, and biocidal agents with excellent bactericidal properties, where the minimal effective concentration over the infected area was 22 mg/m^2.

References

Aliev AE, Oh J, Kozlov ME, Kuznetsov AA, Fang S, Fonseca AF, Ovalle R et al (2009) Giant-stroke, superelastic carbon nanotube aerogel muscles. Science 323(5921):1575–1578
Arnáiz FJ, Aguado R, Arnáiz S (1998) Microscale thermite reactions. J Chem Educ 75(12):1630
Atwater JE, Sauer RL, Schultz JR (1996) Numerical simulation of iodine speciation in relation to water disinfection aboard manned spacecraft I. Equilibria. J Environ Sci Health A 31 (8):1965–1979
Baughman RH, Cui C, Zakhidov AA, Iqbal Z, Barisci JN, Spinks GM, Wallace GG et al (1999) Carbon nanotube actuators. Science 284(5418):1340–1344
Chaalane A, Chemam R, Houabes M, Yahiaoui R, Metatla A, Ouari B, Metatla N, Mahi D, Dkhissi A, Esteve D (2015) A MEMS-based solid propellant microthruster array for space and military applications. J Phys Conf Ser 660(1):012137. IOP Publishing
Cheng JL, Hng HH, Ng HY, Soon PC, Lee YW (2010) Synthesis and characterization of self-assembled nanoenergetic Al Fe_2O_3 thermite system. J Phys Chem Solids 71(2):90–94
Clark BR, Pantoya ML (2010) The aluminium and iodine pentoxide reaction for the destruction of spore forming bacteria. Phys Chem Chem Phys 12(39):12653–12657
Comet M, Pichot V, Siegert B, Schnell F, Ciszek F, Spitzer D (2010) Phosphorus-based nanothermites: a new generation of energetic materials. J Phys Chem Solids 71(2):64–68
Ding Y, Ma L, Xu Z, Li M, Huo H, Tan S, Gou X et al (2015) Progress on the use of satellite technology for gravity exploration. Geodesy Geodyn 6(4):299–306

Dlott DD (2006) Thinking big (and small) about energetic materials. Mater Sci Technol 22 (4):463–473

Dreizin EL (2009) Metal-based reactive nanomaterials. Prog Energy Combust Sci 35(2):141–167

Fut'ko SI, Ermolaeva EM, Dobrego KV, Bondarenko VP, Dolgii LN (2011) Thermodynamic analysis of solid-fuel mixtures glycidyl azide polymer (GAP)/RDX for miniengines of microelectromechanical systems. J Eng Phys Thermophys 84(5):1068–1073

Gottardi W (1991) Iodine and iodine compounds. In: Block SS (ed) Disinfection, sterilization, and preservation, 4th edn. Lea & Febiger, Philadelphia, PA, pp 152–166

Greiner W, Neise L, Stöcker H (1995) Thermodynamics and statistical mechanics. Springer-Verlag, p 101

Haines CS, Lima MD, Li N, Spinks GM, Foroughi J, Madden JDW, Kim SH et al (2014) Artificial muscles from fishing line and sewing thread. Science 343(6173):868–872

Hobosyan MA, Martirosyan KS (2014) Consolidation of lunar regolith simulant by activated thermite reactions. J Aerosp Eng 28(4):04014105

Hobosyan MA, Martirosyan KS (2016) Tuning of lunar regolith thermal insulation properties utilizing reactive consolidation by activated thermites. In: Lunar and planetary science conference, vol. 47, p 1035

Hobosyan MA, Martirosyan KS (2017) Iodine pentoxide nano-rods for high density energetic materials. Propellants Explos Pyrotech 42(5):506–513

Hobosyan M, Kazansky A, Martirosyan KS (2012) Nanoenergetic composite based on I_2O_5/Al for biological agent defeat. In: Technical proceeding of the 2012 NSTI nanotechnology conference and expo, pp 599–602

Hobosyan MA, Kirakosyan KG, Kharatyan SL, Martirosyan KS (2015) PTFE–Al_2O_3 reactive interaction at high heating rates. J Therm Anal Calorim 119(1):245–251

Hobosyan MA, Yolchinyan SA, Martirosyan KS (2016) A novel nano-energetic system based on bismuth hydroxide. RSC Adv 6(71):66564–66570

Hobosyan MA, Martinez PM, Zakhidov AA, Haines CS, Baughman RH, Martirosyan KS (2017) Laminar composite structures for high power actuators. Appl Phys Lett 110(20):203101

Hobosyan M, Lyshevski SE, Martirosyan KS (2018) Design and evaluations of 3D-printed microthrusters with nanothermite propellants. In: 2018 IEEE 38th international conference on electronics and nanotechnology (ELNANO). IEEE (accepted)

Johnson CE, Fallis S, Chafin AP, Groshens TJ, Higa KT, Ismail IMK, Hawkins TW (2007) Characterization of nanometer-to micron-sized aluminum powders: size distribution from thermogravimetric analysis. J Propul Power 23(4):669

Kim SH, Zachariah MR (2004) Enhancing the rate of energy release from nanoenergetic materials by electrostatically enhanced assembly. Adv Mater 16(20):1821–1825

Lima MD, Fang S, Lepró X, Lewis C, Ovalle-Robles R, Carretero-González J, Castillo-Martínez E et al (2011) Biscrolling nanotube sheets and functional guests into yarns. Science 331 (6013):51–55

Liu X, Li T, Li Z, Ma H, Fang S (2015) Design, fabrication and test of a solid propellant microthruster array by conventional precision machining. Sens Actuators A 236:214–227

Madden JDW, Vandesteeg NA, Anquetil PA, Madden PGA, Takshi A, Pytel RZ, Lafontaine SR, Wieringa PA, Hunter IW (2004) Artificial muscle technology: physical principles and naval prospects. IEEE J Ocean Eng 29(3):706–728

Martirosyan KS (2011) Nanoenergetic gas generators, principle and applications. J Mater Chem 21:9400–9405

Martirosyan KS (2012) High-density nanoenergetic gas generators. In: Handbook of nanoscience, engineering, and technology, 3rd edn. CRC Press, pp 739–758

Martirosyan KS, Wang L, Luss D (2009a) Novel nanoenergetic system based on iodine pentoxide. Chem Phys Lett 483(1):107–110

Martirosyan KS, Wang L, Vicent A, Luss D (2009b) Nanoenergetic gas generators: design and performance. Propellants Explos Pyrotech 34(6):532–538

Martirosyan KS, Wang L, Vicent A, Luss D (2009c) Synthesis and performance of bismuth trioxide nanoparticles for high energy gas generator use. Nanotechnology 20(40):405609

Martirosyan KS, Hobosyan M, Lyshevski SE (2012) Enabling nanoenergetic materials with integrated microelectronics and MEMS platforms. In: 2012 12th IEEE conference on nanotechnology (IEEE-NANO). IEEE, pp 1–5
McDonnell G, Russell AD (2001) Antiseptics and disinfectants: activity, action, and resistance. Clin Microbiol Rev 14(1):227
Miziolek A (2002) Nanoenergetics: an emerging technology area of national importance. Amptiac Q 6(1):43–48
Patel VK, Ganguli A, Kant R, Bhattacharya S (2015a) Micropatterning of nanoenergetic films of Bi_2O_3/Al for pyrotechnics. RSC Adv 5(20):14967–14973
Patel VK, Saurav JR, Gangopadhyay K, Gangopadhyay S, Bhattacharya S (2015b) Combustion characterization and modeling of novel nanoenergetic composites of Co_3O_4/nAl. RSC Adv 5 (28):21471–21479
Perry WL, Smith BL, Bulian CJ, Busse JR, Macomber CS, Dye RC, Son SF (2004) Nano-scale tungsten oxides for metastable intermolecular composites. Propellants Explos Pyrotech 29 (2):99–105
Puchades I, Hobosyan M, Fuller LF, Liu F, Thakur S, Martirosyan KS, Lyshevski SE (2014) MEMS microthrusters with nanoenergetic solid propellants. In: 2014 IEEE 14th international conference on nanotechnology (IEEE-NANO). IEEE, pp 83–86
Puchades I, Fuller LF, Lyshevski SE, Hobosyan M, Ting L, Martirosyan KS (2017) MEMS and 3D-printing microthrusters technology integrated with hydroxide-based nanoenergetic propellants. In: 2017 IEEE 37th international conference on electronics and nanotechnology (ELNANO). IEEE, pp 67–70
Puszynski JA (2009) Processing and characterization of aluminum-based nanothermites. J Therm Anal Calorim 96(3):677–685
Puszynski JA, Bulian CJ, Swiatkiewicz JJ (2007) Processing and ignition characteristics of aluminum-bismuth trioxide nanothermite system. J Propul Power 23(4):698–706
Rogachev AS, Mukasyan AS (2010) Combustion of heterogeneous nanostructural systems. Combust Explos Shock Waves 46(3):243–266
Rossi C, Larangot B, Lagrange D, Chaalane A (2005) Final characterizations of MEMS-based pyrotechnical microthrusters. Sens Actuators A 121(2):508–514
Rossi C, Zhang K, Esteve D, Alphonse P, Tailhades P, Vahlas C (2007) Nanoenergetic materials for MEMS: a review. IEEE/ASME J Microelectromech Syst 16(4):919–931
Ru C, Wang F, Xu J, Dai J, Shen Y, Ye Y, Zhu P, Shen R (2017) Superior performance of a MEMS-based solid propellant microthruster (SPM) array with nanothermites. Microsyst Technol 23(8):3161–3174
Sathiyanathan K, Lee R, Chesser H, Dubois C, Stowe R, Farinaccio R, Ringuette S (2011) Solid propellant microthruster design for nanosatellite applications. J Propul Power 27(6):1288–1294
Shiryaev AA (1995) Thermodynamics of SHS processes: advanced approach. Int J Self-Propag High-Temp Synth 4(4):351–362
Staley CS, Raymond KE, Thiruvengadathan R, Apperson SJ, Gangopadhyay K, Swaszek SM, Taylor RJ, Gangopadhyay S (2013) Fast-impulse nanothermite solid-propellant miniaturized thrusters. J Propul Power
Sullivan K, Zachariah M (2010) Simultaneous pressure and optical measurements of nanoaluminum thermites: investigating the reaction mechanism. J Propul Power 26(3):467–472
Sullivan KT, Piekiel NW, Chowdhury S, Wu C, Zachariah MR, Johnson CE (2010) Ignition and combustion characteristics of nanoscale Al/$AgIO_3$: a potential energetic biocidal system. Combust Sci Technol 183(3):285–302
Umbrajkar SM, Schoenitz M, Dreizin EL (2006) Control of structural refinement and composition in Al-MoO_3 Nanocomposites prepared by arrested reactive milling. Propellants Explos Pyrotech 31(5):382–389
Varma A, Rogachev AS, Mukasyan AS, Hwang S (1998) Combustion synthesis of advanced materials: principles and applications. Adv Chem Eng 24:79–226
Wang Y, Jiang W, Cheng Z, Chen W, An C, Song X, Li F (2007) Thermite reactions of Al/Cu core-shell nanocomposites with WO_3. Thermochim Acta 463(1):69–76

Wang L, Luss D, Martirosyan KS (2011) The behavior of nanothermite reaction based on Bi_2O_3/Al. J Appl Phys 110(7):074311

Xu X, Li X, Zhou J, Zhang B, Xiao D, Huang Y, Wu X (2017) Numerical and experimental analysis of cold gas microthruster geometric parameters by univariate and orthogonal method. Microsyst Technol 1–14

Yolchinyan SA, Hobosyan MA, Martirosyan KS (2018) Tailoring bismuth oxide flower-, bowtie- and brushwood-like structures through microfluidic synthesis. Mater Chem Phys 207:330–336

Zhang M, Atkinson KR, Baughman RH (2004) Multifunctional carbon nanotube yarns by downsizing an ancient technology. Science 306(5700):1358–1361

Zhang Y, Evans JRG, Yang S (2011) Corrected values for boiling points and enthalpies of vaporization of elements in handbooks. J Chem Eng Data 56(2):328–337

Zhou X, Xu D, Yang G, Zhang Q, Shen J, Lu J, Zhang K (2014) Highly exothermic and superhydrophobic Mg/fluorocarbon core/shell nanoenergetic arrays. ACS Appl Mater Interfaces 6(13):10497–10505

Part IV
Nano-energetic Materials: The Emerging Paradigm

Chapter 10
Recent Advancement in the Fabrication of Energy Storage Devices for Miniaturized Electronics

Poonam Sundriyal, Megha Sahu, Om Prakash and Shantanu Bhattacharya

Abstract The rapidly increasing need of the energy and the requirement of the current and further generation compact electronic devices have emerged the development of micro-scaled energy storage devices. These energy storage devices should be efficient enough to store the sufficient energy in a limited area. The micro-supercapacitors have reported as the best alternative to power the miniaturized electronic devices. A lot of energy storage materials, fabrication methods, and the electrode design have been explored to achieve the high performance of the micro-scaled energy storage devices. This review focuses on the current progress to fabricate and improve the performance of the micro-supercapacitor devices for their potential application in the miniaturized electronic devices.

Keywords Energy storage · Micro-supercapacitors · Fabrication methods Electrode design

10.1 Introduction

The rapidly increasing demand for energy and the limited supply from the conventional energy sources has emerged the urgent need of exploring new approaches for energy generation, storage, and its management (Beidaghi and Gogotsi 2014; Kyeremateng et al. 2017). The portable, wireless, and miniaturized electronic devices have recently emerged as the potential candidates for the current and future

---

P. Sundriyal (✉) · S. Bhattacharya
Department of Mechanical Engineering, Indian Institute of Technology, Kanpur 208016, Uttar Pradesh, India
e-mail: poonams@iitk.ac.in

P. Sundriyal · S. Bhattacharya
Microsystems Fabrication Laboratory, Indian Institute of Technology, Kanpur 208016, India

M. Sahu · O. Prakash
Boeing India Corporation, New Delhi, India

S. Bhattacharya et al. (eds.), *Nano-Energetic Materials*, Energy, Environment and Sustainability, https://doi.org/10.1007/978-981-13-3269-2_10

smart electronics. The applications of these smart devices range from the electronics used in our daily lives (smartphones, wearable devices, laptop, etc.) to the medical (wearable healthcare devices and medical implants) and industrial applications (electrical vehicles, nanorobots, microelectromechanical systems, wireless sensors, etc.) (Choi et al. 2016). These applications require a micro-scaled energy storage component integrated with the miniaturized electronic devices. To date, a lot of research is focused on the miniaturization of the energy storage devices to power the portable electronics (Pech et al. 2010; El-Kady and Kaner 2013; El-Kady et al. 2015; Hyun et al. 2017; Li et al. 2017). However, to achieve the good electrochemical performance of the energy storage unit at the small scales, their integration with the other electronic components and the application of these components in the real-life applications are the major challenges (Beidaghi and Gogotsi 2014; Tyagi et al. 2015).

Out of the different energy storage devices such as batteries, supercapacitors, and the electrolytic capacitors, the micro-supercapacitors have displayed the huge potential for the miniaturized electronic applications, providing high rate capability, long lifetime, and easy integration (Chmiola et al. 2010). Recently, several micro-supercapacitors have been developed by using the different micro-fabrication methods. The widely used fabrication techniques are printing, lithography, laser scribing, chemical vapor deposition, and electrochemical/electrophoretic deposition (Pech et al. 2010; El-Kady et al. 2012; El-Kady and Kaner 2013; Su et al. 2014b; Wu et al. 2014; Xu et al. 2014; Choi et al. 2016; Sundriyal and Bhattacharya 2017a, b). The performance and the scalability of the micro-supercapacitors also depend on the selection of the fabrication method. The current research focus is on the development of a proper guideline to select a suitable fabrication process that can fabricate a high-performance micro-supercapacitor with a simple, cost-effective, environment-friendly, and time-saving manner. Out of the different techniques, the laser scribing and the inkjet printing are the trending methods for making micro-supercapacitors. However, most of the developed devices are at the research stage, and further focus is required to make them suitable or the practical applications (Beidaghi and Gogotsi 2014; Tyagi et al. 2015).

The electrode design and the device architecture are the key elements to get the high electrochemical performance and the proper integration of the micro-supercapacitors with the electronic devices (Beidaghi and Gogotsi 2014). The in-plane and the 3-D electrode designs are favorable approaches to get the high electrochemical performance of the micro-supercapacitor devices (Kyeremateng et al. 2017). Also, they promote the flexibility and easy mounting of these devices to the other components.

The choice of the electrode material is one of the main parameters which greatly affects the overall performance of an energy storage device (Gogotsi 2014; Yu et al. 2015). The device stability, lifetime, power capabilities depend on the electrode material, its morphology, and the crystal structure. Therefore, the tremendous literature is focused on the material's aspects of the energy storage devices. Several materials including carbon-based materials, conducting polymers, transition metal oxides, and their hybrids are extensively explored as the advanced energy storage materials. However, their utilization for the micro-supercapacitors still suffers from the processing and scaling issues which need further attention to make them efficient for the real-life applications. This review focuses on the recent progress in the area of micro-supercapacitor fabrication, electrode design, and the material's development for application in the microelectronic devices.

10.2 Fabrication Methods for Constructing the Energy Storage Devices

The output performance of the MSCs depends on the electrode materials, electrode design, and the device architecture. The selection of the efficient materials and a suitable manufacturing technique are the crucial points for constructing the high-performance energy storage device. The thin-film electrodes are widely used for fabrication of the energy storage devices for on-chip electronics. The potential fabrication techniques for constructing micro-supercapacitors are printing technique, laser scribing, lithography, vapor deposition, electrochemical deposition, and stamping. Figure 10.1 shows the schematic of the different fabrication processes for constructing the micro-supercapacitors.

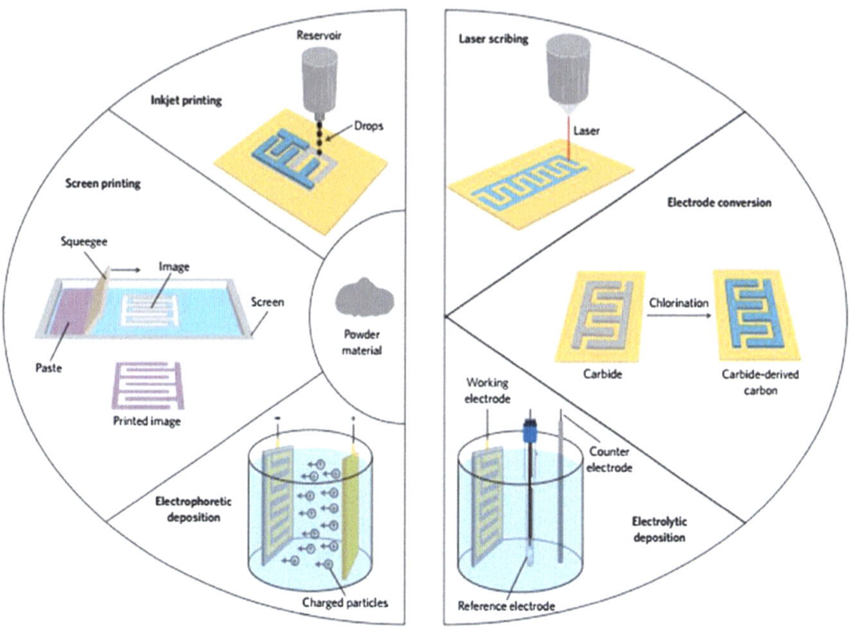

Fig. 10.1 Schematic illustration of the different fabrication processes for making thin-film micro-supercapacitors. Reproduced from Kyeremateng et al. (2017) with permission from Nature

10.2.1 Printing Techniques

The printing methods are recently explored for fabricating the thin-film micro-scaled energy storage devices (Wang et al. 2015; Choi et al. 2016; Sundriyal and Bhattacharya 2017a, b). These methods have gained much acceptance for the shape and size variable electronics devices as demanded by the flexible and miniaturized electronic devices. The main attractive feature of this method is the user-defined variation in the printed patterns, which provides full control over the shape, size, thickness, design, and positioning of the electrodes. Out of the different printing techniques, inkjet printing has been greatly used for the construction of the micro-scaled energy storage devices.

Inkjet printing is a direct and non-contact material deposition method which emerged as a promising fabrication technique for thin-film devices due to its low cost, environment friendliness, simplicity, and industrial maturity (Ervin et al. 2014). Due to its favorable nature and compatibility with different flexible substrates, it is prominently used for making the flexible and thin-film supercapacitor and batteries. Choi et al. (2016) have developed a solid-state flexible supercapacitor on an A4 paper using the desktop inkjet printer. The supercapacitors with different shapes and sizes were successfully fabricated by selecting the commercial activated carbon/carbon nanotube as the electrode material, ionic gel as an electrolyte, and silver nanowires as the conducting medium. A thin layer of cellulose nanofibril (CNF) was printed over the paper substrate to promote high resolution of the printed patterns. Figure 10.2a shows schematic of the stepwise construction of different components of the supercapacitor device and photograph of the printed supercapacitor device. Figure 10.2b–e displays the electrochemical performance of the printed supercapacitor device. Moreover, the developed device exhibits negligible variation in the output performance under mechanical deformation (bending) condition, indicating the excellent flexibility of the device. The practical feasibility of the printed device was checked by powering a LED from it, which also worked well at the high temperature (as shown in Fig. 10.2f, g).

Our group has also reported an ultra-low-cost paper-based supercapacitor with excellent flexibility and high electrochemical performance utilizing an inkjet printer (Sundriyal and Bhattacharya 2017a, b). The rheological properties of the used inks, the concentration of the materials in the inks, and printing layers of electrode materials were properly optimized to develop the high-performance micro-supercapacitor. The device showed excellent electrochemical performance and displayed 89.6% capacitance retention after 9000 charge–discharge cycles. The maximum current density of 22 mWh/cm^3 was achieved at a power density of 99 mW/cm^3. Also, the device displayed a negligible change in the electrochemical performance under different mechanical deformation conditions, which indicated its high flexibility. Figure 10.3 shows the schematic of the printed electrodes on the cellulose substrate, assembly of the device components, and the cyclic voltammograms curves of the developed paper-based micro-supercapacitor (inset shows an LED powered by two devices in the series connection).

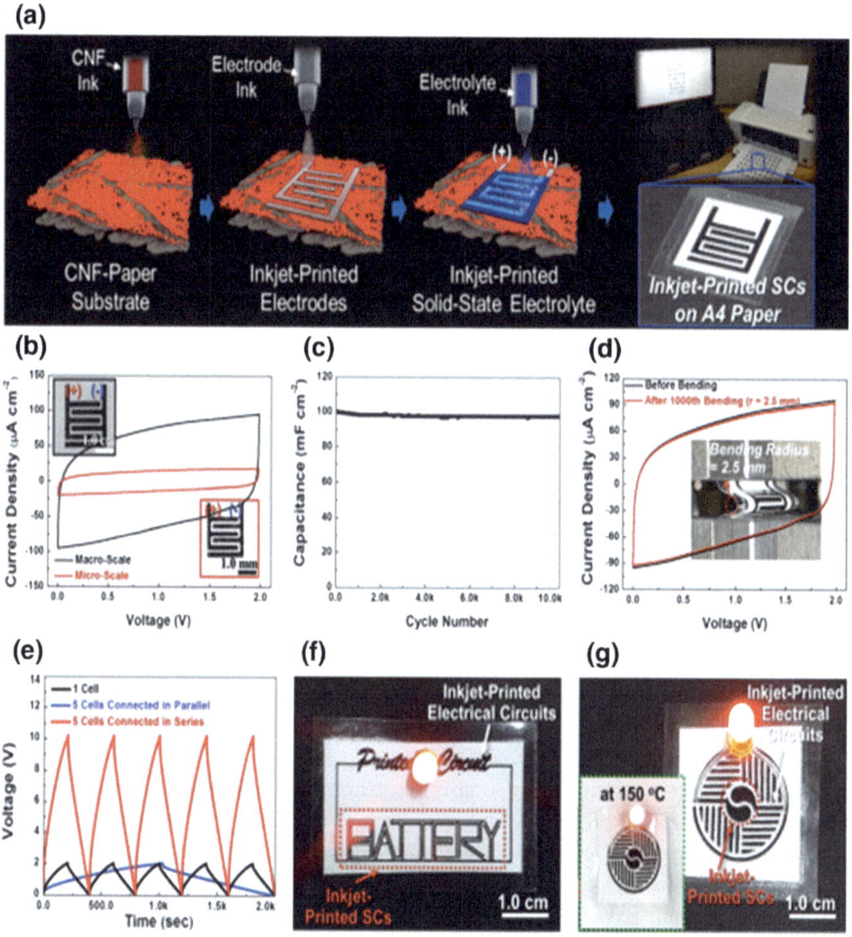

Fig. 10.2 **a** Schematic of the printing steps to develop a micro-supercapacitor, **b–e** electrochemical performance of the developed device, and **f**, **g** LED powered by the printed micro-supercapacitors. Reproduced from Choi et al. (2016) with permission from the Royal Society of Chemistry

Lin et al. have used commercial Dimatrix printer method for constructing interdigitated electrode-based architecture of the in-plane micro-supercapacitors. They reported a maximum areal capacitance of 52.9 mF/cm^2 with a nanocoral structure of the Ni@MnO_2 electrode material. The developed micro-supercapacitors were unitized to a wearable system, and the practical realization of the micro-supercapacitor was demonstrated by powering a self-powered wearable device.

Hyun et al. (2017) have also reported graphene-based micro-supercapacitors on the plastic substrate using a capillary-assisted printing process. This capillary

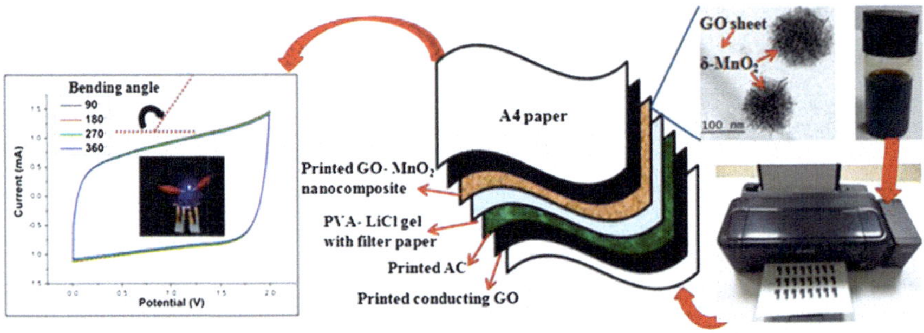

Fig. 10.3 Inkjet printing of the electrode materials, the assembly of the supercapacitor components, and the electrochemical performance of the device under different bending conditions. Reproduced from Sundriyal and Bhattacharya (2017a, b) with permission from the American Chemical Society

flow-based deposition method provided an excellent print resolution of 20 μm (width of the line) and displayed a maximum specific capacitance of 269 $\mu F/cm^2$ with the graphene material's electrode. The presented method was highly efficient to fabricate the energy storage device for large-area electronics; however, this method needs much improvement to be commercialized for the wide applications. Li et al. have also developed inkjet-printed graphene-based micro-supercapacitors on the variety of the substrates. They reported large-scale micro-supercapacitor arrays using 100 devices in series and parallel connection. These arrays displayed a remarkable high voltage of 12 V and stability up to 8 months.

Although inkjet printing is becoming a good solution for the micro-scaled energy storage devices, some inherent limitation of this method hinders its complete utilization. The major problem with the inkjet printing method is the formation of printable inks with the suitable rheological properties. Therefore, only solution processed and small particle-sized materials can be used for the inkjet printing.

10.2.2 Laser Scribing

Laser scribing is another direct fabrication method for the fabrication of thin-film electrodes for energy storage devices. El-Kady et al. and the group have recently presented laser writing as an efficient method for fabricating graphene-based thin-film micro-supercapacitors (El-Kady et al. 2012, 2015; El-Kady and Kaner 2013). They performed laser scribing of graphene oxide-coated DVD optical drives and obtained highly conducting (electrical conductivity of 1738 S/m) graphene patterns. The resulted graphene films were directly used as electrodes for the micro-supercapacitors. The developed micro-supercapacitors exhibited excellent electrochemical performance (El-Kady et al. 2012). In another approach, the laser-scribed graphene-based micro-supercapacitors were realized for the flexible electronic applications (El-Kady and Kaner 2013). Figure 10.4a–e shows the schematic of the single-device fabrication and the mass fabrication using a laser scribing method. The digital and optical image of a single device is shown in Fig. 10.4f, h. The optical image shows well-defined patterns with no short-circuiting, and SEM image (Fig. 10.4g) shows clear identification between the graphene oxide and the laser-reduced graphene side of the patterns. The

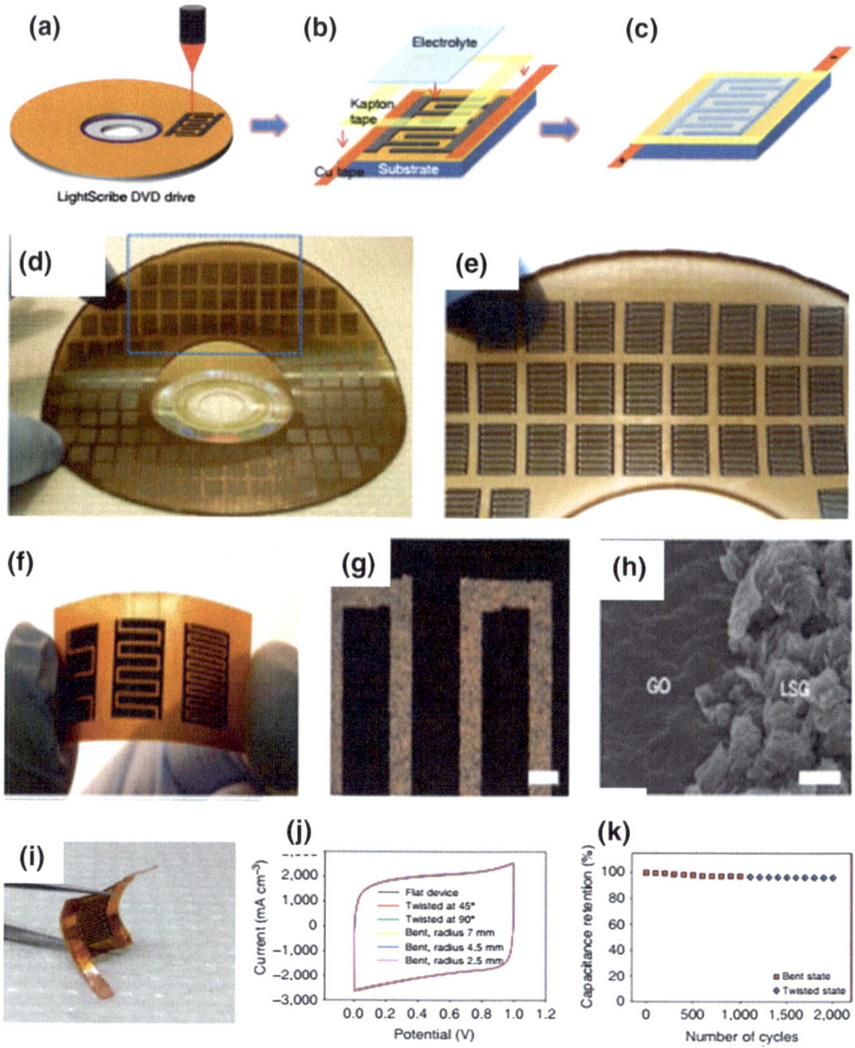

Fig. 10.4 **a–e** Fabrication of the micro-supercapacitors using laser scribing method, **f**, **i** digital photographs of the laser-scribed micro-supercapacitors, **g**, **h** optical and FESEM image of the developed patterns, **j**, **k** electrochemical performance of the device. Reproduced from El-Kady and Kaner (2013) with permission from the American Chemical Society

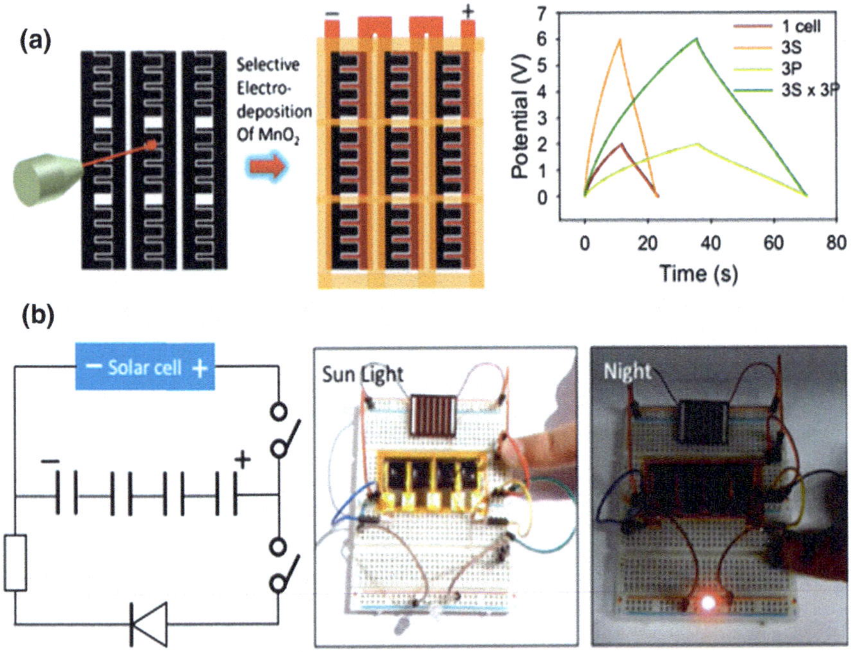

Fig. 10.5 **a** Fabrication of the micro-supercapacitor arrays using laser and electrodeposition method, and charge–discharge curves of the device arrays and **b** integration of micro-supercapacitor with a solar cell device. Reproduced from El-Kady et al. (2015) with permission from Proceedings of the National Academy of Sciences

photograph of the flexible device and its electrochemical performance under different flexibility conditions is displayed in Fig. 10.4i–k.

The same group also reported a combined system of micro-supercapacitor and a solar cell for integrated energy harvesting–energy storage systems (El-Kady et al. 2015). The micro-supercapacitor was developed with the laser scribing of graphene oxide followed by the MnO_2 deposition. Figure 10.5a displays the fabrication steps of the in-plane micro-supercapacitor device and its electrochemical performance with different series–parallel connections. Figure 10.5b shows the schematic and photographs of the integrated solar cell and micro-supercapacitor device.

10.2.3 Lithography

Lithography is a micro-fabrication method to construct the thin-film patterns by using a pre-patterned mask. The patterns are transferred to the photoresist using a light source. It is a multi-step process and involves a lot of chemical treatment process to get the final patterns. This method is widely used to fabricate the

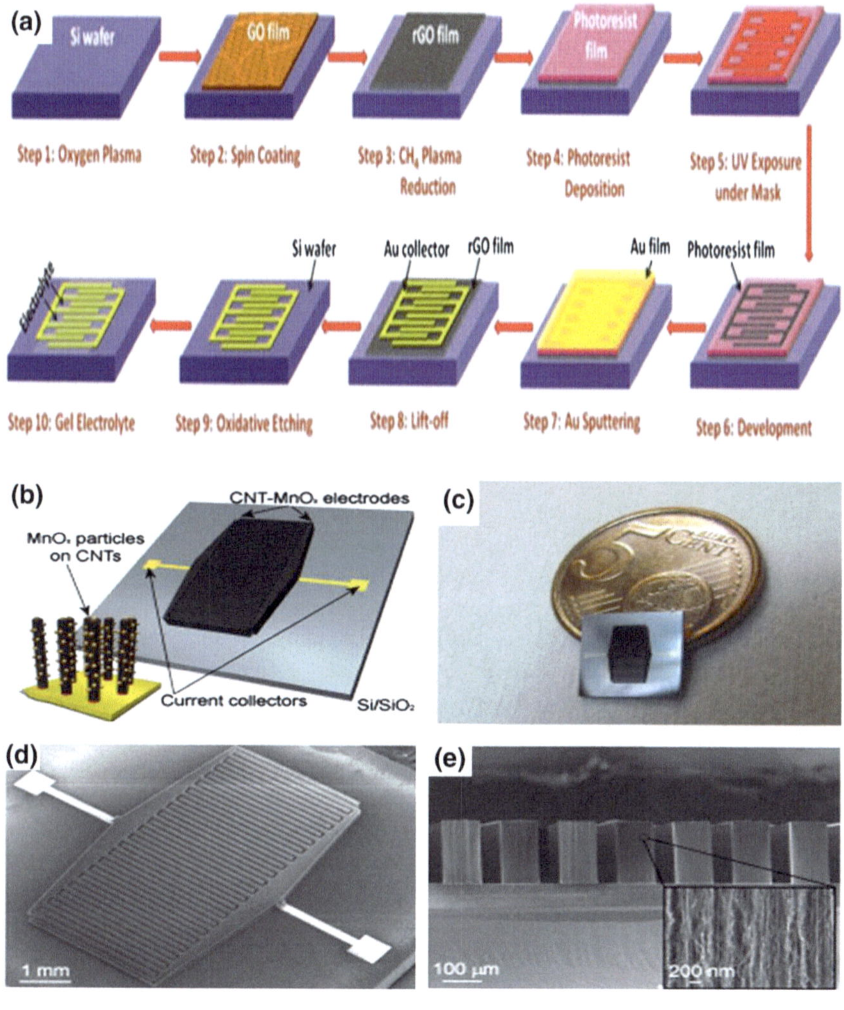

Fig. 10.6 **a** Steps involved in the micro-supercapacitor fabrication using lithography method, **b**, **c** schematic and a digital image of the CNT–MnO_2 micro-supercapacitor, and **d–e** FESEM images of the prepared patterns. Reproduced from Wu et al. (2014) and Pitkänen et al. (2017) with permission from the Royal Society of Chemistry and Nature

thin-film patterns for miniaturized applications. It is also reported as one of the effective methods to construct the electrodes for in-plane micro-supercapacitors (Wu et al. 2014; Bhatt et al. 2017; Pitkänen et al. 2017). Figure 10.6a shows the steps involved in the fabrication of thin-film interdigitated electrodes of graphene oxide using this method. Micro-supercapacitors with a different number of the electrode finger were fabricated which were used to find the correlation between the electrode fingers with the output performance. The developed device displayed outstanding rate capability up to 2000 V/s scan rate and excellent cyclic stability up to 50,000 cycles (Wu et al. 2014).

Pitkanen et al. (2017) have also developed a micro-supercapacitor for on-chip energy storage purpose using a photolithography method. The Pt and Mo layers were used as a current collector. The selected electrode material was CNT–MnO_x with 3-D architecture. They achieved a maximum areal capacitance of 37 mF/cm^2 with in-plane electrode design. Figure 10.6b shows the schematic of the in-plane CNT–MnO_x electrode design, and Fig. 10.6c shows the optical camera image of the patterned device. The FESEM image of the patterned device is shown in Fig. 10.6d, and cross-sectional FESEM images of the patterns are displayed in Fig. 10.6e.

10.2.4 Chemical Vapor Deposition

Chemical vapor deposition (CVD) is an efficient process to fabricate thin-film devices used in all applications of the semiconducting devices. Yoo et al. (2011) reported 1–2 layer films of the graphene using CVD technique, which exhibited a specific capacitance of 0.08 F/cm^2. Miller et al. (2011) have also developed vertically aligned graphene films on the nickel foil using the CVD method. The prepared micro-supercapacitor device displayed outstanding frequency response with a time constant of 200 ms and a good power output.

Lee et al. (2013) have demonstrated a porous graphene nanoball film for supercapacitor application. The surface area of the films was 508 m^2/g with an average pore diameter of 4.27 nm. The developed supercapacitor shows a specific capacitance of 206 F/g and excellent performance up to 10,000 charge–discharge cycles. Xu et al. (2014) presented an interesting approach to fabricate CVD-grown transparent graphene films for micro-supercapacitors with excellent electrochemical performance, flexibility, and stretchability. They reported 73% transmittance and good electrochemical performance up to 40% stretching of the device. Figure 10.7a–d shows FESEM images of the CVD-grown graphene films showing the surface ripples. Figure 10.7e displays AFM image of these films. These surface ripples are favorable for high performance of the supercapacitors. Figure 10.8 shows the electrochemical performance of the developed device with a various stretching.

10.2.5 Electrochemical and Electrophoretic Depositions

The electrochemical and electrophoretic techniques have widely gained much attention for fabricating thin-film electrodes for micro-supercapacitor devices. The widespread acceptance of this method is due to its low-cost, large-area fabrication capability, and easily operated instrument. Recently, a lot of micro-scaled energy storage devices with electrode materials such as graphene, CNT, carbon onion,

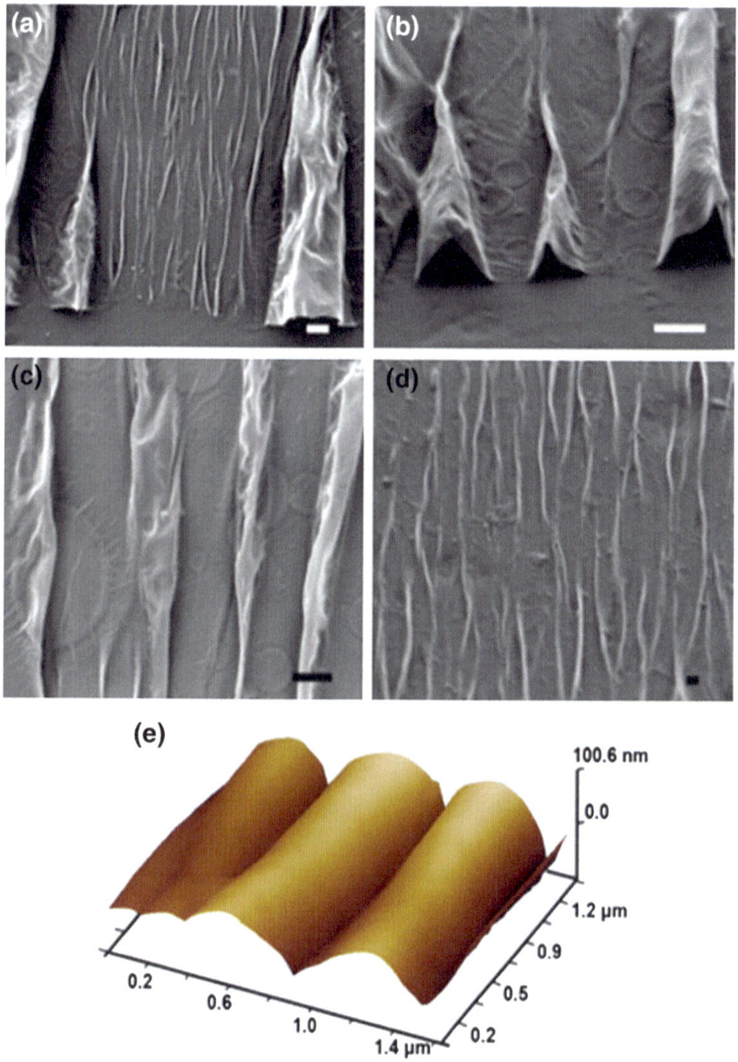

Fig. 10.7 **a–d** FESEM images of the CVD-grown graphene films and **e** atomic force microscopy image of the surface. Reproduced from Xu et al. (2014) with permission from the American Chemical Society

activated carbon, transition metal oxides, and conducting polymers have been fabricated by these methods. One of the important literature reported by Pech et al. (2010) is the development of the carbon onions by using an electrophoretic deposition technique. The developed electrodes were the nanoscaled carbon onions (with 6–7-nm diameter), which displayed excellent rate performance as compared to the commercial activated carbon-based supercapacitors. Figure 10.9a–e shows

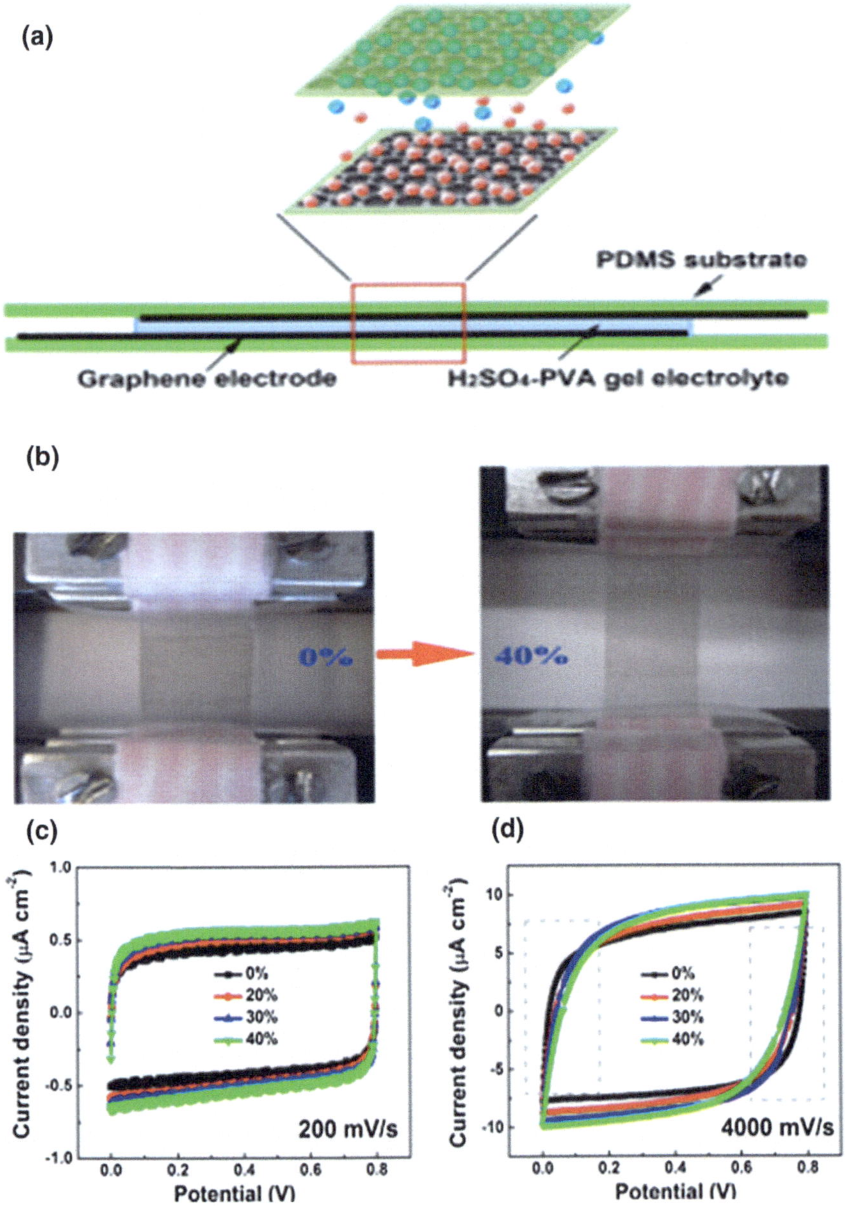

Fig. 10.8 **a** Schematic of the micro-supercapacitor components, **b**, **c** device in the stretched condition, **d** electrochemical performance of the device with different stretching conditions. Reproduced from Xu et al. (2014) with permission from the American Chemical Society

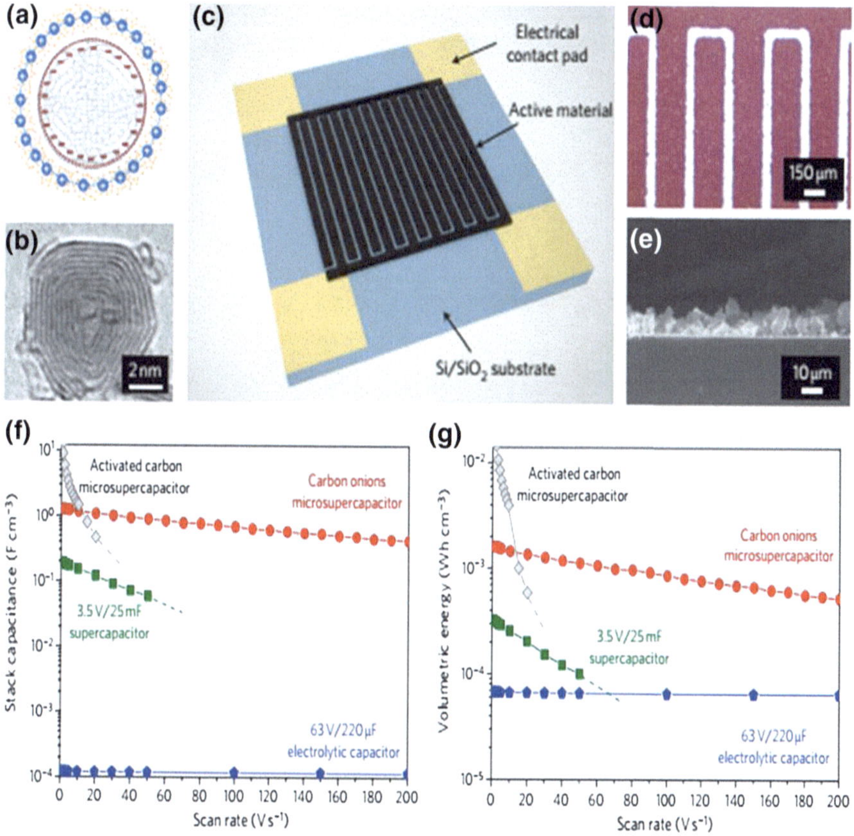

Fig. 10.9 **a–e** Design of the micro-supercapacitors with the onion-like carbon and **f**, **g** comparison of the electrochemical performance of the onion-like carbon with the commercial devices. Reproduced from Pech et al. (2010) with permission from Nature

the construction of the micro-supercapacitor with zero-dimensional carbon onion electrode, and Fig. 10.9f, g shows the superior electrochemical performance of the carbon onion electrode-based micro-supercapacitor device as compared to the activated carbon supercapacitor, commercial supercapacitors, and the commercial electrolytic capacitor.

In another approach, Su et al. have deposited MnO_2 on a 3-D nickel nanocone arrays (as shown in Fig. 10.10). It formed a 3-mm-thin self-supported electrode films and displayed an excellent electrochemical performance with activated carbon as a counter electrode. The thin-film device displayed a good lifetime of 20,000 cycles with 95.3% capacitance retention and maximum energy density of 52.2 Wh/kg (Su et al. 2014b).

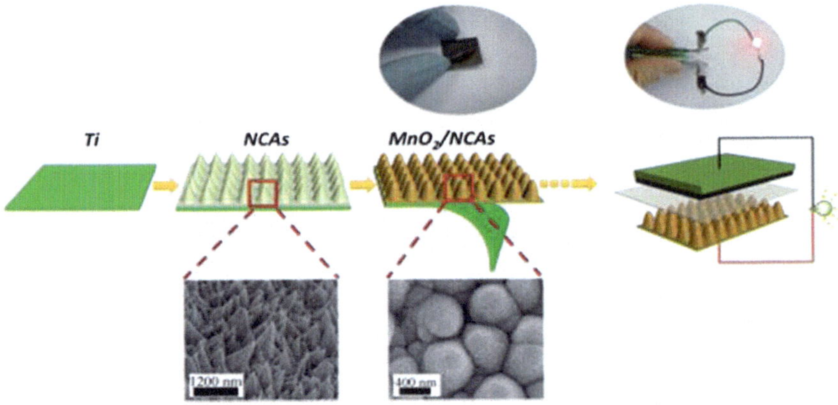

Fig. 10.10 Schematic of the MnO_2 deposition on the Ni nanocones to make the three-dimensional micro-supercapacitor electrodes. Reproduced from Su et al. (2014a) with permission from the Royal Society of Chemistry

10.3 Current Advancement in the Electrode Design for Micro-supercapacitors

A lot of research work is focused on the improvement of the electrode design and the device architecture to achieve the high performance of the micro-supercapacitor devices for miniaturized or on-chip electronic. The electrode design significantly affects the output performance of the energy storage devices (Kyeremateng et al. 2017). Out of the various electrode configurations, the interdigital in-plane design and the 3-D architecture of the electrodes significantly improved the performance of the micro-supercapacitors to fulfill the energy requirements of the miniaturized electronic devices.

10.3.1 In-Plane Electrode Design

The in-plane electrode configuration is an efficient design to get the high output performance of the micro-supercapacitor devices. It provides much higher power capabilities than the conventional sandwich structure due to the shorter diffusion paths and fast electrolyte ion transfer provided by the planar architecture (Yoo et al. 2011).

It provides the significant benefit to achieve the high capacitance, high-frequency response, good rate capability, and excellent cyclic lifetime of the device (Liu et al. 2015). The planer design also reduces the weight and thickness of the energy storage component and provides more flexibility to integrate with the compact and portable electronics. The in-plane micro-supercapacitors can be fabricated by the micro-fabrication techniques already reported in Sect. 10.2. Recently, the solid-state in-plane micro-supercapacitors were introduced to eliminate the electrolyte leakage problem and for the efficient operation of the energy storage components in practical applications (Gao and Lian 2014). Most of the solid-state electrolytes are prepared from the mixture of the polymers and the standard electrolytes in the water solution. However, due to the poor ionic conductivity and incompatibility of these electrolytes at the high temperatures, some ionic gels were reported to overcome these problems (Le Bideau et al. 2011). Figure 10.11 shows a schematic of the in-plane and conventional sandwich design of the graphene-based supercapacitors. The planer design is benefitted from the easy and fast ion transfer, which results in fast charging–discharging of the device and provides a huge gain as compared to the sandwich structure. Most of the potential materials for energy storage applications such as carbon-based material (graphene, CNT, activated carbon), metal oxides (RuO_2, MnO_2, NiO, V_2O_5, etc.), and conducting polymers have been utilized for making in-plane micro-supercapacitors. Although significant research is already conducted to both regarding fabrication and the power performance of the planer design micro-supercapacitors, however their proper integration with the other electronic components, achieving the required energy density for practical applications and scaling the device size without shorting the anode and cathode parts, is still a big challenge.

10.3.2 3-D Electrodes Design

Despite the several advantages, the thin-film micro-supercapacitors suffer from the low power densities which hinder its use for the real-life applications (Kyeremateng et al. 2017). This problem arises from the less volume of the electroactive material which is responsible for the less charge storage. Therefore, there should be a sufficient volume of the electrode material per unit area. One approach is to make the thicker electrodes. However, they may be problematic for the performance of the micro-supercapacitors. Another method is to make three-dimensional electrode designs instead of the planer surface of the electrodes. Figure 10.12 shows the schematic of the interdigitated micro-supercapacitors with 3-D electrode design.

In recent years, several studies have focused on the development of the 3-D electrode design with nanocone, nanowires, nanosheets, and other morphologies. Qui et al. (2014) have presented a 3-D electrode design of the micro-supercapacitors using the soft-printing method. The MnO_2 electrode material was deposited on the nanocone-shaped arrays of the current collector to form the 3-D architecture of the electrodes (as shown in Fig. 10.13). The developed device shows maximum aerial capacitance of 88.2 mF/cm^2 at 2 A/g current density and 96.5% capacitance retention after 2000 charge–discharge cycles.

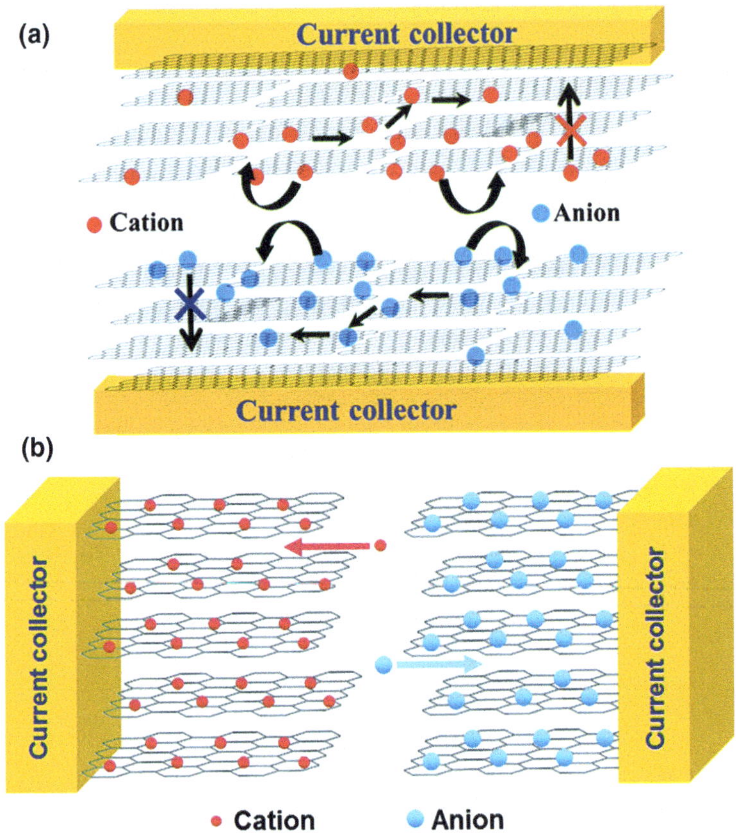

Fig. 10.11 **a** Schematic of the conventional sandwich design shows the obstructs in the ion transfer path due to the stacked layers which results in the incomplete utilization of the active material and may reduce the output performance, and **b** schematic of the in-plane design shows the full accessibility of all sides of the active electrode layers and easy transportation of the ions, which is beneficial to get the high electrochemical performance. Reproduced from Yoo et al. (2011) with permission from the American Chemical Society

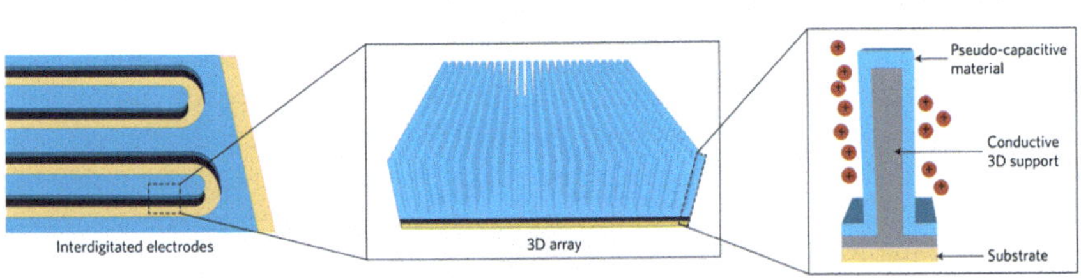

Fig. 10.12 Schematic illustration of a planer micro-supercapacitor with three-dimensional electrode design. Reproduced from Kyeremateng et al. (2017) with permission from Nature

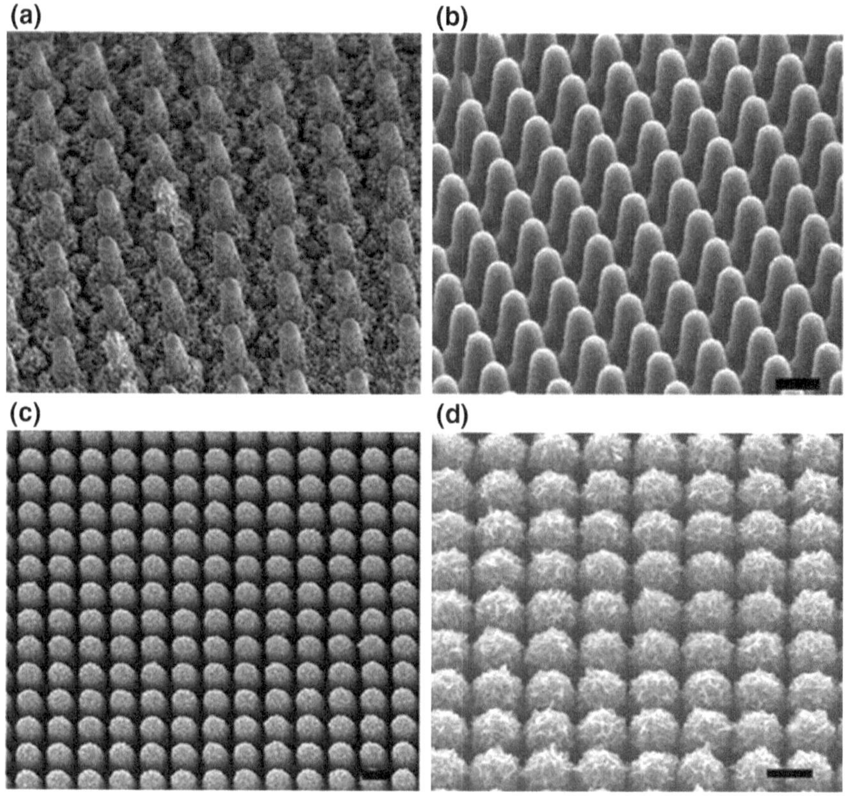

Fig. 10.13 FESEM images of the MnO_2 deposited nanocone arrays at different deposition instants: **a** 70 s, **b** 175 s, **c** 350 s, and **d** 700 s (scale bar: 1 mm). Reproduced from Qiu et al. (2014) with permission from the Royal Society of Chemistry

10.4 Advancement in the Material's Development for Improved Energy Storage Devices

The selection of electrode material is an important part of the energy storage devices as it directly affects the performance of the device (Yu et al. 2015). The required parameters of an energy storage device such as specific capacitance, cyclic stability, power density, energy density, conductivity, and rate performance depend on the type of electrode material. Therefore, the material research community has devoted much research attention to develop the efficient energy storage materials. To date, the different morphologies and structures of the carbon-based materials, metal oxides/sulfides/nitrides/hydroxides, and conducting polymers have widely explored as the electrode materials.

10.4.1 Carbon-Based Materials

The carbon-based materials have extensively used in the energy storage devices due to their availability, conducting nature, high surface area, good electrochemical stability, and easy fabrication process. Various carbon-based materials, such as graphene, carbon nanotubes, carbon onions, activated carbon with different morphologies, have been reported for the micro-supercapacitors (Xiong et al. 2014). Also, one of the recent focuses is on the conversion of bio-waste products into the efficient carbon-based electrodes in an eco-friendly way (Sundriyal and Bhattacharya 2017a, b). The major limitation of the carbon-based material is their low capacitance; therefore, significant efforts are still required to improve their electrochemical performance.

10.4.2 Metal Oxides

The advancement in the metal oxides/sulfides/nitrides/hydroxides has greatly boosted the output performance of the energy storage units. The widely explored metal oxides are RuO_2, MnO_2, NiO, V_2O_5, CoO, ZnO, and others (Chauhan and Bhattacharya 2016, 2018; Chauhan et al. 2017; Patel et al. 2017).

10.4.2.1 Two-Dimensional Morphology

Two-dimensional transition metal oxides have emerged as a potential high-efficiency electrode material for micro-supercapacitors. The thin structure of these materials provides a shorter ion diffusion path and the larger surface area which lead to better ion diffusion and higher conductivity compared to their bulk counterparts. Many research groups have explored 2-D nanostructures of $CoCO_3$ (Jiang et al. 2016), MnO_2 (He et al. 2012; Shi et al. 2013; Gao et al. 2017; Kumbhar and Kim 2018), NiO (Zhi et al. 2016), MoO_3 (Zhu et al. 2017), etc., for micro-supercapacitors. The thin and flexible nature of these materials also provides scope for the development of flexible and portable electronic devices (Zhi et al. 2016). The efficient methods for synthesis of the 2-D metal oxides are milling and microwave exposure. Figure 10.14 shows the FESEM and TEM images of the 2-D MnO_2 using the milling method.

Another transition metal oxide Co_3O_4 has attractive benefits as they are economical and have high theoretical specific capacitance. In nanostructured form, the specific capacitance is even higher due to the larger surface area of the material. Ge et al. have achieved a specific capacitance of 238.4 F/g with mesoporous structure of the Co_3O_4, Jing et al. have reported a specific capacitance of 288 F/g for the porous Co_3O_4 sheets, and Zaun et al. have reported a specific capacitance of 1862 F/g for the 2-D morphology of the Co_3O_4.

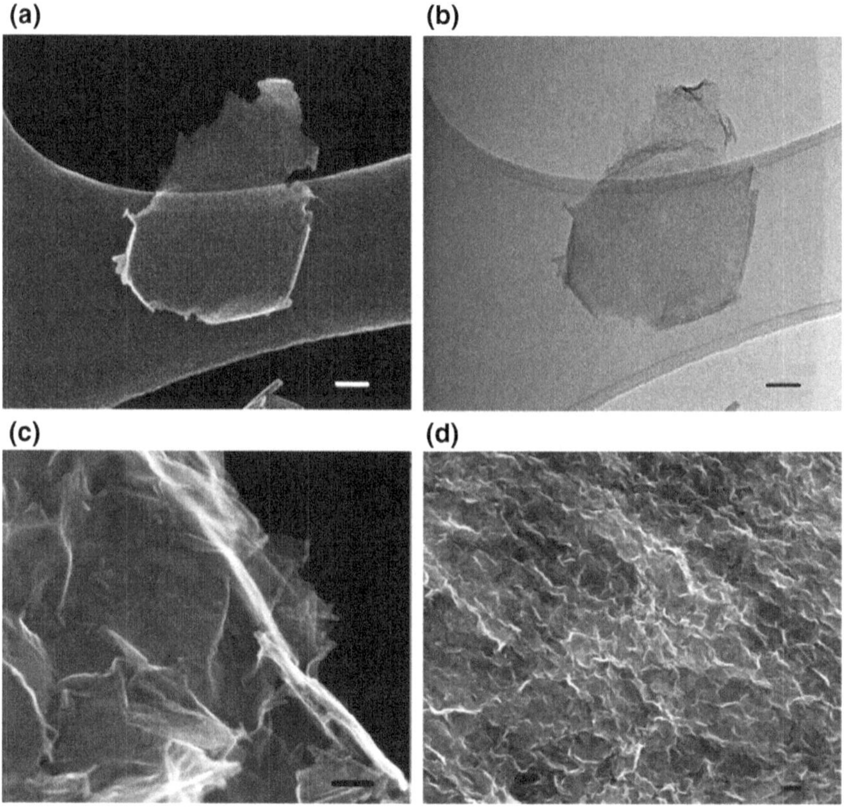

Fig. 10.14 Electron microscopy images of the 2-D MnO_2: **a**, **c**, and **d** the FESEM images and **b** bright-field TEM image [scale bar 50 nm (**a**–**c**) and 500 nm (**d**)]. Reproduced from Gao et al. (2017) with permission from Nature

10.4.2.2 Core/Shell Nanohybrid Structures

In order to improve the performance of the supercapacitor electrodes, one of the methods that have been adopted extensively is the use of core/shell morphology of metal oxide/metal oxide, metal oxide/metal hydroxide, etc. The materials for the core/shell morphology are selected to improve the electrochemical stability, to enlarge the electroactive sites, and to promote the redox reaction with anions and cations (Liu et al. 2011). The core/shell structure enables in achieving better electrostatic charge storage capacity for supercapacitors. Many core/shell morphologies in the form of needle, nanowires, and ribbon have been tried using various metal oxides such as Co_3O_4 nanowire@MnO_2 (Liu et al. 2011; Kong et al. 2014), $NiCo_2O_4$@$NiWO_4$ (Chen et al. 2016), CoO@MnO_2 (Li et al. 2016), $ZnFe_2O_4$@MnO_2 (Liu et al. 2017), Co_3O_4@NiO (Han et al. 2017). All these research works involved the use of nanostructures to enhance the surface area and

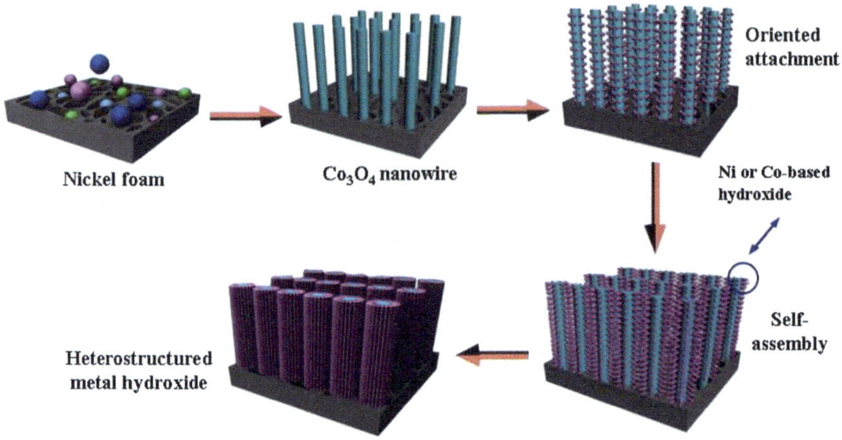

Fig. 10.15 Schematic illustration of the formation of core/shell structure of Co_3O_4@NiO nanowire@nanosheet arrays. Reproduced from Han et al. (2017) with permission from the Elsevier

provide a shorter diffusion path for better capacitive properties. Figure 10.15 displays the schematic of the CO_3O_4@NiO nanowire@nanosheet arrays on a nickel foam substrate, and Table 10.1 summarizes the electrochemical performance of the recently reported core/shell arrays. Figure 10.16 displays the FESEM images of the CoO@MnO_2 core/shell nanostructures, and Fig. 10.17 also shows TEM images of the $Ni(OH)_2$@MnO_2 core/shell structures.

10.4.2.3 Ternary Transition Metal Oxides/Mixed Transition Metal Oxides (MTMOs)

Mixed transition metal oxides are being studied enormously to fulfill the need for an electrode with high power and energy density. The pseudocapacitance of MTMOs is due to fast redox reaction at the electrode/electrolyte interface. Spinel structures of MTMOs have two TMOs. The MTMOs have attracted substantial attention as they have superior electrochemical properties, they are easy to synthesize and economical in nature. MTMO-based electrodes have much higher specific capacitance compared to graphite-based electrodes (Yuan et al. 2014).

Table 10.1 Electrochemical performance of the recently reported core/shell arrays

Sr. No.	Metal oxide core/shell material	Specific capacitance (F g^{-1})	Current density (A g^{-1})	Synthesis method	References
1	Co_3O_4 nanowire@MnO_2	480	2.67	Hydrothermal	Liu et al. (2011)
2	Co_3O_4@MnO_2 nanoneedles	1693.2	1	Hydrothermal	Kong et al. (2014)
3	$NiCo_2O_4$@$NiWO_4$ nanowire	1384	1	Hydrothermal annealing	Chen et al. (2016)
4	CoO@MnO_2 nanonetwork	1835	1	Hydrothermal annealing	Li et al. (2016)
5	$NiCo_2O_4$@NiCoAl double hydroxide nanoforest	1814	1	Hydrothermal annealing	He et al. (2017)
6	Co_3O_4/graphene quantum dots	2435	1	Solution–thermal treatment	Shim et al. (2017)
7	$MnCo_2O_4$@$MnMoO_4$ nanosheets			Hydrothermal annealing	Lv et al. (2018)
8	Phosphorus-doped Ni $(OH)_2/MnO_2$	911	1	Hydrothermal	Li et al. (2018)

10.4.3 Conducting Polymers

The several conducting polymers such as polyaniline, poly (3,4-ethylene dioxythiophene), polythiophene have been reported as the potential energy storage electrode materials due to their good electrical conductivity and multistate redox reactions (Wang et al. 2014). A lot of research attention is focused on the utilization of these conducting polymers to improve the electrochemical performance of the existing supercapacitor electrode materials. The application of these electrode materials in the micro-supercapacitors may greatly boost the electrochemical performance of the thin-film energy storage devices. Although a lot of research has carried out to develop the efficient energy storage materials, however their practical use in the micro-supercapacitor application is still a challenge.

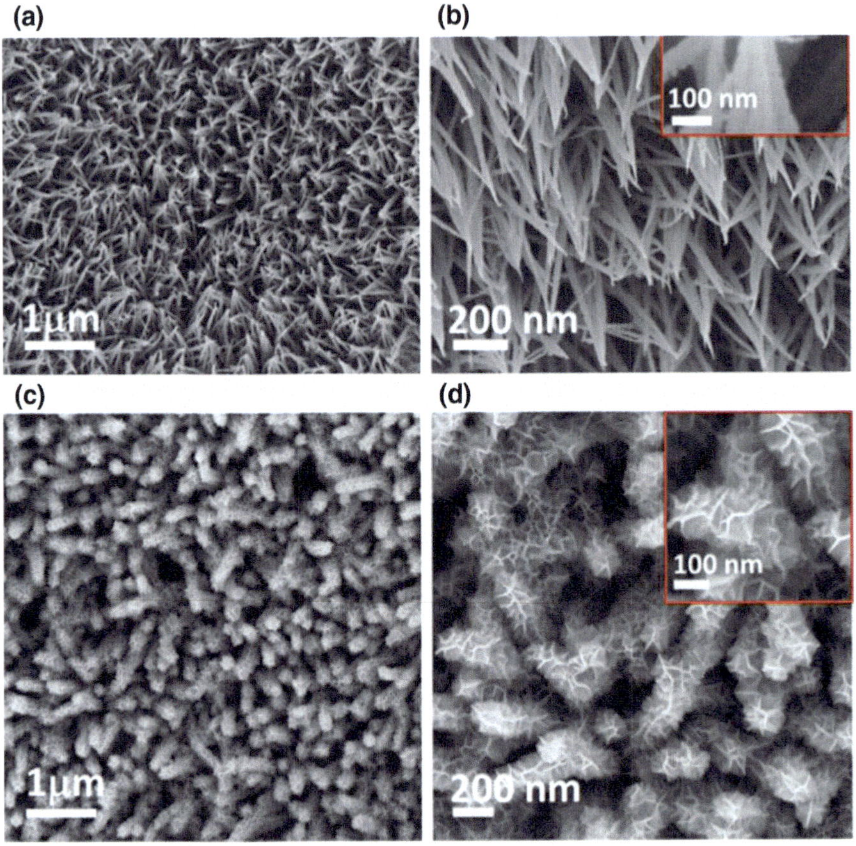

Fig. 10.16 Low and high magnification FESEM images of **a**, **b** CoO nanowires and **c**, **d** 3-D CoO@MnO_2 core/shell nanohybrid. Reproduced from Li et al. (2016) with permission from the Royal Society of Chemistry

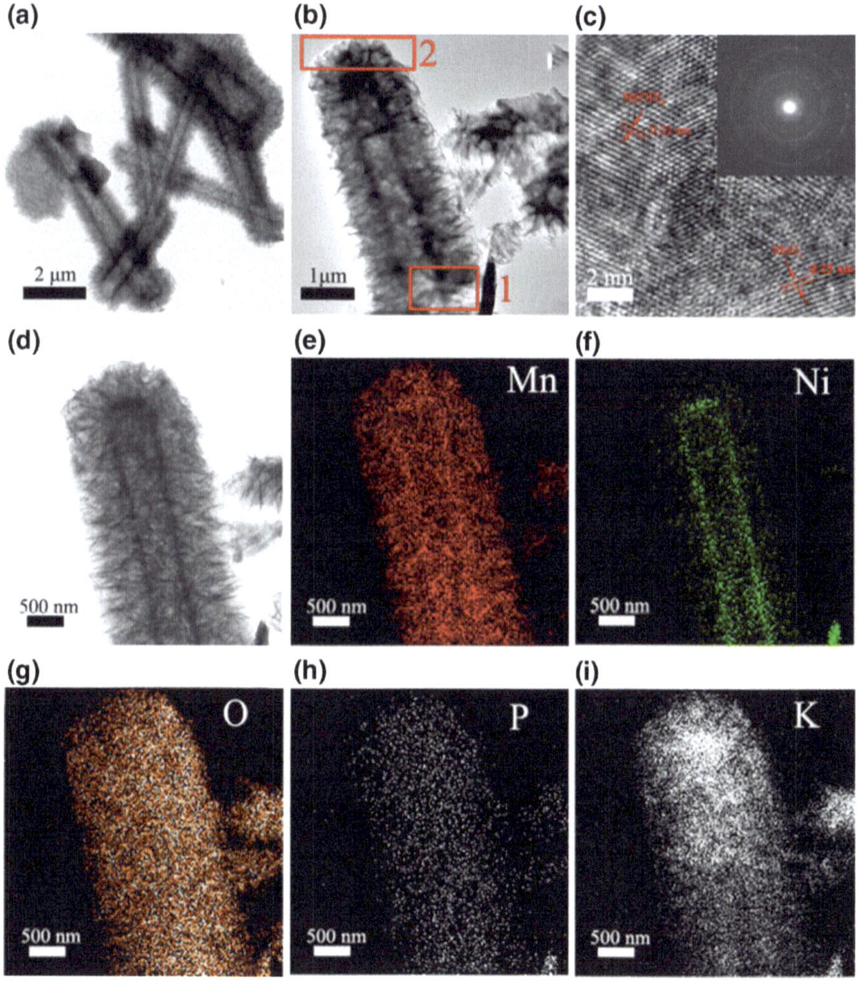

Fig. 10.17 **a** and **b** TEM images of the porous $Ni(OH)_2@MnO_2$ core/shell nanostructures, **c** HR-TEM image of the marked region in (**b**) (the inset shows the SEAD pattern), and **d–i** EDS elemental mapping of the porous $Ni(OH)_2@MnO_2$ core/shell nanostructure. Reproduced from Li et al. (2018) with permission from the Royal Society of Chemistry

10.5 Conclusion

In summary, the micro-supercapacitors are the promising energy storage devices to power the portable and miniaturized electronic systems and they are expected to fulfill the needs of the smart electronics of the future. A lot of research attention is already devoted to the fabrication, design, and the material's improvements. However, further research is required for their integration in the miniaturized electronic devices and their realization for the real-life applications.

References

Beidaghi M, Gogotsi Y (2014) Capacitive energy storage in micro-scale devices: recent advances in design and fabrication of micro-supercapacitors. Energy Environ Sci 7(3):867–884

Bhatt G, Kant R, Mishra K, Yadav K, Singh D, Gurunath R, Bhattacharya S (2017) Impact of surface roughness on dielectrophoretically assisted concentration of microorganisms over PCB based platforms. Biomed Microdevice 19(2):28

Chauhan PS, Bhattacharya S (2016) Vanadium pentoxide nanostructures for sensitive detection of hydrogen gas at room temperature. J Energy Environ Sustain 2:69–74

Chauhan PS, Bhattacharya S (2018) Highly sensitive $V_2O_5 \cdot 1.6\ H_2O$ nanostructures for sensing of helium gas at room temperature. Mater Lett 217:83–87

Chauhan PS, Rai A, Gupta A, Bhattacharya S (2017) Enhanced photocatalytic performance of vertically grown ZnO nanorods decorated with metals (Al, Ag, Au, and Au–Pd) for degradation of industrial dye. Mater Res Express 4(5):055004

Chen S, Yang G, Jia Y, Zheng H (2016) Three-dimensional $NiCo_2O_4$@$NiWO_4$ core–shell nanowire arrays for high performance supercapacitors. J Mater Chem A 5(3):1028–1034

Chmiola J, Largeot C, Taberna P-L, Simon P, Gogotsi Y (2010) Monolithic carbide-derived carbon films for micro-supercapacitors. Science 328(5977):480–483

Choi K-H, Yoo J, Lee CK, Lee S-Y (2016) All-inkjet-printed, solid-state flexible supercapacitors on paper. Energy Environ Sci 9(9):2812–2821

El-Kady MF, Kaner RB (2013) Scalable fabrication of high-power graphene micro-supercapacitors for flexible and on-chip energy storage. Nat Commun 4: ncomms2446

El-Kady MF, Strong V, Dubin S, Kaner RB (2012) Laser scribing of high-performance and flexible graphene-based electrochemical capacitors. Science 335(6074):1326–1330

El-Kady MF, Ihns M, Li M, Hwang JY, Mousavi MF, Chaney L, Lech AT, Kaner RB (2015) Engineering three-dimensional hybrid supercapacitors and microsupercapacitors for high-performance integrated energy storage. Proc Nat Acad Sci 112(14):4233–4238

Ervin MH, Le LT, Lee WY (2014) Inkjet-printed flexible graphene-based supercapacitor. Electrochim Acta 147:610–616

Gao H, Lian K (2014) Proton-conducting polymer electrolytes and their applications in solid supercapacitors: a review. RSC Adv 4(62):33091–33113

Gao P, Metz P, Hey T, Gong Y, Liu D, Edwards DD, Howe JY, Huang R, Misture ST (2017) The critical role of point defects in improving the specific capacitance of δ-MnO_2 nanosheets. Nat Commun 8:14559

Gogotsi Y (2014) Materials science: energy storage wrapped up. Nature 509(7502):568

Han D, Jing X, Wang J, Ding Y, Cheng Z, Dang H, Xu P (2017) Three-dimensional Co_3O_4 nanowire@NiO nanosheet core-shell construction arrays as electrodes for low charge transfer resistance. Electrochim Acta 241:220–228

He Y, Chen W, Li X, Zhang Z, Fu J, Zhao C, Xie E (2012) Freestanding three-dimensional graphene/MnO_2 composite networks as ultralight and flexible supercapacitor electrodes. ACS Nano 7(1):174–182

He X, Liu Q, Liu J, Li R, Zhang H, Chen R, Wang J (2017) Hierarchical $NiCo_2O_4$@NiCoAl-layered double hydroxide core/shell nanoforest arrays as advanced electrodes for high-performance asymmetric supercapacitors. J Alloy Compd 724:130–138

Hyun WJ, Secor EB, Kim CH, Hersam MC, Francis LF, Frisbie CD (2017) Scalable, self-aligned printing of flexible graphene micro-supercapacitors. Adv Energ Mater 7(17):1700285

Jiang Y, Chen L, Zhang H, Zhang Q, Chen W, Zhu J, Song D (2016) Two-dimensional Co_3O_4 thin sheets assembled by 3D interconnected nanoflake array framework structures with enhanced supercapacitor performance derived from coordination complexes. Chem Eng J 292:1–12

Kong D, Luo J, Wang Y, Ren W, Yu T, Luo Y, Yang Y, Cheng C (2014) Three-dimensional Co_3O_4@MnO_2 hierarchical nanoneedle arrays: morphology control and electrochemical energy storage. Adv Func Mater 24(24):3815–3826

Kumbhar VS, Kim D-H (2018) Hierarchical coating of MnO_2 nanosheets on $ZnCo_2O_4$ nanoflakes for enhanced electrochemical performance of asymmetric supercapacitors. Electrochim Acta 271:284–296

Kyeremateng NA, Brousse T, Pech D (2017) Microsupercapacitors as miniaturized energy-storage components for on-chip electronics. Nat Nanotechnol 12(1):7

Le Bideau J, Viau L, Vioux A (2011) Ionogels, ionic liquid based hybrid materials. Chem Soc Rev 40(2):907–925

Lee J-S, Kim S-I, Yoon J-C, Jang J-H (2013) Chemical vapor deposition of mesoporous graphene nanoballs for supercapacitor. ACS Nano 7(7):6047–6055

Li C, Balamurugan J, Thanh T, Kim N, Lee J (2016) 3D hierarchical CoO@MnO_2 core–shell nanohybrid for high-energy solid state asymmetric supercapacitors. J Mater Chem A 5(1): 397–408

Li J, Sollami Delekta S, Zhang P, Yang S, Lohe MR, Zhuang X, Feng X, Östling M (2017) Scalable fabrication and integration of graphene microsupercapacitors through full inkjet printing. ACS Nano 11(8):8249–8256

Li K, Li S, Huang F, Yu X-Y, Lu Y, Wang L, Chen H, Zhang H (2018) Hierarchical core–shell structures of P-$Ni(OH)_2$ rods@MnO_2 nanosheets as high-performance cathode materials for asymmetric supercapacitors. Nanoscale 10(5):2524–2532

Liu J, Jiang J, Cheng C, Li H, Zhang J, Gong H, Fan H (2011) Co_3O_4 nanowire@MnO_2 ultrathin nanosheet core/shell arrays: a new class of high-performance pseudocapacitive materials. Adv Mater 23(18):2076–2081

Liu W, Lu C, Wang X, Tay RY, Tay BK (2015) High-performance microsupercapacitors based on two-dimensional graphene/manganese dioxide/silver nanowire ternary hybrid film. ACS Nano 9(2):1528–1542

Liu C, Peng T, Wang C, Lu Y, Yan H, Luo Y (2017) Three-dimensional $ZnFe_2O_4$@MnO_2 hierarchical core/shell nanosheet arrays as high-performance battery-type electrode materials. J Alloy Compd 720:86–94

Lv Y, Liu A, Che H, Mu J, Guo Z, Zhang X, Bai Y, Zhang Z, Wang G, Pei Z (2018) Three-dimensional interconnected $MnCo_2O_4$ nanosheets@$MnMoO_4$ nanosheets core-shell nanoarrays on Ni foam for high-performance supercapacitors. Chem Eng J 336:64–73

Miller JR, Outlaw R, Holloway B (2011) Graphene electric double layer capacitor with ultra-high-power performance. Electrochim Acta 56(28):10443–10449

Patel V, Sundriyal P, Bhattacharya S (2017) Aloe vera vs. poly (ethylene) glycol-based synthesis and relative catalytic activity investigations of ZnO nanorods in thermal decomposition of potassium perchlorate. Part Sci Technol 35(3):361–368

Pech D, Brunet M, Durou H, Huang P, Mochalin V, Gogotsi Y, Taberna P-L, Simon P (2010) Ultrahigh-power micrometre-sized supercapacitors based on onion-like carbon. Nat Nanotechnol 5(9):651

Pitkänen O, Järvinen T, Cheng H, Lorite G, Dombovari A, Rieppo L, Talapatra S, Duong H, Tóth G, Juhász KL (2017) On-chip integrated vertically aligned carbon nanotube based super-and pseudocapacitors. Sci Rep 7(1):16594

Qiu Y, Zhao Y, Yang X, Li W, Wei Z, Xiao J, Leung S-F, Lin Q, Wu H, Zhang Y (2014) Three-dimensional metal/oxide nanocone arrays for high-performance electrochemical pseudocapacitors. Nanoscale 6(7):3626–3631

Shi S, Xu C, Yang C, Chen Y, Liu J, Kang F (2013) Flexible asymmetric supercapacitors based on ultrathin two-dimensional nanosheets with outstanding electrochemical performance and aesthetic property. Sci Rep 3:2598

Shim J, Ko Y, Lee K, Partha K, Lee C-H, Yu K, Koo H, Lee K-T, Seo W-S, Son D (2017) Conductive Co_3O_4/graphene (core/shell) quantum dots as electrode materials for electrochemical pseudocapacitor applications. Compos B Eng 130:230–235

Su Y-Z, Xiao K, Li N, Liu Z-Q, Qiao S-Z (2014a) Amorphous $Ni(OH)_2$@three-dimensional Ni core–shell nanostructures for high capacitance pseudocapacitors and asymmetric supercapacitors. J Mater Chem A 2(34):13845–13853

Su Z, Yang C, Xie B, Lin Z, Zhang Z, Liu J, Li B, Kang F, Wong CP (2014b) Scalable fabrication of MnO_2 nanostructure deposited on free-standing Ni nanocone arrays for ultrathin, flexible, high-performance micro-supercapacitor. Energy Environ Sci 7(8):2652–2659

Sundriyal P, Bhattacharya S (2017a) Inkjet-printed electrodes on A4 paper substrates for low-cost, disposable, and flexible asymmetric supercapacitors. ACS Appl Mater Interfaces 9(44):38507–38521

Sundriyal P, Bhattacharya S (2017b) Polyaniline silver nanoparticle coffee waste extracted porous graphene oxide nanocomposite structures as novel electrode material for rechargeable batteries. Mater Res Express 4(3):035501

Tyagi A, Tripathi KM, Gupta RK (2015) Recent progress in micro-scale energy storage devices and future aspects. J Mater Chem A 3(45):22507–22541

Wang K, Wu H, Meng Y, Wei Z (2014) Conducting polymer nanowire arrays for high performance supercapacitors. Small 10(1):14–31

Wang S, Liu N, Yang C, Liu W, Su J, Li L, Yang C, Gao Y (2015) Fully screen printed highly conductive electrodes on various flexible substrates for asymmetric supercapacitors. RSC Adv 5(104):85799–85805

Wu Z-S, Parvez K, Feng X, Müllen K (2014) Photolithographic fabrication of high-performance all-solid-state graphene-based planar micro-supercapacitors with different interdigital fingers. J Mater Chem A 2(22):8288–8293

Xiong G, Meng C, Reifenberger RG, Irazoqui PP, Fisher TS (2014) A review of graphene-based electrochemical microsupercapacitors. Electroanalysis 26(1):30–51

Xu P, Kang J, Choi J-B, Suhr J, Yu J, Li F, Byun J-H, Kim B-S, Chou T-W (2014) Laminated ultrathin chemical vapor deposition graphene films based stretchable and transparent high-rate supercapacitor. ACS Nano 8(9):9437–9445

Yoo JJ, Balakrishnan K, Huang J, Meunier V, Sumpter BG, Srivastava A, Conway M, Mohana Reddy AL, Yu J, Vajtai R (2011) Ultrathin planar graphene supercapacitors. Nano Lett 11 (4):1423–1427

Yu Z, Tetard L, Zhai L, Thomas J (2015) Supercapacitor electrode materials: nanostructures from 0 to 3 dimensions. Energy Environ Sci 8(3):702–730

Yuan C, Wu H, Xie Y, Lou X (2014) Mixed transition-metal oxides: design, synthesis, and energy-related applications. Angew Chem Int Ed 53(6):1488–1504

Zhi J, Yang C, Lin T, Cui H, Wang Z, Zhang H, Huang F (2016) Flexible all solid state supercapacitor with high energy density employing black Titania nanoparticles as a conductive agent. Nanoscale 8(7):4054–4062

Zhu L, Nuo Peh CK, Zhu T, Lim Y-F, Ho GW (2017) Bifunctional 2D-on-2D MoO_3 nanobelt/Ni $(OH)_2$ nanosheets for supercapacitor-driven electrochromic energy storage. J Mater Chem A 5 (18):8343–8351

Chapter 11
Solid Energetic Materials-Based Microthrusters for Space Applications

Vinay K. Patel, Jitendra Kumar Katiyar and Shantanu Bhattacharya

Abstract In this global scenario, the current state-of-the-art technologies in energy policy and management systems involve the integration of solid propellants/ energetic materials into microelectromechanical system (MEMS) to exploit the onboard thermal, mechanical, and chemical energy for civilian and defense needs. The solid propellants are recognized as attractive onboard energy sources owing to contain high energy density and rapid energy release and high actuation pressure as demanded in micro-propulsion. Microthrusters are used to propel and guide the missiles, shells, and also to orient and propel the satellites and to launch the rockets. This chapter details the technological developments and advancements made in the design, fabrication, and modeling of solid energetic materials (propellants and nano-thermites)-based microthrusters and their characterization in terms of propulsion performance as applied for space applications.

Keywords Solid propellant · Microelectromechanical system (MEMS) Micro-spacecraft · Micro-propulsion · Microthrusters · Microigniter Nano-thermites · Nano-energetic materials

V. K. Patel
Department of Mechanical Engineering, Govind Ballabh Pant Institute of Engineering and Technology Ghurdauri, Pauri-Garhwal 246194, Uttarakhand, India

J. K. Katiyar
Department of Mechanical Engineering, SRM Institute of Science and Technology, Kattankulathur Campus, Kattankulathur, Chennai 603203, Tamil Nadu, India

S. Bhattacharya (✉)
Department of Mechanical Engineering, Indian Institute of Technology, Kanpur 208016, Uttar Pradesh, India
e-mail: bhattacs@iitk.ac.in

S. Bhattacharya et al. (eds.), *Nano-Energetic Materials*, Energy, Environment and Sustainability, https://doi.org/10.1007/978-981-13-3269-2_11

11.1 Introduction

The advancements in Microelectromechanical systems (MEMS) or microsystems technology have evolved micromachines capable to the extent to be applied for miniaturization of the space industry with the aim to reduce the size of parts, mass, fabrication costs, and launch costs of satellites along with enhancements in reliability, flexibility, and performance of micro-propulsion devices such as satellite, rocket. In early 1995, the US space agency National Astronautics and Space Administration implemented "New Millennium Program" to dramatically miniaturize spacecraft by developing an under 10 kg micro-spacecraft. For such micro-propulsion system, most of the parts, involving thrusters for altitude control, were targeted to be miniaturized using the MEMS technology. Propulsion plays a pivotal role in miniaturization of spacecraft because micro-/nano-spacecrafts require very low (smaller than the Newton scale) and precise force to attain stabilization, station keeping, and pointing (Rossi 2002). The necessary level of impulse precision, space, degree of thrust (Helvejian 1999), and weight necessities cannot be fulfilled with conventional propulsion units, and for this space requirement, a new micro-product of microthruster has been evolved using MEMS technology. The microthrusters are basically applied for mainly propulsion and altitude control of micro-spacecraft, micro-satellites (10–100 kg), nano-satellites (1–10 kg), and pico-satellites (0.1–1 kg).[1]

Chemical propulsion microsystems convert the chemical energy available in a liquid/solid/gaseous fuel substance into thrust by accelerating the burning gases through a nozzle. Chemical propulsion systems mainly constitute of a tank to store propellant, an ignition device, and a converging–diverging portion to accelerate the combusted gases. The main limitations of liquid energetic materials are the power input essential for pumps and valves, and the essential cooling system to avoid overheating. Additional precautions are exercised for handling toxic liquids, storage, and leakage, making the solid energetic materials alternatively more reliable (See Footnote 1). Solid energetic materials are mainly classified into three groups, i.e., propellants, explosives, and pyrotechnics. Propellants/pyrotechnics release their energy through relatively slow deflagration processes (combustion), whereas the explosives detonate to release their energy (Rossi et al. 2007). The theory of solid propellant microthruster is realized by combustion of solid energetic fuels in a micro-chamber and ultra-fast ejection of gaseous streams producing thrust. The solid propellant microthruster generates short impulses and has the advantages of simple in design and construction, reliable, and economical and can produce relatively accurate specific impulses (100–200 s) (Rossi 2002). The solid propellant microthruster has the following major advantages (See Footnote 1; Rossi et al. 2002; Tanaka et al. 2003).

[1]Technology challenges in solid energetic materials for micro propulsion applications.

- There is less number of components and great overall compactness.
- The frictional forces are eliminated due to the absence of moving components.
- There is no chance of leakage with solid propellant as inherent with liquid fuel, and thus the solid propellants remain stable with time.
- The gas pressurization rate in the micro-chamber is not as extent as that in micro-gas turbine thus prevents the mechanical failures.
- It is capable to exert thrust in very accurate and of predetermined amounts.
- The system is highly redundant.

11.2 Survey of Design, Development, and Performance of Microthrusters

The advantages of solid energetic materials enable the microthruster applicable for the simple altitude control of a sub 10 kg class micro-spacecraft (Tanaka et al. 2003). In recent years, the pyrotechnics including nano-thermite has been evolved as advanced functional materials having properties such as ultra-rapid combustion reactivity and energy release, excellent pressurization rate resulting in pulse power which make these materials direct/indirect deployment in many civilian, defense, and micro-manufacturing applications (Zhou et al. 2014; Patel and Bhattacharya 2013; Patel et al. 2015, 2018). In this chapter, it is aimed to critically explore the research and development held in design, fabrication, modeling and performance characterization of microthrusters driven by the solid propellants and nano-thermites.

11.2.1 Solid Propellants-Based Microthruster

Rossi et al. (2001) introduced the concept of miniaturized, low-cost microthruster with high level of integration which had high responsiveness to the spatial constraints such as volume constraints. They fabricated an array of 36 independent microthrusters in the same substrate with novelty of the use of single solid propellant (glycidyl azide polymer GAP) loaded in a small micro-chamber fabricated in ceramic, glass, or silicon chip. The microthruster's design consisted of a sandwich of three micromachined structures: igniters consisting of patterned poly-silicon resistor, propellant tanks, and nozzles on top of the structure. Two types of micro-tank arrays were fabricated with the first one being 1 mm diameter hole created in a 3 mm ceramic substrate with spacing between each thruster of 0.5 mm and the second one being 1 mm^2 holes machined into a Si-substrate (thickness: 1 mm) by deep reactive ion etching (DRIE). They claimed that this device was able to produce precise impulses useful orbit moving, accurate station keeping, and altitude control of micro-satellite. The flexibility of loading of various categories/types of propellants in individual thrusters provided it a great maneuverability and flexibility to the device. There was also a plenty of possibilities due to flexibility in the design such as flexibility in designing the chamber and throat dimensions to obtain programmable thrust delivery and thus capable to produce thrust forces ranging from a few μN to a few tens mN, useful for wide range applications. Figure 11.1 displays the design of a single microthruster which is able to produce total impulse in the range of 0.1 mN s–1 N s when arrayed (Rossi et al. 2002). Orieux et al. (2002) predicted the performance of a solid propellant micro-rocket (throat diameter—108 μm, chamber diameter—850 μm, chamber length—1500 mm, convergent length—500 μm, divergent length—500 μm) based on a lumped parameter theory. The model predicted the chamber pressure of about 5 bar and a thrust value of about 3 mN, at steady state. The model illustrated that the divergent length had influenced the thrust force and therefore must be optimized in case the external pressure was very near to vacuum, whereas the divergent length had no influence on thrust value when the external pressure was atmospheric.

Tanaka et al. (2003) developed and tested a prototype (10 × 10 φ0.8 mm solid fuel micro-scale rockets arrayed at a pitch of 1.2 mm on a 20 × 22 mm) substrate for simple altitude control of a 10 kg class micro-spacecraft like penetrator. The penetrator required a velocity increment of 1 m/s to direct its nose toward the lunar surface by exerting a total impulse thrust of 10 Ns. The microthruster comprised of three layers. The first one (Si-layer) housed nozzles and ignition heaters machined on diaphragms. The diaphragm was used to insulate the ignition heater and also to seal the solid propellant. The second glass layer constituted the solid energetic propellants and electrical feed-throughs. The electrical feed-through also assisted in connecting directly the control circuit to the backside of the microthruster. This reduced the dimension of the system and protected the control circuit from damage

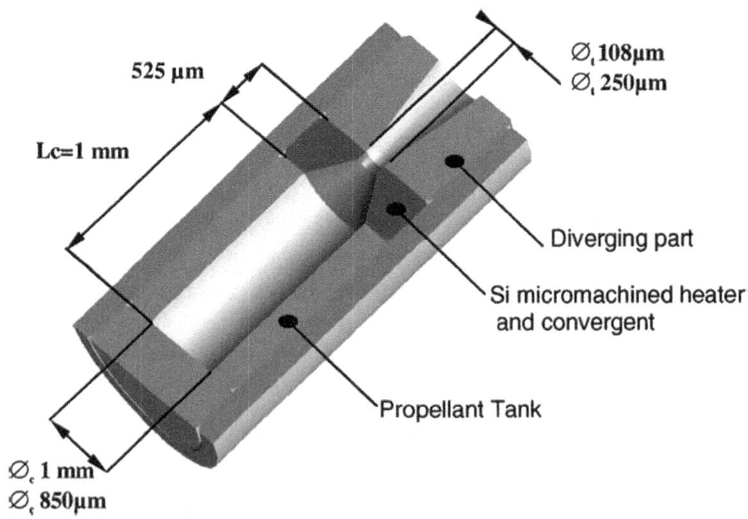

Fig. 11.1 Schematic representation of a single microthruster. Reproduced from Rossi et al. (2002) with permission from Elsevier

caused by exposure to cosmic rays. The impulse thrusts were found to range from 0.2×10^{-4} to 3×10^{-4} Ns after compensating the air damping which was below the required magnitude of 1 mNs.

Zhang et al. (2004) designed, micro-fabricated, and tested a solid propellant single microthruster. The depth of wafer trench was kept 0.35 mm and nozzle divergence angle was machined as 12°. The nozzle throat was designed to be a 100 μm plane to avoid sharp edges and to ease the micro-fabrication steps. Continuous combustion of the solid fuels of gunpowder and potassium perchlorate generated a total impulse ranging $(0.91–1.52) \times 10^{-4}$ Ns at sea level. Zhang et al. (2005) further reported the design, micro-fabrication, and experimentation of low-temperature co-fired ceramic (LTCFC) microthrusters. The gunpowder-based LTCFC microthruster produced a total impulse $(0.38–1.3) \times 10^{-4}$ N s and a specific impulse of 5.6–14.4 s at sea level; and a total impulse of $(1.3–2.8) \times 10^{-4}$ N s and specific impulse 19.1–31.6 s in vacuum. The LTCFC solid propellant microthruster was found more advantageous than that of former Si-based microthrusters, like superior suitability for batch production and high-scale integration, better ignition and thermal properties, and a better design flexibility. The performance of LTCFC microthrusters was found improved with a wire igniter than the former Si-based microthruster. Zhang et al. (2005) reported the improvement in performance of microthruster with Au/Ti igniter patterned by using standard micro-fabrication technologies. They demonstrated that the solid propellant microthruster with Au/Ti igniter delivered the same total impulse, but a greater specific impulse than the microthruster ignited by a wire igniter. Figure 11.2 represents the front view and side view of thruster array clearly reflecting the assembly of igniter and microthruster.

The computational fluid dynamics analysis (Zhang et al. 2006) revealed the great impact of chamber-area-to-throat-area ratio and exit-to-throat-area ratio on the jet flow phenomenon. The simulation revealed that the total impulse of microthruster

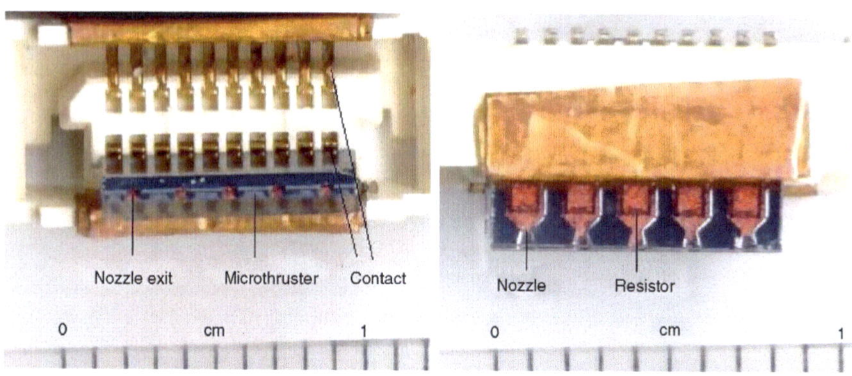

Fig. 11.2 Front (left) and side (right) views of the microthrusters installed a micro-connector showing microthruster and igniter. Reproduced from Zhang et al. (2005) with permission from Elsevier

had been changed by 5–7% due to wall heat loss. Slip wall boundary-layer effect was recognized a significant factor owing to rarefaction effects playing significant role in microthruster. According to the simulation, the microthruster total impulse was found to differ up to 11.3% due to slip wall boundary-layer effect. Zhang et al. (2007) performed finite element analysis of electrothermal characteristics to study the transient propellant ignition process of Au/Ti igniter applied in 3-D solid propellant microthruster. They demonstrated that the SiO_2 insulation layer had increased the ignition efficiency by reducing the leakage current.

Rossi et al. (2005) built an array (16 Ø1.5 mm × 1.5 mm rockets on 200 mm^2) of pyrotechnical microthruster to validate its working concept. The structural assembly of microthruster (Fig. 11.3) housed one Si nozzle part (throat: 250 μm), one igniter part containing zirconium perchlorate potassium (ZPP), one empty reservoir of 0.8 mm^3, and one Foturan reservoir containing GAP/ammonium perchlorate/Zr material. The utilization of ZPP as igniter instead of GAP propellant had allowed reaching 100% of ignition accomplishment with only 100 mW. The usage of Foturan for chamber had enabled a better thermal insulation between the arrays of thrusters and thus prevented the thermal cross talk between each thruster. Without nozzle on top, the combustion rate was found to be 2.8 mm/s, and with a nozzle on top, it reached 3.2 mm/s. Thrust had the range from 0.3 to 2 mN.

Rossi et al. (2006) fabricated and modeled an array (16) of solid propellant microthrusters with each microthruster constituting four main components: nozzle, heater, chamber, and seal. The simulation of the microthruster (chamber area: 2.25 mm^2, varied throat diameters: 160, 250, and 500 μm) revealed the generation of thrusts in the range of 0.3–30 mN to meet station keeping as required for

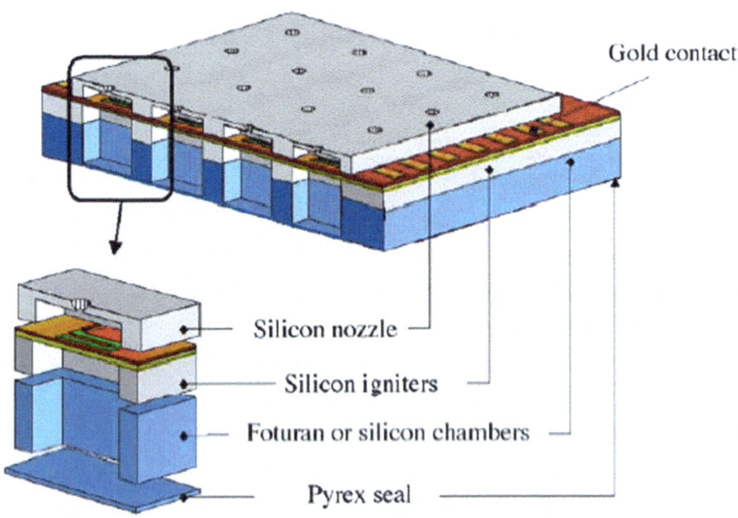

Fig. 11.3 Structure of an array of pyrotechnical thrusters (without pyrotechnical materials). Reproduced from Rossi et al. (2005) with permission from Elsevier

micro-satellites. The ignition power ranged from 80 to 150 mW depending on the energetic material utilized as igniter. The ignition attainment had reached 100% with ZPP propellant in a similar investigation (Rossi et al. 2005).

Lee and Kim (2013) fabricated and evaluated the performance of a microigniter integrated with an improved membrane for solid propellant (lead styphnate) thruster array. They fabricated the microigniter by anisotropic etching of a photosensitive glass membrane (thickness: 35 μm) and applying platinum as a heating element. The glass membrane was able to withstand ~3 times more pressure, i.e., 1.53 MPa than that withstood by the dielectric membrane. The proposed microigniter delivered the sufficient heat to ignite the solid propellant for a given electric power. The glass membrane had high structural stability to allow the solid propellant to be loaded more easily into the micro-chamber. The maximum thrust and total impulse were measured to be 3619 mN and 0.38 mN s, respectively.

Liu et al. (2015) fabricated a 10 × 10 scale and a 100 × 100 scale solid propellant (ammonium perchlorate-hydroxyl-terminated polybutadiene) microthruster array in multilayer structure (ignition circuit structure, ignition powder structure, propellant structure, cavity structure, and nozzle structure). The thermal infrared imager measured the ignition temperature to be 261.7 °C, and the result was verified by transient heat transfer numerical modeling. The lowest power that enables the total ignition successful was found to be 0.72 W. They measured the mean specific impulse to be 117.67 N s/kg of the 10 × 10 array thruster.

Zhang et al. (2016) studied the influence of wall (materials and thickness) heat transfer properties on the performance of solid propellant (AP/HTPB)-based microthruster. They demonstrated that employing a wall of smaller thermal conductivity and larger heat capacity could result in enhancements of thruster performance. Wall thickness was found to affect the combustion processes significantly, and a thinner wall resulted in a gradual increase of temperature for a shorter combustion duration.

11.2.2 Nano-thermites-Based Microthruster

Apperson et al. (2009) explored CuO/Al nano-thermites as great potential energy source to generate short-duration and large-amplitude thrust for microthruster applications. They fabricated the thrust motors by boring out (inner diameter of chamber: 1.59 mm) stainless steel bolts. They tested the nano-thermite composite in the range of 9–38 mg by varying the packing densities in two separate kinds of thrust motors: one without any nozzle and another with a convergent–divergent nozzle. They reported that at low packing density, the combustion occurred in fast-reaction regime producing small-duration large-amplitude thrust pulses (thrust forces: ~75 N generating within a time of less than 50 μs at full width at half maximum). At high packing density, the combustion occurred at relatively slow reaction regime producing lower thrust forces value of 3–5 N for a time duration of 1.5–3 ms. The both regimes generated the similar specific impulse of 20–25 s.

They recommended no particular benefits of employing convergent–divergent nozzle in the fast-reaction regime, due to significant deviations from ideal rocket performance conditions.

Staley et al. (2013) fabricated miniaturized thrusters from stainless steel-SS304 in three-piece designs without using converging/diverging nozzle and measured the thrust performance with two different nano-thermite composites of Bi_2O_3/Al and CuO/Al. The Bi_2O_3/Al nano-thermites exerted 46.1 N average thrusts for a time duration of 1.7 ms with a specific impulse of 41.4 s, whereas CuO/Al produced a thrust of 4.6 N (time duration: 5.1 ms) and a specific impulse of 20.2 s. Thus, Bi_2O_3/Al nano-thermites outshined CuO/Al in properties such as specific impulse, volumetric impulse, and energy conversion efficiency owing to its higher gas release rate, and theoretical mean density. They further found that adding nitrocellulose binder to nano-thermites decreased the average thrusts and extended burn times as a function of weight concentration. Nitrocellulose tends to decompose into gaseous products during combustion which enhances the thrust output. The specific and volumetric impulses of all the nano-thermites were enhanced by nitrocellulose up to a 2.5% weight concentration. The unique performance advantages of nano-thermite-based thrusters were attributed to high-reactivity combustion kinetics coupled with large theoretical mean density. Nitrocellulose additives were also demonstrated to substantially increase the high-g acceleration launch tolerance with regard to ignition sensitivity which was attributed to nitrocellulose acting as a passivation coating and cushioning the nano-thermites during high-g loading (Staley et al. 2014).

Ru et al. (2017) proposed a 10 × 10 solid propellant (CuO/nano-Al) microthruster array to meet the precise propulsion requirements micro-/nano-satellites. Similar to Staley et al. (2013), the group found that the nitrocellulose as additives into the CuO/nano-Al nano-thermites enhanced the thrust performance. The specific impulse and total impulse of nitrocellulose-free CuO/*n*Al were found to be 10.2 s and 155.9 μN s, which were increased to 27.2 s and 346.9 μN s, respectively by adding nitrocellulose binder.

11.3 Conclusion

This chapter imparts the important details of fabrication, assembly, experimentation, and modeling of solid propellant and nano-thermite-based microthruster. The solid propellant-based microthruster generally contains three main components/layers: the first Si-layer of micro-nozzles; second Si-layer as heating elements; the third Si-layer as propellant reservoirs. An intermediate Si-chamber is attached between the igniter and the propellant to allow achieving 100% of ignition accomplishment and a reliable flame transition between the igniter and propellant. The Au/Ti igniter was demonstrated to deliver same total impulse, but a greater specific impulse than wire igniter when applied in solid propellant microthruster. The glass membrane used as platform for microigniter has the ability to withstand ~ 3 times higher fracture pressure than that of dielectric membrane. A wall material of smaller thermal conductivity and larger heat capacity should be preferably selected result to achieve better thruster performance. For fast combustion reaction regime, a convergent–divergent nozzle is not required due to considerable deviations from the conditions necessary for ideal propulsion performance. Bi_2O_3/Al nano-thermites outclassed the CuO/Al with regard to specific impulse, volumetric impulse, and energy conversion efficiency due to higher gas evolution rate during combustion and higher theoretical mean density. Nitrocellulose tends to decompose into gaseous products during combustion and therefore is used as additives into basic nano-thermite composite to enhance the thrust output. Thus, this chapter systematically reports the progress and technological advancements in solid energetic materials (propellants and nano-thermites)-based microthrusters for space applications.

References

Apperson SJ, Bezmelnitsyn AV, Thiruvengadathan R, Gangopadhyay K, Gangopadhyay S, Balas WA, Anderson PE, Nicolich SM (2009) Characterization of nanothermite material for solid-fuel microthruster applications. J Propul Power 25:1086–1091

Helvejian H (1999) Microengineering aerospace systems. AIAA

Lee J, Kim T (2013) MEMS solid propellant thruster array with micro membrane igniter. Sens Actuators, A 190:52–60

Liu X, Li T, Li Z, Ma H, Fang S (2015) Design, fabrication and test of a solid propellant microthruster array by conventional precision machining. Sens Actuators, A 236:214–227

Orieux S, Rossi C, Esteve D (2002) Compact model based on a lumped parameter approach for the prediction of solid propellant micro-rocket performance. Sens Actuators, A 101:383–391

Patel VK, Bhattacharya S (2013) High-performance nanothermite composites based on aloe-vera-directed CuO nanorods. ACS Appl Mat Interfaces 5:13364–13374

Patel VK, Ganguli A, Kant R, Bhattacharya S (2015) Micropatterning of nanoenergetic films of Bi_2O_3/Al for pyrotechnics. RSC Adv 5:14967–14973

Patel VK, Kant R, Choudhary A, Painuly M, Bhattacharya S (2018) Performance characterization of Bi_2O_3/Al nanoenergetics blasted micro-forming system. Defence Technol. https://doi.org/10.1016/j.dt.2018.07.005

Rossi C (2002) Micropropulsion for space—a survey of MEMS-based micro thrusters and their solid propellant technology. Sens Update 10:257–292

Rossi C, Do Conto T, Esteve D, Larangot B (2001) Design, fabrication and modelling of MEMS-based microthrusters for space application. Smart Mater Struct 10:1156

Rossi C, Orieux S, Larangot B, Do Conto T, Esteve D (2002) Design, fabrication and modeling of solid propellant microrocket-application to micropropulsion. Sens Actuators, A 99:125–133

Rossi C, Larangot B, Lagrange D, Chaalane A (2005) Final characterizations of MEMS-based pyrotechnical microthrusters. Sens Actuators, A 121:508–514

Rossi C, Larangot B, Pham PQ, Briand D, de Rooij NF, Puig-Vidal M, Samitier J (2006) Solid propellant microthrusters on silicon: design, modeling, fabrication, and testing. J Microelectromech Syst 15:1805–1815

Rossi C, Zhang K, Estève D, Alphonse P, Tailhades P, Vahlas C (2007) Nanoenergetic materials for MEMS: a review. J Microelectromech Syst 16:919–931

Ru C, Wang F, Xu J, Dai J, Shen Y, Ye Y, Zhu P, Shen R (2017) Superior performance of a MEMS-based solid propellant microthruster (SPM) array with nanothermites. Microsyst Technol 23:3161–3174

Staley CS, Raymond KE, Thiruvengadathan R, Apperson SJ, Gangopadhyay K, Swaszek SM, Taylor RJ, Gangopadhyay S (2013) Fast-impulse nanothermite solid-propellant miniaturized thrusters. J Propul Power 29:1400–1409

Staley CS, Raymond KE, Thiruvengadathan R, Herbst JJ, Swaszek SM, Taylor RJ, Gangopadhyay K, Gangopadhyay S (2014) Effect of nitrocellulose gasifying binder on thrust performance and high-g launch tolerance of miniaturized nanothermite thrusters. Propellants Explos Pyrotech 39:374–382

Tanaka S, Hosokawa R, Tokudome SI, Hori K, Saito H, Watanabe M, Esashi M (2003) MEMS-based solid propellant rocket array thruster with electrical feedthroughs. Trans Jpn Soc Aeronaut Space Sci 46:47–51

Zhang KL, Chou SK, Ang SS (2004) Development of a solid propellant microthruster with chamber and nozzle etched on a wafer surface. J Micromech Microeng 14:785

Zhang KL, Chou SK, Ang SS (2005a) Development of a low-temperature co-fired ceramic solid propellant microthruster. J Micromech Microeng 15:944

Zhang KL, Chou SK, Ang SS, Tang XS (2005b) A MEMS-based solid propellant microthruster with Au/Ti igniter. Sens Actuators, A 122:113–123

Zhang KL, Chou SK, Ang SS (2006) Performance prediction of a novel solid-propellant microthruster. J Propul Power 22:56–63

Zhang KL, Chou SK, Ang SS (2007) Investigation on the ignition of a MEMS solid propellant microthruster before propellant combustion. J Micromech Microeng 17:322

Zhang T, Li GX, Chen J, Yu YS, Liu XH (2016) Effect of wall heat transfer characteristic on the micro solid thruster based on the AP/HTPB aerospace propellant. Vacuum 134:9–19

Zhou X, Torabi M, Lu J, Shen R, Zhang K (2014) Nanostructured energetic composites: synthesis, ignition/combustion modeling, and applications. ACS Appl Mat Interfaces 6:3058–3074

Chapter 12
Nanomaterials for Hydrogen Production Through Photocatalysis

Ahmed M. A. El Naggar, Mohamed S. A. Darwish and Asmaa S. Morshedy

Abstract In the last few decades, nanostructured materials have been of great interest worldwide due to their unique characteristics and their sub-driven reactivity. Furthermore, the unlimited applications of such materials in different fields and their associated success had added extra value for their importance. The combination between nanomaterials and photocatalytic processes has been recently given a great attention in different applications. It may enhance the viability of the nanotechnology principles. One of these applications is the usage of nanophotocatalytic materials in hydrogen production via water-splitting reaction. This chapter will cover the main concepts of photocatalysis and its associated terms. The main features of efficient photocatalysts and the ways of measuring such properties are illustrated in this chapter. Also, a brief presentation for the current methods that are utilized in the preparation of these catalysts is provided. An overview of the different types of semiconductors that are employed in the domain of photo-based hydrogen generation via splitting of water is introduced through this chapter too. A new approach in the water-splitting process by introducing noble metals attached magnetic nanoparticle (core/shell structure) as promising photomaterials is also described in this chapter. However, these materials are used in different fields such as bio-medical processes, water treatment, and energy storage. They are expected to be of high significance in field of catalysis, since they can be easily separated and recovered, due to their magnetic character, for reuse in further reactions. Implementation of a suitable magnetic force can enhance the hydrogen productivity during the water-splitting process. These materials are possessing magnetic properties which could reveal a new approach to their photocatalytic activity via quenching the radiation scattering.

Keywords Photocatalysis · Water-splitting processes hydrogen energy Magnetic materials · Semiconductors

---

A. M. A. El Naggar (✉) · M. S. A. Darwish · A. S. Morshedy
Refining Department, Egyptian Petroleum Research Institute, Cairo, Egypt
e-mail: drmeto1979@yahoo.com

S. Bhattacharya et al. (eds.), *Nano-Energetic Materials*, Energy, Environment and Sustainability, https://doi.org/10.1007/978-981-13-3269-2_12

Chapter Outline
This first part of this chapter illustrates the important definitions and terms of photocatalysis as well as the application of photo-based processes in various applications. The classifications and essential optical/ electronic characteristics of photocatalysts are reported in this part of the book. The relation between catalysts properties and proper irradiation sources is also discussed in details. An expanded overview of photocatalysis in water-splitting processes via two different routes is introduced in the second part of this chapter. The most common and efficient photocatalysts to generate hydrogen by splitting of water are also mentioned in this part.

12.1 Photocatalysis

Photocatalysis is a field of research that is developing fast with a great potentiality for use in various industrial applications, which include disinfection of air and water, mineralizing the organic contaminates, renewable fuels making, and organic creation. Photocatalysis is a word that was first known in Greece and consists of two pieces: photo (means: radiance) and the catalysis (partial break or decomposition). This term can be generally used to describe processes that are based on the light to activate a substance. The photocatalyst is a major item in such processes. It has been utilized to modify the rate of a chemical response without being involved in the chemical alteration. Therefore, conventional thermal catalyst is mainly different from a photocatalyst in the way of activation. Particularly, the first can be made under thermal effect, while second may be induced through light photons that have suitable power.

12.1.1 Types of Photocatalysis

Photocatalytic processes may occur in homogeneous or heterogeneous routes; however, heterogeneous photocatalysis has been rigorously investigated in past decades since it is highly likely to be used in versatile applications that are in relevance to environment, energy, and fabrication of organics.

12.1.1.1 Homogeneous Photocatalysis

In homogeneous photocatalysis, reacting species and catalysts are present in the same phase. Homogeneous photocatalysts that are of most common use includes photo-Fenton system (Fe^{2+} and Fe^{2+}/H_2O_2). Reactive species in such media is $^{\cdot}OH$

radicals (Wu and Chang 2006) that are generated according to the subsequent mechanism (as shown in Eqs. (12.1)–(12.5)) (Ciesla et al. 2004).

$$Fe^{2+} + H_2O_2 \rightarrow {}^{\cdot}OH + Fe^{3+} + OH^- \quad (12.1)$$

$$Fe^{3+} + H_2O_2 \rightarrow HO_2^{\cdot} + Fe^{2+} + H^+ \quad (12.2)$$

$$Fe^{2} + {}^{\cdot}OH \rightarrow Fe^{3+} + OH^- \quad (12.3)$$

In case of photo-Fenton course of action, extra supply of $^{\cdot}OH$ radicals should be taken into account. Specifically, hydroxyl radicals ($^{\cdot}OH$) are released through H_2O_2, photolysis, and consequent reduction of Fe^{+3} ions by the influence of UV irradiation; see Eqs. (12.4) and (12.5)

$$H_2O_2 + h\upsilon \rightarrow 2\ {}^{\cdot}OH \quad (12.4)$$

$$Fe^{3+} + H_2O_2 + h\upsilon \rightarrow OH + Fe^{2+} + H^+ \quad (12.5)$$

The effectiveness of Fenton-based processes may be affected by various operational factors such as hydrogen peroxide strength, pH, and radiation intensity (Wu and Chang 2006; Ciesla et al. 2004). The chief benefit of homogeneous photo-dependent methods is the possibility of using sunlight with radiation sensitivity around 450 nm; therefore, elevated expenditures of UV lamps and electricity can be avoided. These reactions could exhibit higher efficiency than diverse photocatalysis. On the other hand, a major drawback of this process is that it has to be carried out at low pH values. This obviously is due to the precipitation of Iron ions at increased pH values.

12.1.1.2 Heterogeneous Photocatalysis

In heterogeneous photocatalysis, the system entails the earlier configuration of a boundary between a solid photocatalyst and a fluid restrains reactants and outputs of the reaction. Interactions which involve lighted-up metal interfaces generally lie in the division of photochemistry.

So, concept of heterogeneous photocatalysis may basically be implemented in the processes that include a light-absorbing semiconductor in connection to liquid or gas phase. Although not all the heterogeneous photocatalysts are semiconductors, this type of solids is characterized of being most representative and broadly studied photoactive substances. So, it has been crucial to acquire some basic insights of physicochemical characteristics that belong to semiconductors and means of their interaction with radiation sources (Ciesla et al. 2004).

12.1.2 Electronic and Optical Properties of Heterogeneous Semiconductors

Semiconductors and insulators electronic individuality is of strong dependency to the band theory. Due to the huge number of atoms and their relevant electrons which act together in solid materials, equivalent levels of energy are spaced intimately forming bands. Each one of these energy band has its own level where the electrons occupy such bands from the level of minimum energy to the highest one. Latter statement is alike to the trend that electrons follow to engage orbitals of a certain atom (Moliton and Hiorns 2004).

The bands, of energy that are equivalent to engaged orbitals (HOMO) of the highest energy of definite molecule, are known as valance bands (V_B). The following band that has higher energy capacity than V_B contains free orbital (LUMO) of this molecule. Therefore, this band is identified as conduction band (C_B). Obviously, it can be said that V_B and C_B have been alienated via gap known as band gap (E_{bg}). Hence, the classification of a solid material type as conductor, semiconductor, or insulator is related to its filled bands, size of its particles and band gap, as given in Fig. 12.1.

Generally, electrons can move within a solid species by applying an electric field if these species have a partially occupied or entirely empty band. On the other hand, electrical insulator has no possibility of electron flow as a result of the full occupation of V_B by electrons and the C_B is too far away in energy to be reached by its electrons. In conductors, V_B interferes with C_B (Park et al. 1995). Therefore, the electrons may transfer freely crossways all atoms of which the solid substance is made. In the case of semiconductor, E_{bg} is less than 4 eV. Thus, it can be bridged

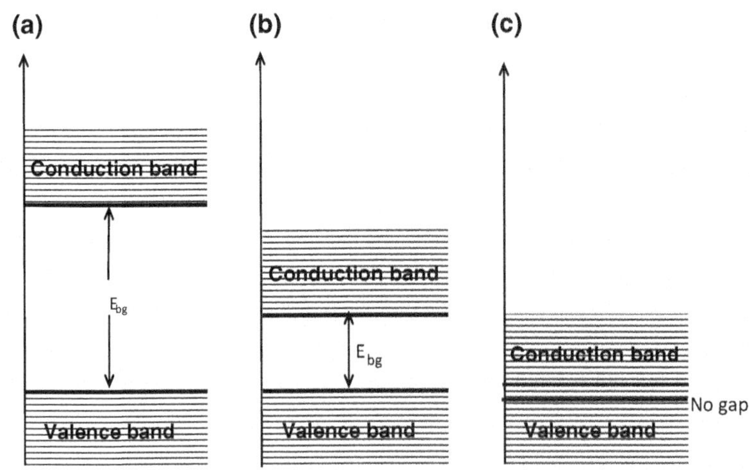

Fig. 12.1 Energy bands in solids: **a** insulator, **b** semiconductor and **c** conductor

either using thermal induction or light since these energy sources can encourage electrons movement from V_B to conduction band.

Semiconductors can be basically divided into two groups: intrinsic and extrinsic. The group of intrinsic semiconductors includes the undoped photoactive materials. Thus, the produced holes in V_B are formed under the effect of thermal excitation of electrons to subsequently reach to C_B. The behavior of such type of semiconductors is strongly dependent on surrounding temperature. For instance, no charge transfer for an intrinsic semiconductor can be attained at absolute zero (°K). Hence, semiconductors are changing to become as an insulator instead (Langot et al. 1996). At a finite temperature, electron (e^-) of V_B may thermally depart to the C_B, creating behind vacant state in V_B, called a hole (h^+). Electrons and holes are known as intrinsic charge carriers, and they usually move in opposite directions.

In intrinsic semiconductors, each electron in C_B has a parallel hole in V_B. Therefore, the number of electrons is always the same as that of holes. This number is symbolized as n_i and is called the intrinsic carrier concentration. An intrinsic semiconductor can be converted to extrinsic one through the insertion of impurity atoms into its crystal via an addition stage known as doping. Doping step makes a change in relative number of electrons and holes in the doped substance. This change has to be of strong dependency to type and inclusive number of doping atom (Perera et al. 1992). Chemical impurities that may involve their electrons to a conduction band are known as donors. Semiconductors that are doped by atoms of these donors can be assigned as n-type semiconductors. Electrons in n-type semiconductors (Shim and Guyot 2000) represent the majority of charge carriers where the number of electrons (n) equals the number of involved atoms by donors. In contrary, holes are major charge carriers in p-type semiconductors (Chang and James 1989). The hole concentration (p) in this type of semiconductors equals the concentration of acceptor atoms.

12.1.3 The Electronic Structure of Semiconductors

12.1.3.1 The Fermi Energy Level

Fermi level, E_f, is of a vital consideration in the band theory. Electrons of solid materials are propagated in existing energy levels according to the statistics of Fermi–Dirac. These statistics describe potentiality of a given energy level (E) to be filled with electron at a certain temperature. Fermi energy has been described as the needed energy to make the probability of occupying an energy level equals 0.5. However, the amount of this energy is strongly configured according to number of electrons in the system (Li et al. 2005). Exhibited Fermi level of intrinsic semiconductors (Fig. 12.2) can be seen as placed halfway between the C_B and V_B to show equal statistical potentiality to get a charge transporter through the two energy bands.

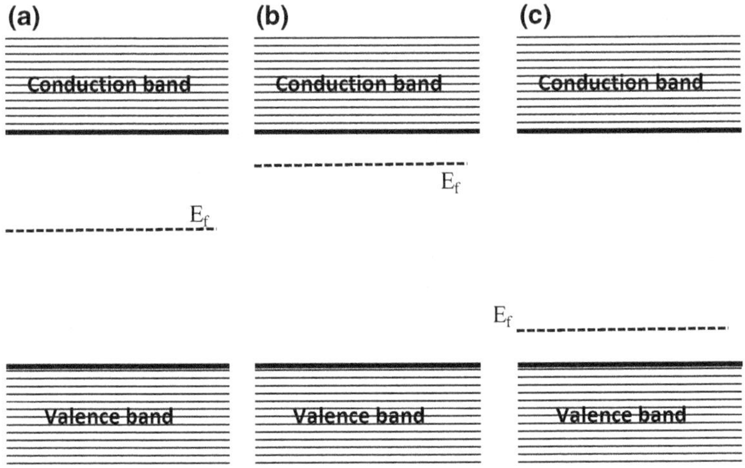

Fig. 12.2 Location of Fermi level in relation to V_B and C_B for **a** intrinsic, **b** n-type, **c** p-type semiconductors

The insertion of donor impurities with the structure of an intrinsic semiconductor will consequently increase the chance of facing free electrons. Therefore, Fermi level makes a gradual move to be closer to C_B than the V_B. On the other hand, doping of an intrinsic semiconductor by acceptor impurities reduces the probability to come across free electrons. Nevertheless, potentiality to meet holes will be enhanced, and thus, the Fermi level reallocates to be nearer to the V_B than C_B.

12.1.3.2 Surface Versus Bulk Properties

Extinction of a semiconductor periodic structure at its free surface may get rid of its symmetry in a right-angle direction to the surface. This in turn may result in the creation of localized electronic state within the surface of semiconductor. This state may then affect the electronic features of surface which may by then play crucial effect through interaction between surface and adsorbed substances. Surface states can be presented as:

(a) Existence of *dangling bonds*, i.e., superficial surface atoms that has no atoms above to bind with,
(b) Surface reconstruction or relaxation, for example, variation of position and/or chemical bonding arrangement of surface atoms; hence, surface energy will be minimized,
(c) Structural defects of the surface,
(d) Adsorption of foreign atoms as impurity on surface,
(e) Bonding involvement of the surface atoms and new phase atoms.

Energy levels that are positioned in the area of banned gap are of particular significance in terms of semiconductor optical properties. The manifestation of surface-localized states carries up movements of charge among core and surface of an atom in order to build up thermal balance between both of them. Thus, the density of charge delivery in environs of the surface moves away from its equilibrium value and makes the development of a non-neutral section in the bulk of semiconductor. This area is often identified as the surface space charge region (SCR). The features of bulk in pure semiconductors are improved at the rear of a few atomic layers in the core of their crystals. However, for doped semiconductors, the induction of surface states may lead to construction of a SCR that can go far (up to thousands angstroms) into the solid. Electrostatic potential, (V), that might be obtained via the space charge region will subsequently make a development in the band configuration close to the surface, named as band bending. For n-type semiconductors, bands go up to the surface, while in p-type semiconductors, bands twist downward; these actions are due to the move of electrons from the surface region of donor to surface states in the first case, while in the second case, electrons transport from the surface states to acceptor. Work function (impact of surface electronic states and bending) of a doped semiconductor is non-relevant to the concentration of doping species. Particularly, doping of a semiconductor with donors inflates Fermi level; thus, work function of such semiconductor decreases. However, this effect can be stopped via elevating the energy spent by a skiving electron to pass over the SCR. This phenomenon in doped semiconductors has been called "pinning of Fermi level" (Walukiewicz 1988).

12.1.4 Ultraviolet and Visible Spectrometry

Absorption of an electromagnetic irradiation normally varies energy status of a molecule. These changes in state of energy include:

- Electronic state ΔE_e = 150–600 kJ/mol
- Vibrational state ΔE_v = 2–60 kJ/mol
- Rotational state $\Delta E_r \approx 3$ kJ/mol.

The relation to the absorbed radiation wavelength is:

$$\Delta E = \Delta E_e + \Delta E_v + \Delta E_r = hv = h.c/\lambda$$

where (Planck's constant) $h = 6.626 \times 10^{-34}$ J s.

The energy gaps between molecular bonding, non-bonding, and anti-bonding orbitals lie in between 125 and 650 kJ per mole. This energy difference is due to the electromagnetic irradiation in both UV and visible regions of a spectrum (Perliski and Solomon 1993; Waymouth 1971; Sommerer 1996).

The UV–Visible spectrum can be split into three distinct regions:

1. Far UV (10–200 nm)
2. Near UV (200–380 nm)
3. Visible light (400–800 nm)

Dependent irradiation to each of the above-mentioned spectrum regions can just pass through certain types of glasswares. Table 12.1 summarizes values of potentially transmitting wavelength through some types of glass; hence, these types can be employed in photocatalytic reactions.

12.1.5 Radiation Sources

Good features of a spectrometric source can be represented through by a stable and high-intensity release. This in turn can provide such source with the capability to covers a broad range of wavelengths. Therefore, there is no specific single source that can be appropriate for all spectral zones. Irradiation sources can be categorized into thermal and electric discharge supply. Thermal radiations are generally resulting from elevating the operational temperature in a certain process. Hereafter, examples for some of the common irradiation sources are illustrated:

- *Deuterium discharge lamp*

This source uses an electrical discharge to dissociate deuterium molecules into atoms (Waymouth 1971). The process is accompanied by the emission of uninterrupted UV irradiation which lies nearly between 160 and 380 nm.

- *Tungsten filament lamp*

The most common source of visible radiation is the ordinary tungsten filament lamp. It is made from a thin spiral of tungsten string sealed inside an evacuated glass bulb. Electrical energy passing through the filament is converted to heat causing it to glow "white hot" (Waymouth 1971). *Mercury* is used under high pressure in Hg discharge tubes. Therefore, such source cannot be of proper use in spectral studies that are carried out with nonstop. This is obviously due to the sharp lines or bands that will be overlaid present on an unremitting background.

- *Xenon discharge lamp*

This lamp operates with a low-voltage DC source similar to that of the linear halogen lamp but at xenon pressures in the range of 10–30 atm. The intensity in the

Table 12.1 Type of glass and their corresponding wavelength cut

Type of glass	Wavelength cutoff (nm)
Pyrex	<275
Corex	<260
Vycor	<220
Quartz	<170

near UV is actually much greater than that of the LHL lamp, but even greater intensity in the visible region may pose potential stray radiation problems (Sommerer 1996).

- *Monochromators*

Monochromator is a source of radiation that can be utilized in order to disperse its chrome in reference to the desired wavelength. Prisms are broadly employed to attain such purpose. Glass, quartz, and fused silica are the most common materials to fabricate prisms are. Among those materials, glass could be counted as the highest of power resolving. This is because glass can firmly scatter light along the visible zone of a spectrum. Nevertheless, glass does not allow radiations that have wavelengths between 350 and 200 nm to transmit through. Thus, glass cannot be exploited at such range of wavelength. On the other hand, quartz usually permits wavelengths in the range of 200–700 nm to penetrate. Therefore, quartz is suitable for usage in UV region.

12.1.6 Photocatalytic Metal Oxides

In terms of molecular configuration, a metal oxide (MO_x) consists of at least one metal atom combined with one or more oxygen atoms. Excluding a few of the lighter inert gases ([He], [Ne], [Ar], and [Kr]), virtually all other elements are capable of oxidation, especially metal elements. Metals tend to be highly conductive due to the presence of delocalized electrons within the material matrix. However, upon oxidation, metal atoms donate their electrons to oxygen species forming positive ions, thereby adopting an electropositive state. Metal oxides are generally classified as semiconductors because their conductivity is often dependent upon the level of external excitation (Wong and Fierro 2006; Cox 1996).

Typical applications of these metal oxide semiconductors (MOS) include transistors, resistors, light-emitting diodes, solar cells, gas sensors, and piezoelectric transducers. While predominantly found within the electrical and computer industry, MOS technology can also be found across several other industries including, chemical, mechanical, environmental, energy, and petroleum (Wong and Fierro 2006). An ideal photocatalyst for photocatalytic applications has to possess the following properties (Carp et al. 2004):

(1) Stability of irradiation photons,
(2) Inert toward chemical and biological substances,
(3) Abundance and of low fabrication costs,
(4) Ability to attract reacting species at appropriate photostimulation energy ($h\upsilon \geq E_g$), where E_g represents energy band gap.

Several photocatalysts are lately modified for the utilization in multi-applications that can be induced by light irradiation. Photocatalysts/semiconductors that are of a wide employment in different applications are generally oxide forms of transition metal. These metals oxides are commonly characterized by their unique features. Semiconductors always possess a void energy zone in which no levels of energy can be provided. Therefore, recombination between the generated electrons and holes via photoactivation in a semiconductor can be effectively prohibited (Ghosh and Rao 2004; Jing and Guo 2006; Selvam et al. 2011). Table 12.2 gives the band gaps values for some of the extensively utilized photocatalysts in various applications.

Cadmium oxide is considered one of astonishing compounds as a photocatalyst and some other application, for instance, its use as one of solar cells component (Salehi et al. 2014), phototransistors (Kondo et al. 1971), photodiodes (Benko and Koffyberg 1986), transparent electrodes (Chang et al. 2007), and gas sensors (Ghosh and Rao 2004). Another example for the cadmium compounds that are well known as efficient photocatalysts is cadmium sulfide. To best of knowledge, CdS has been counted as one of the most photoactive materials and is widely used in photocatalytic processes (El Naggar et al. 2013). On the other hand, although CdO may be taken into account as valuable photocatalyst, it has not been yet of extensive use. Cadmium oxide posses an exceptional feature based on owning a direct E_{bg} equal to 2.3 eV and indirect E_{bg} of 1.36 eV (Dou et al. 1998). CdO also holds a special catalytic aspect which makes it capable of photodecomposing various organic structures such as dyes and part of environmental contaminations (Nezamzadeh-Ejhieh and Banan 2011; Karunakaran and Dhanalakshmi 2009; Karunakaran et al. 2010).

Another example of efficient photocatalytic oxides (that is of extensive use) is zinc oxide. Zn contains two electrons in its outer most valence shell. Due to its ability to form protective oxides, Zn is frequently utilized in rust preventive coatings. Zn is also commonly combined with other metals to form alloys such as brass when paired with copper (Taylor 1964; Bertini 2007).

ZnO is the most stable form; this inorganic compound can be found naturally in form of a yellowish mineral, known as zincite. It is considered a wide E_{bg} semiconductor at 3.37 eV and has a high excitation binding energy of 60 eV (Fan and Lu 2005; Ashfold et al. 2007; Kim et al. 2010). This semiconductor has been widely used for manufacturing white paints in pigment industry. ZnO can also

Table 12.2 Example for excitation wavelengths of semiconductors and their matched energy of band gaps

Examples for semiconductors	Band gap energy (eV)	Wavelength (nm)
Titanium dioxide (rutile)	3	413
Titanium dioxide (anatase)	3.2	388
Zinc oxide	3.2	388
Zinc sulfide	3.6	335
Cadmium sulfide	2.4	516
Cadmium oxide	2.1	550

promote the catalytic fabrication of rubber stuff. With a moderately high refractive index ($n = 2.0$), it is also added to polymers serving as a protectant from UV radiation (Zhang et al. 2004; Jacobson 2005; Hewitt and Jackson 2008).

Another example for one of the most common semiconductors is titanium oxide. Ti metal has been considered the ninth most plentiful constituent in earth crust at 6600 ppm by weight. Same as Zn, Ti forms a thin protective oxide layer when it is exposed to air in most environmental conditions (Taylor 1964; Bertini 2007). The most stable and most predominant form occurs as TiO_2, which is an inorganic chemical compound found naturally occurring as a white crystalline powder within magmatic rock and hydrothermal veins (Fu et al. 2009).

There are three main structural forms of TiO_2. Rutile and anatase have been the most widely spread phases having tetragonal crystal arrangements. Brookite, remaining less common, but still representative, exists in an orthorhombic crystal form. Ranging from 3.00 to 3.20 eV (depending on the wave factor), TiO_2 is considered a moderately wide E_{bg} semiconductor with a reported binding energy between 45 and 55 eV (Hashimoto et al. 2005; Fujishima et al. 2008).

Due to its high refractive index ($n = 2.7$), TiO_2 is widely used as a whitening pigment and also provides the physical blocker in most sunscreens (Jacobson 2005). TiO_2, while highly effective at generating free radicals, is limited to activation energies of relatively high intensity (only found within the UV spectrum). Within the past few decades, significant progressions in metal oxide photocatalyst development have been made (Miyauchi et al. 2002; Li et al. 2008; Brezesinski et al. 2010). However, titanium oxide maintains the vanguard of research interest as it continues to demonstrate the greatest photocatalytic promise. By hybridizing these two well-established metal oxides in the formation of heterogeneous nanostructures, it is anticipated that photocatalytic advancements can be realized (Moradi et al. 2015).

12.1.7 Importance of Photocatalytic Efficiency of Metal Oxides by Nanostructure Growth Techniques

The desire to fit more technology into smaller packages continues to push the scientific community to develop new and improved material designs. This trend is sustained through continued research in nanostructured materials. A material constrained to nanoscale dimensions offers unique properties that often behave starkly different from the same material in bulk form. As the feature's size shrinks, the effects of quantum mechanics begin to emerge. Vast increases in surface area show considerable potential for technological advancements, especially in energy applications. Understanding the behavior of metal oxide semiconductors at the atomic level continues to hold great interest within the research community. Several methodologies were improved through the past time to produce new as well as innovative MOS nanostructures including nanotubes, nanowires, nanorods, nanoribbons, nanopropellers, and nanocoils (Fan and Lu 2005; Song et al. 2006; Tong et al. 2006). The most popular methods for producing these nanostructures include vapor–liquid–solid phase technique (VLS), pulsed laser deposition, hydrothermal growth (HG), and sputtering methods (Ashfold et al. 2007; Tak et al. 2009; Akhtar et al. 2010; Ottone et al. 2014; Smith et al. 2015). The precipitation synthesis can be considered one of the most potential techniques for creating uniform self-assembling growth patterns because it is inexpensive and effective. This method is promising for producing of a new photocatalytically predominant metal oxide nanostructure.

12.2 Hydrogen Production

Hydrogen as scalar quantity has been paid great attention lately. This attributed to its enormous content of energy as well as of being a clean source of energy generation. Several techniques have been currently utilized for hydrogen production such as pyrolysis and gasification of biomass wastes, catalytic decomposition of hydrocarbons, steam and dry reforming of ethanol and methane, respectively, and photocatalytic water splitting. However, taking into account the simplicity of production procedures and low energy consumption, the photocatalytic generation of hydrogen via cleavage of water molecules is favorable over the other methodologies. The hydrogen production by photocatalysis is often done through either the photocatalytic or photoelectrochemical crack of water.

12.2.1 Photocatalytic Water Splitting

Cleavage of water via photocatalytic route has been recently a perfect trend to obtain hydrogen especially through the utilization of two abundant renewable sources, specifically water and sunlight (Ni et al. 2007; Zheng et al. 2009). Usage of a proper photocatalyst in such process has to be an outstanding route due to the provided benefits which are as follows: Catalysts are introduced as solid phase, low costs of fabrication, no operational risks and against deactivation (Naik et al. 2011; Pradhan et al. 2011).

Improvement of a system for photo-induced breakup of water molecules that has the ability to use the zone of visible light in spectrum of sunlight can be carried out via various routes. One of these efficient routes has been dye sensitization. In this method, a dye sensitizer is inserted into a photocatalytic reactor. The sensitization is then excited due to the effect of visible irradiation. The molecules of dye sensitizer at excited state bring in electrons to the nearby C_B of a photocatalyst. These electrons transport afterward along the whole of photocatalyst crystalline structure.

These electrons will next reach to the catalyst surface and consequently attain the step of reducing water molecules to release hydrogen (Dhanalakshmi et al. 2001; Sreethawong and Yoshikawa 2012).

12.2.1.1 Semiconductors

Well-organized transfer of electrons in order to be received by molecules of a certain substrate is of a high relevance to the physicochemical features of semiconductors, such as specific surface area as well as crystalline and porous built-up. Generation of hydrogen via water cleavage through the usage of semiconductors has grabbed remarkable attention since the earlier work that had been done by Fujishima and Honda (1972). These scientists are the two who firstly explored that water molecule can be cracked down into hydrogen and oxygen via photoelectrochemical route. To achieve such process, a semiconductor electrode, namely TiO_2, had been utilized under the effect of UV rays. A great number of metals oxides and sulfides such as titanium dioxide (Karakitsou and Verykios 1995), WO_3 (Abe et al. 2005), strontium titanate (Zou and Liu 2006), zinc oxide (Hoffman et al. 1992), cadmium and zinc sulfide (Maeda et al. 2005; Guan et al. 2004), niobates (Izumi et al. 1987), and tantalates (Furube et al. 2002) had been afterward presented as photocatalysts for obtaining hydrogen by splitting of water. In the last decade, scientists are paying attention to modify photocatalysts to enhance their capability toward the usage of irradiations that possess wavelengths ranged between 1400 and 700 nm. The modification stage includes doping of these catalysts by some of transition metal as platinum (Ikeda et al. 2006), chromium (Kim et al. 2005), and vanadium (Wang et al. 2006). The doping can also be carried out using non-metallic component, for example, nitrogen (Gu et al. 2007), sulfur (Yin et al. 2007; Murakami et al. 2007) as well as various carbon species (Yu et al. 2005; Liu et al. 2007). The modified photocatalysts could be by then efficient candidates for the generation of H_2 via photocatalysis.

For an identical photoprocess to produce hydrogen, the potentiality of recombining the segregated charges during the photoreaction is of a strong dependence to the additives of water solution. Loading of metals onto the structure of semiconductor is another factor that can control the integration between separated charges. As soon as step of charge separation is completed, electrons in C_B can be confined via existing metals particles at the surface of semiconductors. Electron arrest at that stage is occurring as a result of the dissimilarity between Fermi energies and work function. In a process of hydrogen evolution from water, a solution contains both water and methanol is utilized. In such solution, the appearance of protons is taking place due to oxidation of either water or methanol via the radiation-produced holes. Protons are then obeying a reduction step, by electrons, at the metallic surface of catalyst in order to generate molecules of hydrogen. Equations (12.6)–(12.9) show the occurring interactions through the photocatalytic process of hydrogen formation (Choi and Kang 2007):

$$h\upsilon \rightarrow e^- + h^+ \quad (12.6)$$

$$4h^+ + 2H_2O \rightarrow O_2 + 4H^+ \quad (12.7)$$

$$2H^+ + 2e^- \rightarrow H_2 \quad (12.8)$$

The overall reaction is

$$4h\upsilon + 2H_2O \rightarrow O_2 + 2H_2 \quad (12.9)$$

Methanol is added into the mixture to work as an oxidation mediator to prohibit the reveal of oxygen gas that may undertake due to adsorption of water molecules at the surface of photocatalyst. In other words, methanol has been employed to efficiently segregate the hole charges that are the reason behind the process of recombining hole–electron pair. Furthermore, methanol performs another role specifically as a hole scavenger. At this stage, methanol could take part in producing hydrogen, as presented in Eqs. (12.10)–(12.12). This subsequently develops the amount of total hydrogen gas productivity. In addition to the generated hydrogen, the occurring side reactions cause the production of carbon dioxide (Kawai and Sakata 1980; Chen et al. 1999).

$$MeOH \leftrightarrow H_2CO + H_2 \quad (12.10)$$

$$H_2CO + H_2O \leftrightarrow H_2CO_2 \quad (12.11)$$

$$H_2CO_2 \leftrightarrow CO_2 + H_2 \quad (12.12)$$

12.2.1.2 Efficient Water-Splitter Catalysts

Cadmium sulfide is an n-type semiconductor that has an energy band gap of 2.4 eV. It is one of the photocatalysts that exhibited a reasonable photocatalytic action toward H_2 generation in the presence of visible light irradiation. However, the use of C_2H_5OH, HS, or SO_3 (Sathish et al. 2006), as sacrificial electron donors, is highly advised to acquire significant amounts of H_2. It is also favorable to overcome the photo-induced corrosion of cadmium sulfide in attendance of O_2. From another perspective, electronic states as well as photoactivity of a semiconductor generally and cadmium sulfide in particular can be adjusted through varying or managing its crystals size with no influence on chemical structure. Hoffman and co-authors had stated an increment, by tenfolds, in efficiency of ZnO nanoparticles toward photocatalytic making of H_2O_2 by decreasing the crystal size of semiconductor from 40 to 23 nm. Another study by Hoffman et al. (1994) had reported the increase in quantum competence in photopolymerization of methyl methacrylate with a parallel decline of crystal size via utilization of size-quantized particles of cadmium sulfide.

Combining two or more semiconductors that have various energy levels can be advantageous to attain successful charge disconnection.

Designation of nanocomposite structures had been of much interest owing to their enhanced response in both UV and visible light zones versus photo-based processes (Morales-Torres et al. 2012, 2013). TiO_2 is well thought-out as the photocatalyst of highest activity. Nevertheless, the usage of TiO_2 in photocatalytic cracking of water is dependent on its redox potential with consideration for the normal hydrogen electrode (NHE) (Pérez-Larios and Gómez 2013). Several investigations were previously carried out in order to elevate the photocatalytic performance of TiO_2 toward the reaction of splitting water molecules. An improved structure of TiO_2 had been obtained through doping with iron, zinc, copper, vanadium, magnesium, and nickel (Tseng and Jeffrey 2004). Another way of modifying titanium dioxide was done by impregnating its particles with precious metals such as platinum, palladium, and gold, as reported in Sreethawong et al. (2005).

The presentation of mixed oxide-based photocatalysts has also grabbed the attention of many scientists. These mixed structures may simultaneously contain two or more metals oxides, for example, the combination between at least two of the following oxides: copper oxide, zinc oxide, nickel oxide, and cerium oxide (Erdóelyi et al. 2006; Yoong et al. 2009; César et al. 2008; Banerjee 2011; Galindo-Hernández and Gómez 2011). Mixed oxide-based photocatalysts had displayed outstanding photocatalytic efficiency and are counted as materials of low fabrication expenses. The influence of combining two oxides or more has been of strong relevance to existing oxygen vacancies in their crystal structure (Nakamura et al. 2000; Ihara et al. 2003). El Naggar and co-authors had demonstrated in a prior study (El Naggar et al. 2013) the photocatalytic cleavage of water to evolute hydrogen from water–methanol mixture in the presence of zinc oxide (E_{bg} of 3.37 eV) and cadmium sulfide (E_{bg} of 2.42 eV) either separately or combined together. It had been explicitly mentioned that CdS and ZnO have distinctive performance toward photocatalytic conversion of water into hydrogen, in the occurrence of either UV or visible spectrums.

12.2.2 Photoelectrochemical Hydrogen Production

In photoelectrochemical process, hydrogen is produced using sunlight through water splitting at a semiconducting material by direct conversion of solar to chemical energy, particularly the hydrogen gas.

The most challenging task in generating hydrogen via photocatalytic cleavage of water molecules is to maximize amount of produced H_2 via a cost-effective process. Improvements are required in cost, efficiency, and immovability for market viability as:

- Efficiency is being improved through better surface catalysis to enhance sunlight absorption.
- Immovability is being amended with rugged and protective surface coatings.
- Cost-effective is being lowered through reduced materials and processing costs.

The arrangement of a semiconductor electrode may have a considerable impact on the performance of photoelectrochemical cracking of water. Different semiconductor compositions, for instance, TiO_2, α-Fe_2O_3, and WO_3 were investigated in hydrogen evolution by photoelectrochemical process (Cha et al. 2011; Kim et al. 2012). Heterojunction electrodes (has at least two or various photocatalysts) suggest better benefits when compared with single semiconductor electrode (Hoshino et al. 2006). CdS/ZnSe nanocore/shell photoelectrodes were prepared recently by Hyun et al. CdS/ZnSe exhibited increased photocatalytic features in comparison to thin film of CdS photoelectrodes. The band structure had stimulated the charge separation of carriers in the semiconductor electrodes and in turn could speed up the water-splitting process at lower and at higher bias voltages compared to CdS thin-film electrodes (Ki and Yun 2017). The photocatalytic hydrogen (H_2) production activities of various CdSe semiconductor nanoparticles were investigated and compared including CdSe and CdSe/CdS quantum dots (QDs), CdSe quantum rods (QRs), and CdSe/CdS dot-in-rods (DIRs). The rate of H_2 production orders as CdSe QDs $\gg$ CdSe QRs > CdSe/CdS QDs > CdSe/CdS DIRs. Calculations of photoexcited surface charge densities are positively correlated with the H_2 production rate and found that the efficiency of H_2 production effected mainly by the size and morphology of the nanoparticle (Fen et al. 2016). In another report, the hydrogen production rate increased more than 4 times after the deposition of Ru on the surface of ZnS-CdS. In addition, the as-prepared ZnS-CdS nanocomposite demonstrated an excellent stability over 50 h (Jingyi 2017).

The use of an embedded structure, where one ingredient is surrounded by another, creates a larger contact area for electrons to be transferred more freely, thereby better realizing a synergistic effect in the heterojunction system. The Si-doped and Ti-doped α-Fe_2O_3 exhibited much higher photoelectrochemical activity than the undoped material. The proposed mechanism for the enhanced photocurrent is reduction of recombination due to an improvement of the charge-transfer rate coefficient at the surface and also possibly passivation of the grain boundaries by the dopants (Glasscock et al. 2007). Carbon-doped In_2CO_3 films were fabricated with spray pyrolysis and estimated for photoelectrocatalysis. Photocurrent densities go up to 1 mA/cm^2 for the C-doped films, with a high contribution from visible light irradiation. An enhanced photocurrent density has been observed for C-doped In_2O_3 than the undoped films, with the same irradiation conditions (Yanping et al. 2008). Recently, Cu-doped ZnS/zeolite composites have been synthesized and subsequently evaluated for their photocatalytic activities based on hydrogen production from an aqueous S^{2-}/SO_3^{2-} solution under UV–visible light irradiation. It was found that the hydrogen production rate over CuZnS/zeolite composites was significantly higher than over the bulk ZnS (Toru and Morio 2017). To date, a variety of nanostructures with large structural voids and surface areas have been fabricated, including nanoparticles, nanorods, nanotubes, and nanowires. The efficiency of silicon for hydrogen production can be improved by developing the antireflective property of nanosilicon substrate. Silicon nanowires coated with an iron sulfur carbonyl catalyst produced much greater photocurrent densities (−17 mA/cm^2) when compared to bare silicon nanowires (−5 mA/cm^2). In addition, the amount of hydrogen gas was produced (315 μmol/h) at low bias potentials for the silicon nanowires coated with an iron sulfur carbonyl catalyst (Soundarrajan et al. 2016). The SiC-TiO_2-Sm_2O_3 was fabricated by solgel process and thermal process at 600 °C with different contents of samarium. It was reported that hydrogen was produced on SiC-TiO_2-Sm_2O_3 from a sulfuric acid solution by applying a bias potential with UV light. The results showed a maximum value of percentage of incident photon conversion efficiency (% IPCE = 4.7) for the sample with 2.0 wt% of Sm^{3+} ions (Isaías et al. 2015). WO_3/$BiVO_4$ heterojunction is considered one of the best pairs for hydrogen production, but its photocurrent density is scanty. The advantage of using helical nanostructures in photoelectrochemical solar water splitting was investigated. Bismuth vanadate-decorated tungsten trioxide helical nanostructures lead to the highest photocurrent density at 1.23 V versus the reversible hydrogen electrode as a result of the combination of effective light scattering, charge separation and transportation, and an enlarged contact surface area with electrolytes (Xinjian et al. 2014). The fabrication of CdSe quantum dot-sensitized photocathodes on NiO-coated indium tin oxide (ITO) electrodes for hydrogen production upon light irradiation was reported. The rainbow photocathodes with forward energetic gradient for charge separation and subsequent electron transfer to a solution-based hydrogen-evolving catalyst show good light harvesting ability and improved photoresponses than the reverse rainbow photocathodes under white LED light illumination. Under minimally optimized conditions, a photocurrent density of as high as 115 μA cm^{-2} and a Faradaic efficiency of 99.5% are achieved; this is among the most effective QD-based photocathode water-splitting systems (Hongjin et al. 2017). A novel configuration of the photoelectrochemical hydrogen production device is based on TiO_2 beads as the primary photoanode material with the addition of a heterostructure of silver nanoparticles/grapheme. The heterostructure not only caters to a great improvement in light harvesting efficiency (LHE) but also enhances the charge collection efficiency. The PCE of the TiO_2 beads/Ag/grapheme cell is improved 2.5 times than for pure P_{25} cell (Chun-Ren et al. 2016). Pt and grapheme (GN) were used to modify TiO_2 nanoparticles. GN/TiO_2 (TG), Pt-TiO_2 (PT), Pt-GN/TiO_2 (PTG) and have been investigated in hydrogen production. The maximum hydrogen production rate was approximately 4.71 mmol h^{-1} g^{-1} when the Pt content was 1.0 wt%. The situation of grapheme is similar to Pt. The highest hydrogen production rate is 6.58 mmol h^{-1} g^{-1} by 1.5 wt% Pt-5 wt% GN/TiO_2 (1.5PTG5), which is about 1.4 and 2.2 times higher than that of Pt-TiO_2 and GN/TiO_2 binary composites, respectively (Nguyen et al. 2018).

Using of noble metals/magnetic nanoparticle (core/shell structure) as photomaterials is promising in water-splitting process. The rate of hydrogen production can be enhanced with magnetic heating of noble-metal-free catalysts. Using nickel/iron carbide core–shell, which are exposed to magnetic heating with high-frequency magnetic field, an over potential (at 20 mA cm^{-2}) declined by 200 mV is needed for oxygen generation in an alkaline water-electrolysis flow-cell, while for hydrogen production it had been decreased by 100 mV. Enhancement of oxygen-release kinetics has corresponded to an increase of cell temperature to $\sim$200 °C; however practically, it had elevated by 5 °C only (Christiane et al. 2018).

12.3 Conclusion

The presented chapter reported at the beginning the definition, importance, and the types of photocatalysis. The electronic and optical characteristics of photocatalytic semiconductors had been then overviewed in details. Energy band gaps of the different semiconductors and their relevant type of irradiation spectrums had been widely discussed. The suitability of the various glass types and their matching to the radiation sources has been also presented through this chapter. The various metals oxide-based semiconductors had been reported as the most efficient and utilized photocatalysts in the versatile applications. Among these metals oxides, the titanium, zinc, and cadmium oxides have been listed as the photocatalysts of the highest activity in general and energy production in particular. One of the distinctive usages for the photocatalysis especially for the subject of energy generation, namely photocatalytic water conversion into hydrogen, was described in this chapter. Importance of water-splitting process in order to generate hydrogen is also illustrated taking into account that hydrogen has been counted as one of remarkable new energy vectors. Two routes of water splitting, photocatalytic and photoelectrochemical, and their dependent photocatalysts are also presented. The efficiency of various semiconductors through the two routes of splitting process is displayed through this chapter.

References

Abe R, Takata T, Sugihara H, Domen K (2005) Photocatalytic overall water splitting under visible light by TaON and WO_3 with an IO3_/I_ shuttle redox mediator. Chem Commun 30: 3829–3831

Akhtar MS, Hyung JH, Kim DJ, Kim TH, Lee SK, Yang O (2010) A comparative photovoltaic study of perforated ZnO nanotube/TiO_2 thin film and ZnO nanowire/TiO_2 thin film electrode-based dye-sensitized solar cells. J Korean Phys Soc 56(3):813–817

Ashfold MNR, Doherty RP, Ndifor-Angwafor NG, Riley DJ, Sun Y (2007) The kinetics of the hydrothermal growth of ZnO nanostructures. Thin Solid Films 515:8679–8683

Banerjee AN (2011) The design, fabrication, and photocatalytic utility of nanostructured semiconductors: focus on TiO_2-based nanostructures. Nanotechnol Sci Appl 4:36–65

Benko FA, Koffyberg FP (1986) Quantum efficiency and optical transitions of CdO photoanodes. Solid State Commun 57:901–903

Bertini I (2007) Biological inorganic chemistry: structure and reactivity, chap. 2. University Science Books, pp 7–30

Brezesinski K, Ostermann R, Hartmann P, Perlich J, Brezesinski T (2010) Exceptional photocatalytic activity of ordered mesoporous TiO_2-Bi_2O_3 thin films and electrospun nanofiber mats. Chem Mater 22:3079–3085

Carp O, Huisman CL, Reller A (2004) Photoinduced reactivity of titanium dioxide. Prog Solid State Chem 32:33–177

César DV, Robertson RF, Resende NS (2008) Characterization of ZnO and TiO_2 catalysts to hydrogen production using thermoprogrammed desorption of methanol. Catal Today 133:136–141

Cha HG et al (2011) Facile preparation of Fe_2O_3 thin film with photoelectrochemical properties. Chem Commun 47:2441–2443

Chang YC, James RB (1989) Saturation of intersubband transitions in p-type semiconductor quantum wells. Phys Rev B 39:12672

Chang J, Mane RS, Ham D, Lee W, Cho BW, Lee JK, Han S-H (2007) Electrochemical capacitive properties of cadmium oxide films. Electrochim Acta 53:695–699

Chen J, Ollis DF, Rulken WH, Bruning H (1999) Photocatalyzed oxidation of alcohols and organochlorides in the presence of native TiO_2 and metallized TiO_2 suspensions. Part (II): photocatalytic mechanisms. Water Res 33:669–676

Choi H-J, Kang M (2007) Hydrogen production from methanol–water decomposition in a liquid photosystem using the anatase structure of Cu loaded TiO_2. Int J Hydrogen Energy 32:3841–3848

Christiane N, Stéphane F, Alexis B, Jonathan D, Marian C, Julian C, Bruno C, Alain R (2018) Improved water electrolysis using magnetic heating of FeC–Ni core–shell nanoparticles. Nat Energy 3:476–483

Chun-Ren K, Jyun-Sheng G, Yen-Hsun S, Jyh-Ming T (2016) The effect of silver nanoparticles/graphene-coupled TiO_2 beads photocatalyst on the photoconversion efficiency of photoelectrochemical hydrogen production. Nanotechnology 27(43)

Ciesla P, Kocot P, Mytych P, Stasicka Z (2004) Homogeneous photocatalysis by transition metal complexes in the environment. J Mol Catal A Chem 224:17–33

Cox PA (1996) The surface science of metal oxides, chap. 5. Cambridge University Press, Cambridge, pp 158–246

Dhanalakshmi KB, Latha S, Anandan S, Maruthamuthu P (2001) Dye sensitized hydrogen evolution from water. Int J Hydrogen Energy 26:669–674

Dou Y, Egdell RG, Walker T, Law DSL, Beamson G (1998) N-type doping in CdO ceramics: a study by EELS and photoemission spectroscopy. Surf Sci 398:241–258

El Naggar AMA, Nassar IM, Gobara HM (2013) Enhanced hydrogen production from water via a photo-catalyzed reaction using chalcogenide d-element nanoparticles induced by UV light. Nanoscale 5:9994–9999

Erdóelyi A, Raskó J, Kecskés T, Tóth M, Dömök M, Baán K (2006) Hydrogen formation in ethanol reforming on supported noble metal catalysts. Catal Today 116:367–376

Fan Z, Lu JG (2005) Zinc oxide nanostructures: synthesis and properties. J Nanosci Nanotechnol 5:1561–1573

Fen Q, Zhiji H, Jeffrey JP, Michael YO, Kelly LS, Todd DK (2016) Photocatalytic hydrogen generation by CdSe/CdS nanoparticles. Nano Lett 16(9):5347–5352

Fu N, Wu Y, Jin Z, Lu G (2009) Structural-dependent photoactivities of TiO_2 nanoribbon for visible-light-induced H_2 evolution: the roles of nanocavities and alternate structures. Langmuir 26:447–455

Fujishima A, Honda K (1972) Electrochemical photolysis of water at a semiconductor electrode. Nature 238:37

Fujishima A, Zhang X, Tryk DA (2008) TiO_2 photocatalysis and related surface phenomena. Surf Sci Rep 63:515–582

Furube A, Shiozawa T, Ishikawa A, Wada A, Domen K, Hirose C (2002) Femtosecond transient absorption spectroscopy on photocatalysts: K4Nb6O17 and Ru(bpy)3 2+ intercalated K4Nb6O17 thin films. J Phys Chem B 106:3065–3072

Galindo-Hernández F, Gómez R (2011) Degradation of the herbicide 2, 4-dichlorophenoxyacetic acid over TiO_2–CeO_2 sol–gel photocatalysts: effect of the annealing temperature on the photoactivity. J Photochem Photobiol A Chem 217:383–388

Ghosh M, Rao CNR (2004) Solvothermal synthesis of CdO and CuO nanocrystals. Chem Phys Lett 393:493–497

Glasscock JA, Barnes PRF, Plumb IC, Savvides N (2007) Enhancement of photoelectrochemical hydrogen production from hematite thin films by the introduction of Ti and Si. J Phys Chem C 111(44):16477–16488

Gu DE, Yang BC, Hu YD (2007) A novel method for preparing V-doped titanium dioxide thin film photocatalysts with high photocatalytic activity under visible light irradiation. Catal Lett 118:254–259

Guan GQ, Kida T, Kusakabe K, Kimura K, Fang XM, Ma TL et al (2004) Photocatalytic H2 evolution under visible light irradiation on CdS/ETS-4 composite. Chem Phys Lett 385: 319–322

Hashimoto K, Irie H, Fujishima A (2005) TiO_2 photocatalysis: a historical overview and future prospects. Jpn J Appl Phys 44:8269–8285

Hewitt CN, Jackson AV (2008) Handbook of atmospheric science: principles and applications, chap. 13. Wiley, pp 339–371

Hoffman AJ, Yee H, Mills G (1992) Photoinitiated polymerization of methyl methacrylate using Q-sized zinc oxide colloids. J Phys Chem 96:5540–5546

Hoffman AJ, Carraway ER, Hoffman MR (1994) Photocatalytic production of H_2O_2 and organic peroxides on quantum-sized semiconductor colloids. Environ Sci Technol 28(5):776–785

Hongjin L, Congcong W, Guocan L, Rebeckah B, Todd DK, Yongli G, Richard E (2017) Semiconductor quantum dot-sensitized rainbow photocathode for effective photoelectrochemical hydrogen generation. PNAS 114(43):11297–11302

Hoshino K, Hirasawa Y, Kim S-K, Saji T, Katano J-I (2006) Bulk heterojunction photoelectrochemical cells consisting of oxotitanyl phthalocyanine nanoporous films and I_3^-/I^- redox couple. J Phys Chem B 110:23321–23328

Ihara T, Miyoshi M, Iriyama Y, Matsumoto O, Sugihara S (2003) Visible-light-active titanium oxide photocatalyst realized by an oxygen-deficient structure and by nitrogen doping. Appl Catal B Environ 42:403–409

Ikeda S, Fubuki M, Takahara YK, Matsumura M (2006) Photocatalytic activity of hydrothermally synthesized tantalite pyrochlores for overall water splitting. Appl Catal A 300:186–190

Isaías J-R, Leticia MT-M, Christian G-S, Juan CB (2015) Photoelectrochemical hydrogen production using SiC-TiO_2-Sm_2O_3 as electrode. J Electrochem Soc 162(4):H287–H293

Izumi N, Junji A, Masayuki I, Ritsuro M, Yoshinori S, Kichiro K (1987) Characterization of the amorphous state in metamict silicates and niobates by EXAFS and XANES analyses. Phys Chem Miner 15:113–124

Jacobson MZ (2005) Fundamentals of atmospheric modeling, chap. 21. Cambridge University Press, Cambridge, pp 681–708

Jing D, Guo L (2006) A novel method for the preparation of a highly stable and active CdS photocatalyst with a special surface nanostructure. J Phys Chem B 110:11139–11145

Jingyi X (2017) Preparation of ZnS-CdS nanocomposite for photoelectrochemical hydrogen production. Int J Electrochem Sci 12:2253–2261

Karakitsou KE, Verykios XE (1995) Definition of the intrinsic rate of photocatalytic cleavage of water over Pt-RuO_2/TiO_2 Catalysts. J Catal 152:360–367

Karunakaran C, Dhanalakshmi R (2009) Selectivity in photocatalysis by particulate semiconductors. Open Chem 7:134–137

Karunakaran C, Dhanalakshmi R, Gomathisankar P, Manikandan G (2010) Enhanced phenol-photodegradation by particulate semiconductor mixtures: interparticle electron-jump. J Hazard Mater 176:799–806

Kawai T, Sakata T (1980) Photocatalytic hydrogen production from liquid methanol and water. J Chem Soc Chem Commun 15:694

Ki HC, Yun MS (2017) Photoelectrochemical performance of CdS/ZnSe core/shell nanorods grown on FTO substrates for hydrogen generation. J Electrochem Soc 164:H382–H388
Kim S, Hwang SJ, Choi WY (2005) Visible light active platinum-ion-doped TiO_2 photocatalyst. J Phys Chem B 109:24260–24267
Kim SJ, Kim HH, Kwon JB, Lee JG, Beom-Hoan O, Lee SG, Lee EH, Park SG (2010) Novel fabrication of various size ZnO nanorods using hydrothermal method. Microelectron Eng 87:1534–1536
Kim JK, Moon JH, Lee T-W, Park JH (2012) Inverse opal tungsten trioxide films with mesoporous skeletons: synthesis and photoelectrochemical responses. Chem Commun 48:11939–11941
Kondo R, Okimura H, Sakai Y (1971) Electrical properties of semiconductor photodiodes with semitransparent films. Jpn J Appl Phys 10:1493–1659
Langot P, Tommasi R, Vallae F (1996) Non-equilibrium hole relaxation dynamics in an intrinsic semiconductor. Phys Rev B 54:1775
Li SX, Yu KM, Wu J, Jones RE, Walukiewicz W, Ager JW, Shan W, Haller EE, Lu H, Schaff WJ (2005) Fermi-level stabilization energy in group III nitrides. Phys Rev B 71:161201–161225
Li HX, Xia RH, Jiang ZW, Chen SS, Chen DZ (2008) Optical absorption property and photo-catalytic activity of tin dioxide-doped titanium dioxides. Chin J Chem 26:1787–1792
Liu H, Imanishi A, Nakato Y (2007) Mechanisms for photooxidation reactions of water and organic compounds on carbon-doped titanium dioxide, as studied by photocurrent measurements. J Phys Chem C 111:8603–8610
Maeda K, Takata T, Hara M, Saito N, Inoue Y, Kobayashi H, Domen K (2005) ZnO solid solution as a photocatalyst for visible-light-driven overall water splitting. J Am Chem Soc 127: 8286–8287
Miyauchi M, Nakajima A, Watanabe T, Hashimoto K (2002) Photocatalysis and photoinduced hydrophilicity of various metal oxide thin films. Chem Mater 14:2812–2816
Moliton A, Hiorns RC (2004) Review of electronic and optical properties of semiconducting of conjugated polymers: applications in optoelectronics. Polym Int 53:1397–1412
Moradi S, Vossoughi M, Feilizadeh M, Zakeri SME, Mohammadi MM, Rashtchian D, Booshehri AY (2015) Photocatalytic degradation of dibenzothiophene using La/ PEG-modified TiO_2 under visible light irradiation. Res Chem Intermed 41:4151–4167
Morales-Torres S, Pastrana-Martinez LM, Figueiredo JL, Faria JL, Silva AMT (2012) Desing of grapheme-based TiO_2 photocatalysts—a review. Environ Sci Pollut Res 19:3676–3687
Morales-Torres S, Pastrana-Martinez LM, Figueiredo JL, Faria JL, Silva AMT (2013) Graphene oxide-P25 photocatalysts for degradation of diphenhy-dramine pharmaceutical and methyl orange dye. Appl Surf Sci. https://doi.org/10.1016/j.apsusc.2012.11.157
Murakami Y, Kasahara B, Nosaka Y (2007) Photoelectrochemical properties of the sulfur-doped TiO_2 film electrodes: characterization of the doped states by means of the photocurrent measurements. Chem Lett 36:330
Naik B, Martha S, Parida KM (2011) Facile fabrication of Bi_2O_3/TiO_2-xNx nanocomposites for excellent visible light driven photocatalytic hydrogen evolution. Int J Hydrogen Energy 36:2794–2802
Nakamura I, Negishi N, Kutsuna S, Ihara T, Sugihara S, Takeuch K (2000) Role of oxygen vacancy in the plasma-treated TiO_2 photocatalyst with visible light activity for NO removal. J Mol Catal A Chem 161:205–212
Nezamzadeh-Ejhieh A, Banan Z (2011) A comparison between the efficiency of CdS nanoparticles/zeolite A and CdO/zeolite A as catalysts in photodecolorization of crystal violet. Desalination 279:146–151
Nguyen N-T, Zheng D-D, Chen S-S, Chang C-T (2018) Preparation and photocatalytic hydrogen production of Pt-graphene/TiO_2 composites from water splitting. J Nanosci Nanotechnol 18 (1):48–55
Ni M, Leung MK, Leung DY, Sumathy K (2007) A review and recent developments in photocatalytic water-splitting using TiO_2 for hydrogen production. Renew Sustain Energy Rev 11:401–425

Ottone C, Laurenti M, Motto P, Stassi S, Demarchi D, Cauda VA (2014) ZnO nanowires: synthesis approaches and electrical properties, nanowires, synthesis, electrical properties and uses in biological systems, chap. 1. Nova Publisher, New York, NY, USA, pp 1–57
Park JH, Domenico T, Dragel G, Clark R (1995) Development of electrical insulator coatings for fusion power applications. Fusion Eng Des 27:682–695
Perera AGU, Sherriff RE, Francombe MH, Devaty RP (1992) Far infrared photoelectric thresholds of extrinsic semiconductor photocathodes. Appl Phys Lett 60:3168–3170
Pérez-Larios A, Gómez R (2013) Hydrogen production using mixed oxides: TiO_2-M (CoO and WO_3). Adv Invest Ing 10:27–34
Perliski LM, Solomon S (1993) On the evaluation of air mass factors for atmospheric near ultraviolet and visible absorption spectroscopy. J Geophys Res Atmos 98:10363–10374
Pradhan AC, Martha S, Mahanta SK, Parida KM (2011) Mesoporous nanocomposite Fe/Al_2O_3-MCM-41: an efficient photocatalyst for hydrogen production under visible light. Int J Hydrogen Energy 36:12753–12760
Salehi B, Mehrabian S, Ahmadi M (2014) Investigation of antibacterial effect of cadmium oxide nanoparticles on *Staphylococcus aureus* bacteria. J NanoBioTechnol 12:12–26
Sathish M, Viswanathan B, Viswanath RP (2006) Alternate synthetic strategy for the preparation of CdS nanoparticles and its exploitation for water splitting. Int J Hydrogen Energy 31: 891–898
Selvam NCS, Kumar RT, Yogeenth K, Kennedy LJ, Sekaran G, Vijaya JJ (2011) Simple and rapid synthesis of cadmium oxide (CdO) nanospheres by a microwave-assisted combustion method. Powder Technol 211:250–255
Shim M, Guyot P (2000) N-type colloidal semiconductor nanocrystals. Nature 407:981–983
Smith NA, Evans JE, Jones DR, Lord AM, Wilks SP (2015) Growth of ZnO nanowire arrays directly onto Si via substrate topographical adjustments using both wet chemical and dry etching methods. Mater Sci Eng B 193:41–48
Sommerer TJ (1996) Model of a weakly ionized, low-pressure xenon dc positive column discharge plasma. J Phys D Appl Phys 29:769–780
Song T, Zhang Z, Chen J, Ring Z, Yang H, Zheng Y (2006) Effect of aromatics on deep hydrodesulfurization of dibenzothiophene and 4, 6-dimethyldibenzothiophene over NiMo/Al_2O_3 catalyst. Energy Fuels 20:2344–2349
Soundarrajan C, Thomas N, Nicolas HV (2016) Silicon nanowire photocathodes for photoelectrochemical hydrogen production. Nanomaterials 6(8):144. https://doi.org/10.3390/nano6080144
Sreethawong T, Yoshikawa S (2012) Impact of Pt loading methods over mesoporous-assembled TiO_2–ZrO_2 mixed oxide nanocrystal on photocatalytic dye-sensitized H_2 production activity. Mater Res Bull 47:1385–1395
Sreethawong S, Suzuki Y, Yoshikawa S (2005) Photocatalytic evolution of hydrogen over mesoporous TiO_2 supported NiO photocatalyst prepared by single step sol–gel process with surfactant template. Int J Hydrogen Energy 30:1053–1062
Tak Y, Hong SJ, Lee JS, Yong K (2009) Solution-based synthesis of a CdS nanoparticle/ZnO nanowire heterostructure array. Cryst Growth Des 9:2627–2632
Taylor SR (1964) Abundance of chemical elements in the continental crust: a new table. Geochim Cosmochim Acta 28:1273–1285
Tong Y, Liu Y, Dong L, Zhao D, Zhang J, Lu Y, Shen D, Fan X (2006) Growth of ZnO nanostructures with different morphologies by using hydrothermal technique. J Phys Chem B 110:20263–20267
Toru K, Morio N (2017) Cu-doped ZnS/zeolite composite photocatalysts for hydrogen production from aqueous S^{2-}/SO_3^{2-} solutions. Chem Lett 46(12):1797–1799
Tseng IH, Jeffrey CSW (2004) Chemical states of metal-loaded titania in the photoreduction of CO_2. Catal Today 97:113–119
Walukiewicz W (1988) Fermi level dependent native defect formation: consequences for metals of semiconductor and semiconductor of semiconductor interfaces. J Vac Sci Technol B 6: 1257–1262

Wang DF, Ye JH, Kako T, Kimura T (2006) Photophysical and photocatalytic properties of $SrTiO_3$ doped with Cr cations on different sites. J Phys Chem B 110:15824–15830

Waymouth JF (1971) Electrical discharge lamps, chap. 14. Cambridge University Press, Cambridge, pp 331–347

Wong MS, Fierro JG (2006) Metal oxides, chemistry and applications, chap. 17. Taylor & Francis, pp 543–568

Wu CH, Chang CL (2006) Decolorization of reactive red 2 by advanced oxidation processes: comparative studies of homogeneous and heterogeneous systems. J Hazard Mater 128:265–272

Xinjian S, Yong C, Kan Z, Jeong K, Dong YK, Ja KL, Sang HO, Jong KK, Jong HP (2014) Efficient photoelectrochemical hydrogen production from bismuth vanadate-decorated tungsten trioxide helix nanostructures. Nat Commun 5. Article number: 4775

Yanping S, Carl JM, Karla RR, Enrique AR, Justin PL, Daniel R (2008) Carbon-doped In_2O_3 films for photoelectrochemical hydrogen production. Int J Hydrogen Energy 33:5967–5974

Yin S, Komatsu M, Zhang QW, Saito F, Sato T (2007) Synthesis of visible-light responsive nitrogen/carbon doped titania photocatalyst by mechanochemical doping. J Mater Sci 42: 2399–2404

Yoong LS, Chong FK, Dutt BK (2009) Development of copper-doped TiO_2 photo-catalyst for hydrogen production under visible light. Energy 34:1652–1661

Yu JC, Ho WK, Yu JG, Yip H, Wong PK, Zhao JC (2005) Efficient visible-light-induced photocatalytic disinfection on sulfur-doped nanocrystalline titania. Environ Sci Technol 39:1175–1179

Zhang M, An T, Hu X, Wang C, Sheng G, Fu J (2004) Preparation and photocatalytic properties of a nanometer ZnO-SnO_2 coupled oxide. Appl Catal A 260:215–222

Zheng XJ, Wei LF, Zhang ZH, Jiang QJ, Wei YJ, Xie B et al (2009) Research on photocatalytic H2 production from acetic acid solution by Pt/TiO_2 nanoparticles under UV irradiation. Int J Hydrogen Energy 34:9033–9041

Zou JJ, Liu C (2006) Preparation of NiO/$SrTiO_3$ with cold plasma treatment for photocatalytic water splitting. Acta Phys Chim Sin 22:926–931

Chapter 13
Interface Mechanical Properties in Energetic Materials Using Nanoscale Impact Experiment and Nanomechanical Raman Spectroscopy

Chandra Prakash, Ayotomi Olokun, I. Emre Gunduz and Vikas Tomar

Abstract Energetic materials are sensitive to mechanical shock and defects caused by a high-velocity impact may result in unwanted detonation due to hot-spot formation. In order to understand the underlying mechanism, characterization of high strain-rate mechanical properties needs to be studied. One of the key factors that can contribute to this type of defect is the failure initiated at the interfaces such as those between Hydroxyl-terminated polybutadiene (HTPB) and Cyclo-tetra-methylene-tetra-nitramine (HMX) (or HTPB and Ammonium Perchlorate (AP)). In this work, interface mechanical properties of HTPB-HMX and HTPB-AP interfaces are characterized using nanoscale dynamic impact experiments at strain rates up to 100 s^{-1}. The binding agent was added to the mixture in order to analyze the effect of chemical composition on the interfacial mechanical properties. For HTPB-AP sample, Tepanol is used as the binding agent and for HTPB-HMX sample, Dantocol was used as the binding agent. The impact response is determined in the bulk HTPB, HMX, and AP as well as at the HTPB-HMX and HTPB-AP interfaces. A strain-rate-dependent viscoplastic power law-based constitutive model was obtained by fitting the experimental stress–strain–strain-rate data. The effect of binding agent on interface level failure properties was studied using an in situ nanomechanical Raman spectroscopy (NMRS) setup.

Keywords Energetic material · MRS · HTPB · HMX · AP

C. Prakash · A. Olokun · V. Tomar (✉)
School of Aeronautics and Astronautics, Purdue University, West Lafayette, IN 47907, USA
e-mail: tomar@purdue.edu

I. Emre Gunduz
School of Mechanical Engineering, Purdue University, West Lafayette, IN 47907, USA

S. Bhattacharya et al. (eds.), *Nano-Energetic Materials*, Energy, Environment and Sustainability, https://doi.org/10.1007/978-981-13-3269-2_13

13.1 Introduction

Energetic compounds are composite materials that have been commonly used in a various defense, aerospace, and civil applications, such as explosive, solid propellants, and pyrotechnic formulations. These materials are sensitive to thermal and mechanical shock, which may be caused by a sudden increase in temperature or by impact at any region within the material. Energetic materials are known to start the chemical reactions that cause detonation, at regions of sufficient size within the microstructure which attain a critical temperature, known as 'hot spots'. The composition of such a material mainly consists of a crystalline oxidizer (e.g., HMX or AP) bounded by a polymeric binder (e.g., HTPB). For example, a solid propellant is a mixture, containing about 60–95% oxidizer in the form of explosives (Benson and Conley 1999) and the remainder of the mixture is usually a polymeric binder (e.g., HTPB) or some combination of binder and metal such as Al etc. Energetic materials, due to their complicated microstructure and various chemical and physical processes occurring at multiple length scales, show very complex mechanical behavior under impact. A combination of several of these processes can occur as a result of accidental impact with bullets or fragments and can result in a sudden change in the temperature profile within the microstructure. This can cause the problem of untimely detonation or explosion in these high-energy explosives and propellants. Understanding the physical properties and the response of these materials under impact or shock is an important task in order to accurately predict the mechanical behavior of an energetic material. Special attention is needed to identify the failure mechanisms in energetic materials which can be broadly characterized as fracture within the particle, failure of various interfaces, and void nucleation and collapse in the particle and binder (Palmer et al. 1993). Both experimental and numerical techniques have been used, to study the failure of these materials and the effect of interface, in literature. Interface is weaker than the bulk material phases, and the particle–matrix debonding has been shown to be a dominant failure mode in polymer-bonded explosives (PBXs) under tensile loading (Palmer et al. 1993), which also indicates that interfaces are the sites where crack initiation occurs. The crack propagation occurs predominantly along the interface of particle and binder in solid propellant (Rae et al. 2002a, b). The particle–matrix interface toughness and the composition of the material, i.e., type of particle, binder, and other constituents, have been shown to increase the overall fracture resistance of such materials (Stacer et al. 1990; Stacer and Husband 1990). Several researchers, such as Rae et al. (2002) and others (Kimura and Oyumi 1998; Khasainov et al. 1997; Drodge et al. 2009) have experimentally shown the effect of particle size on the overall mechanical behavior of energetic materials. It has been shown by Drodge et al. (2009) that increasing the size of the particle in a PBX decreases the yield strain. Chen et al. (2011) showed that the interface failure in PBX-9501 is the dominant failure behavior by measuring the strain fields using a digital image correlation technique. The mechanical behavior of PBX composites has been shown by Yeager (2011) to be affected by the interface/interphase between the constituents and the corresponding microstructure. It was also shown that the interfacial structure in the composite can be altered by mixing the binder with a plasticizer which can inhibit the creation of a large interface/interphase within the microstructure. The addition of a plasticizer was shown to alter the interface structure, as well as a significant effect on the crack formation and propagation as well as the explosive sensitivity in such materials.

In order to understand the complex mechanical behavior, several numerical schemes, such as finite element method, mesh-free methods, and other such techniques, have been developed to simulate the impact-induced failure behavior of composites. In order to identify failure criterion in these methods, experimental observation of failure behavior is required. The cohesive zone model-based finite element method (CFEM) is one such technique used to quantitatively analyze the fracture behavior of a material through explicit simulation of fracture processes and has been used widely to model crack growth in different materials and analyze interfacial fracture. The advantage of using CFEM is the ability to work in the microscale. Cohesive zone models (CZMs) have been used by many researchers, (Barua and Zhou 2011; Barua et al. 2012; Tan 2012; Tan et al. 2005a), to simulate the interface debonding. For the particle–matrix debonding in energetic materials, Tan et al. (2005b) used a nonlinear three-stage cohesive law. In their work, Tan et al. used the principle of minimization of potential energy with respect to the jump in the displacement at the interface. It has also been shown that in energetic materials, the presence of smaller particles contributes to the hardening behavior while larger particles soften the composite. Debonding of large particles was shown to result in an unstable crack propagation which leads to catastrophic failure. The effect of strain rate on the delamination of the interface, between an elastic particle and a viscoelastic matrix of a PBX sample, hasbeen shown to affect the overall mechanical behavior of the composite (Tan et al. 2008). All of these studies are based on the assumption of a cohesive law for particle–matrix interfaces using macroscopic failure behavior of material. Interface separation, however, is a microscopic phenomenon which cannot be accurately represented by the macroscopic observations.

The present work focuses on effect of interface chemistry on the strain-rate-dependent constitutive model as well as cohesive fracture properties of energetic material. In this work, the effect of binding agent of the energetic material on the cohesive strength of the interfaces and the strain rate on the constitutive model and failure properties in a model HTPB-HMX (and HTPB-AP) composites are analyzed. A strain-rate-dependent viscoplastic constitutive model is developed based on nanoscale dynamic impact experiments performed on the analyzed particle–matrix interfaces. Cohesive fracture separation properties including interface strength and cohesive fracture energy in the analyzed samples were obtained using an *in situ* nanomechanical Raman spectroscopy (NMRS) setup (Gan and Tomar 2014). The measured local cohesive parameters accurately capture the failure behavior of these interfaces as well as the effect of interface chemistry variation on such parameters.

The remaining of the chapter is organized in the following manner. Section 2 describes the experimental details for nanoscale dynamic impact experiment and the viscoplastic constitutive model. Section 3 explains the mechanical Raman spectroscopy and analyzes the effects of interface chemistry on the interface delamination. Finally, in Sect. 4, a summary of the results obtained from the experiments and concluding remarks are presented.

13.2 Strain-Rate-Dependent Constitutive Model

In this section, experimental methods used to obtain a strain-rate-dependent viscoplastic constitutive model are explained in detail. First, the steps required in the fabrication of samples starting from the liquid HTPB and AP crystals are explained. In the next subsection, nanoscale dynamic impact experiment used is discussed. The scheme to obtain the model parameters from the experimental data and the effect of binding agent on these parameters, as outlined by Prakash et al. (2018a), is discussed next.

13.2.1 Sample Preparation

The samples used in the experiments described below are single-particle samples which consist of particles embedded in HTPB binder. In order to create samples for interface delamination experiments where failure remains symmetric around the interface, nearly spherical particles (HMX and AP obtained from Firefox Enterprise Inc.) were manually selected. For all sample types, in selecting the particles (from the available size range 600–1000 μm), care was taken so that visual inspections of the particle and the particle–matrix interface could be performed. HTPB was first fabricated by mixing of liquid polybutadiene (R-45M from Firefox Enterprise Inc.) and isophorone diisocyanate (IPDI) such that an OH index ratio of 1.05 is maintained. In order to change the interface chemical composition, a surface binding agent, Dantocol for HTPB-HMX sample and Tepanol for HTPB-AP sample, was added to the binder at a mass ratio of 0.5 to so that the surface adhesion of particle-binder increases, while keeping the same OH index ratio. The liquid polybutadiene and IPDI were manually mixed and put in vacuum for 30 min so that the trapped gasses could be avoided as best as possible. They were poured into 1-mm-deep and 50-mm-square PTFE molds and the particles were added. The mixture is again degassed in the vacuum for 30 min. The samples were cured for 7 days at a constant temperature of 60 °C in a convection oven. After the cure, the solidified material was peeled off from the molds and several single-particle samples were cut into desired dimensions for testing.

13.2.2 Nanoscale Dynamic Impact Experiment

It has been shown earlier by Wiegand (1961) that the mechanical properties vary with the size and type of oxidizer used in fabricating the solid propellants. Since the particle and the interface sizes of these materials are in the order of microns, experimental techniques having sufficient (of the order of nanometer) spatial resolution are needed to properly characterize their mechanical properties such as constitutive model and failure strength. In the current work, the mechanical property measurement was done using a dynamic impact setup to measure strain-rate-dependent viscoplastic properties. The dynamic impact experiments for bulk materials (HTPB, HMX, and AP) as well as interfaces within these materials were performed using a modified high strain-rate impact module of Micro Materials, UK, (Verma et al. 2017a, b; Prakash et al. 2017). Several different materials have been successfully characterized for site-specific high strain-rate compressive behavior, such as glass–epoxy interface (Verma et al. 2016), using the above-mentioned impact technique. The novelty lies in the fact that the precision range of the experiments is in the few nanometers and that the impacts are precisely conducted at the interfaces. The energy of the impacts is precisely transmitted to the point of impacts which insures that the deformation is confined to the region of interests, such as the interfacial region. A spherical impactor of radius 1 μm was chosen for all impact tests so that the interface impact measurements could be performed with precise position as desired. The experimental setup consists of a vertical pendulum, on which the impactor is mounted and a 3D stage that allows the sample to move in x, y, and z directions. The vertical pendulum hangs on a frictionless spring so that it can move freely during impact. As shown in Fig. 13.1a, an electromagnet is used to apply the force on the pendulum. A pair of capacitance plates are attached at the back of the pendulum to measure the depth of indents by measuring the change in the capacitance of the plates when the impactor moves. It allows for the nanometer resolution in the depth measurements, which is measured as the movement of the impactor. In order to provide additional force required for high-velocity movements of the impactor, a solenoid is attached at the lower part of the pendulum. Before the start of the experiment, setup is calibrated for both load and depth. The entire setup was kept inside a thermally stable chamber with a vibration isolation table so as to avoid the external mechanical and thermal noise to affect the measurements. As mentioned earlier, the force is applied on the pendulum through both the electromagnets and the solenoid. This means, before the start of the impact, pendulum is pulled at both ends first and then at the instant of impact, the solenoid is turned off which releases the pendulum and the impactor moves freely and hits the sample at a certain velocity. The load with which the dynamic impact occurs is selected beforehand in the experiment which determines the impact velocity. From the time when solenoid pulls the pendulum toward itself, the position of the impactor is continuously recorded as a function of time t, for the entire duration of the impact which includes the initial impact trajectory and subsequent rebound after the impactor hits the material surface. For a single-impact

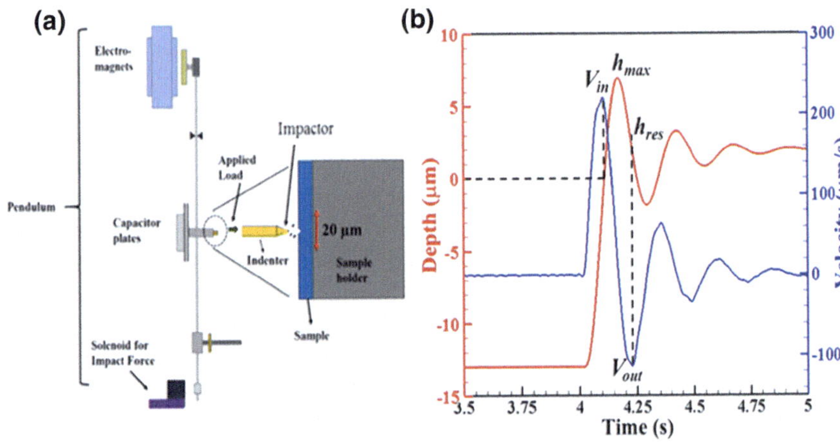

Fig. 13.1 **a** Experimental setup for the nanoscale dynamic impact experiment and **b** a depth versus time data obtained from a single-impact test. Reprinted with permission from Prakash et al. (2017). Copyright (2018) Springer Nature

experiment on HTPB, a representative depth (impactor position) with time including rebound and subsequent motion till impactor comes to a rest is shown in Fig. 13.1b.

The penetration depth ($h_{\max}$) which the impactor reaches at a given load, the contact velocity (V_{in}) with which the impactor first approaches the surface, the velocity (V_{out}) with which the impactor retracts, and the depth (h_{res}) which remains after the impactor leaves the sample surface are calculated from this data as shown in Fig. 13.1b. The time rate of change of strain, $\dot{\varepsilon}$, with which the impact occurs, changes with the depth of impact. In order to define a nominal strain rate, an average strain rate is defined by using the approximation by the expression,

$$\dot{\varepsilon} \approx \frac{V_{\text{in}}}{h_{\max}}, \tag{1}$$

The strain and stress are given by Prakash et al. (2017), Verma et al. (2016),

$$\begin{aligned} \bar{\varepsilon} &= \frac{h_{\text{res}}^2}{h_{\max}^2} \\ \bar{\sigma} &= \frac{P}{\pi h_{\max}^2}, \end{aligned} \tag{2}$$

where, h_{res} and $h_{\max}$ are defined in Fig. 13.1b. For this experimental work, the effective stress–effective viscoplastic strain curve is assumed to be represented by a power law model (Prakash et al. 2018a):

$$\bar{\sigma} = F_0 \left(\frac{\dot{\bar{\varepsilon}}^{vp}}{\dot{\bar{\varepsilon}}^{0}} \right)^m (\bar{\varepsilon}^{vp})^n \tag{3}$$

where, $\dot{\bar{\varepsilon}}^{vp}$ is the effective plastic strain rate, $\dot{\bar{\varepsilon}}^{0}$ is a reference strain which is assumed to be 1 s^{-1}, and F_0, m, and n are the viscoplastic material parameters. The viscoplastic material parameters were obtained by performing multiple nanoscale impact experiments at different positions in the samples following the method outlined by Prakash et al. (2018b).

13.2.3 Viscoplastic Model Parameter Evaluation

In order to obtain the viscoplastic model parameters in Eq. (3), it is rewritten as

$$\bar{\sigma} = A(\bar{\varepsilon}^{vp})^n. \tag{4}$$

where,

$$A = F_0 \left(\frac{\dot{\bar{\varepsilon}}^{vp}}{\dot{\bar{\varepsilon}}^{0}} \right)^m.$$

The parameter A is then plotted as a function of effective viscoplastic strain rate $\dot{\bar{\varepsilon}}^{vp}$ on the log–log scale. Figure 13.2a shows an example plot, for the HTPB binder, of amplitude A versus the effective viscoplastic strain rate. Using the log–log plot

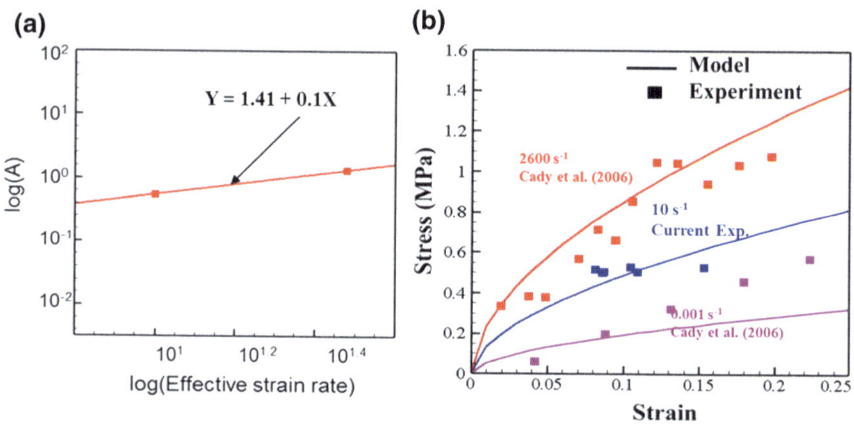

Fig. 13.2 **a** Log–log plot of the parameter A versus the strain rate for the power law model parameters and **b** an example stress–strain curve fitted using these parameters for HTPB for current experimental data. Reprinted with permission from Prakash et al. (2018a). Copyright (2018) Elsevier and comparison with the data obtained by Cady et al. (2006) using SHPB test

shown in Fig. 13.2a, the parameters χ and m can be taken to be the intercept and the slope, respectively. Similarly, n can be obtained by plotting the stress versus strain for a given strain rate.

Once the parameters m, n, and χ are calculated, this power law model can be used to predict the constitutive behavior of the material at higher strain rates. In Fig. 13.2b, we show the effectiveness of the current model by fitting the experimental data obtained from the Split-Hopkinson Bar Experiment (SHPB) test performed by Cady et al. (2006) for HTPB at high strain rate.

Using the similar fitting procedure on the experimentally obtained stress and strain data for other strain rates as well as different samples, we obtain the stress–strain curves for bulk (HTPB, HMX, and AP) as well as interfaces (HTPB-HMX and HTPB-AP with and without the binding agent). Figure 13.3 shows the stress–strain curves fitted using the above procedure on the data obtained from impact experiment for individual constituents of energetic material such as AP, HMX, and the particle–matrix interfaces. As shown in Fig. 13.3, the effect of interface chemistry is most significant on the stress–strain curve for interfaces.

The values for the viscoplastic model parameters for different constituents are presented in Table 13.1. Parameters in Table 13.1 show the effect of chemistry on the interface constitutive law. These parameters are then used to model shock and impact behavior of energetic materials at high strain rates. These parameters then can be used in numerical techniques, such as finite element methods, to simulate the high strain-rate behavior of energetic materials.

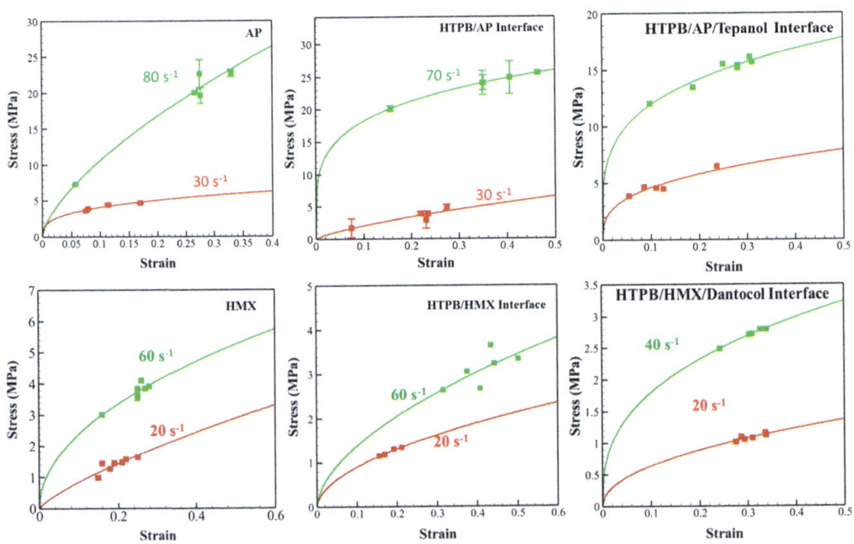

Fig. 13.3 Stress–strain curve fitting using the power law model for AP, HTPB-AP interface, HTPB/AP/Tepanol interface, HMX, HTPB-HMX interface and HTPB/HMX/Dantocol interface. Reprinted with permission from Olokun et al. (2018). Copyright (2018) Springer Nature

Table 13.1 Viscoplastic material parameters obtained for AP, HMX, HTPB, and interfaces

Parameter	F_0 (MPa)	m	n
HTPB	1.41	0.1	0.56
HMX	0.95	0.5	0.63
AP	0.016	1.67	0.17
HTPB/HMX interface	0.78	0.46	0.53
HTPB/HMX/Dantocol interface	0.1	1.14	0.4
HTPB/AP interface	0.01	2.01	0.41
HTPB/AP/Tepanol interface	0.011	2.16	0.5

13.3 Interface Failure Properties Measurement

As was discussed earlier, in order to model the failure behavior of the composite energetic materials, a failure model such as cohesive zone model is required. The experimental methods used in literature to obtain cohesive zone model parameters use macroscopic failure observation which are not suitable for such models. The cohesive zone models are based on local fracture behavior, and the model parameters need to be obtained from local failure consideration. Nanomechanical Raman spectroscopy (NMRS) is one such method that can be used to obtain the cohesive zone model parameters. In order to create a complete cohesive zone model, one needs to obtain the cohesive strength, critical displacement, or the cohesive energy. Cohesive energy can be calculated based on the assumption that the cohesive energy is equal to the energy spent in creating the delaminated surfaces. Cohesive strength, however, needs to be obtained from the local deformation. In this work, an in situ mechanical Raman Spectroscopy test, as proposed by Prakash et al. (2018a), was performed in order to obtain the cohesive zone parameters.

13.3.1 In Situ Nanomechanical Raman Spectroscopy

Mechanical Raman Spectroscopy (MRS) is a technique used to measure a change of frequency of incident light, or 'Raman shift', and ultimately stress when a material is loaded. The stresses near the crack tip area during crack propagation are calculated using NMRS, following the procedure outlined by Prakash et al. (2018a). Anastassakis et al. (1970) were among the very first researchers who investigated the effect of externally applied stresses on silicon Raman modes. From then on, a large number of researchers have used Raman scattering measurements to measure local stresses in the presence of remote-applied stress on different kinds of materials (De Wolf 1996). The shift in the Raman vibration modes is typically reported in wavenumbers. The wavenumber has units of inverse length (usually in cm^{-1}) and is directly related to energy (Fig. 13.4).

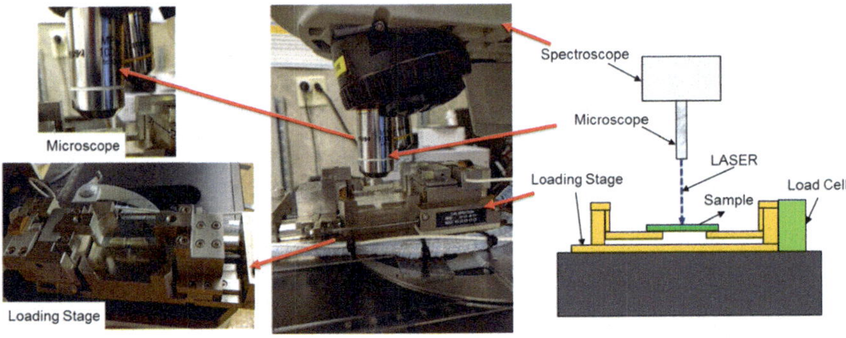

Fig. 13.4 Experimental setup for the mechanical Raman spectroscopy. Reprinted with permission from Prakash et al. (2018). Copyright (2018) Elsevier

A large number of researches have been devoted to the study of Raman vibrational spectroscopy of Polybutadiene and AP and have been reported in the literature (Buback and Schulz 1976; Chakraborty et al. 1986; Nallasamy et al. 2002; Peiris et al. 2000). The effect of externally applied pressure on the internal mode frequency in AP was studied in Peiris et al. (2000), Brill and Goetz (1976), and it was found that the shift in the Raman frequency of both NH_4^+ and ClO_4^- increases as the pressure is increased. However, this was not the case with the N–H stretching mode and Raman shift was found to be decreasing with increasing pressure (Lin et al. 2008). A similar dependence of Raman active modes with temperature was also been studied (Chakraborty et al. 1986), in the single-crystal AP. Raman as well the infrared spectroscopy of cis- and trans-1,4-polybutadiene were performed by Nallasamy et al. (2002), and they identified and assigned the individual modes of vibration to each frequency. Several different types of energetic material and propellant have been characterized by Fell et al. (1995) using the Fourier transform Raman spectroscopy. Raman spectra of these different energetic materials were found to be in the range of 100–3000 cm^{-1}.

13.3.2 Raman Shift Versus Stress Calibration

In the NMRS approach (Gan and Tomar 2014; Prakash et al. 2018a), Raman shifts in a material are measured in situ during loading. These local Raman shifts are then converted to stresses based on a calibration performed under identical loading and boundary conditions. In the current study, the stress versus Raman shift calibration is obtained by using a HTPB sample under tension. A representative Raman spectra and the corresponding vibration modes of each peaks, as reported in (Nallasamy et al. 2002), are also indicated on the spectra, in Fig. 13.5a. A tensile/compressive loading stage (Deben UK Ltd.) was used to apply tensile load on the sample. The load was applied in a quasi-static manner (0.1 mm/min) until the complete failure.

A Raman spectrometer (HORIBA Xplora Plus from HORIBA Ltd.) was used to record the Raman spectra at multiple points in scan area, shown in Fig. 13.5b, of the sample material. An excitation laser of wavelength 532 nm and a 2400-grating size was used for this purpose. The spectral resolution of the Raman spectrometer, using the above-mentioned laser and grating with microscope objective of 100× and 0.95 N.A., near the CH_2-stretching zone was found to be 1.4 cm^{-1}. The spatial resolution of this setup was 1 μm. The shift in the above Raman vibration mode (CH_2-stretching zone) was observed and recorded continuously, Fig. 13.5a, as the applied load was increased quasi-statically. Raman shift versus known stress at a given cross section, shown in Fig. 13.5b, was plotted. Representing the shift in wavenumber due to the applied external load by Δw and the stress by σ, stress can then be evaluated using (Wu et al. 2007),

$$\sigma = C\Delta w, \tag{5}$$

where C is a calibration constant. As shown in Fig. 13.5b, this calibration constant is the inverse of the slope of the Raman shift versus stress curve. In this work, the value of the constant C from the fitted line was found to be 0.23 cm^{-1}/MPa. The calibration constant, obtained in this way, is then used for the stress evaluation in samples under similar boundary conditions. In the current work, above calibration is used for the stress around the interface in a tensile edge-crack sample of a one-particle HTPB-HMX and HTPB-AP samples.

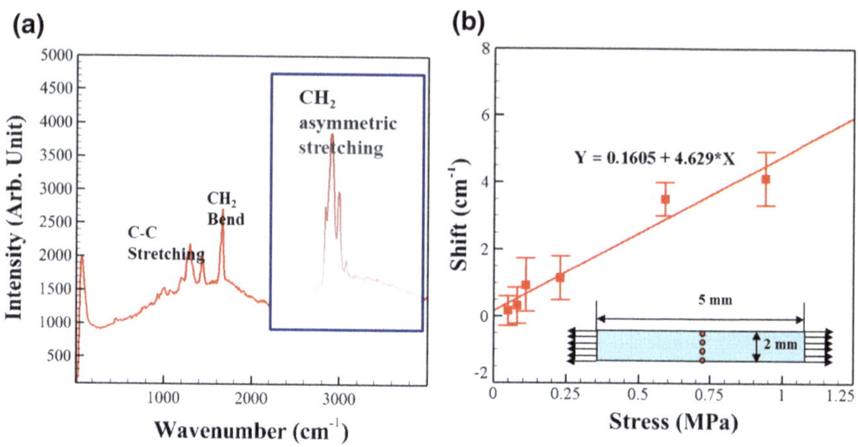

Fig. 13.5 **a** Raman spectra of HTPB observed in the sample and **b** the stress versus shift calibration curve. Reprinted with permission from Prakash et al. (2018a). Copyright (2018) Elsevier

13.3.3 Cohesive Zone Model Parameter Evaluation

The bilinear cohesive zone model parameters, such as interface strength and cohesive energy, for particle–matrix interfaces are obtained in a particle–matrix delamination experiment at particle–matrix interfaces in the analyzed single-particle samples, Fig. 13.6a. The combination of local stresses at the interface and energy spent in completely delaminating the particle from the matrix is used to obtain a bilinear cohesive zone model parameter. NMRS measurements, as explained in the previous subsection, are used to obtain stress near the interface during delamination of particle–matrix interface. The experiment is performed in a single-particle sample with an initial edge crack, which extends very close to the interfaces, by loading the sample in tension, Fig. 13.6b. As increasing load is applied at the sample ends, the crack further propagates and reaches the interface and the interface separation start immediately. The energy required in extending the crack can be obtained from the load–displacement curve, as shown by the shaded region in Fig. 13.6c.

From the load versus displacement data, Fig. 13.6c shows the location when crack after propagating reaches the interface and the location where the interface completely fails. The area under the load–displacement curve, between these two points, is considered to be the energy needed for interface failure (Prakash et al. 2017), which is equal to the cohesive energy. The stresses, in the sample, at the crack tip near the interface region during crack propagation are calculated using NMRS.

A combination of the above cohesive energy and stress measurements is then applied to derive a bilinear cohesive zone law parameter. Here, it has been assumed that once the crack reaches the interface, no other dissipation mechanism takes place except delamination. The assumption here is valid because of the fact that the sample is under a quasi-static loading. The HTPB-AP sample delaminated at around 0.7 N, whereas HTPB-HMX sample failed at 0.6 N. The crack extension were

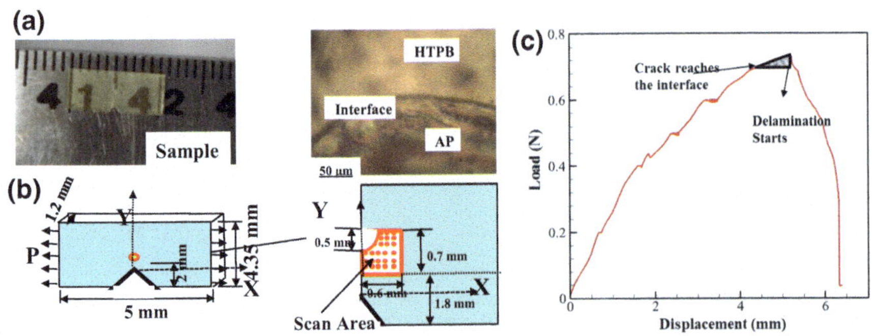

Fig. 13.6 **a** A representative sample used for the NMRS-based experiments, **b** sample dimensions, loading conditions, and the Raman scan region, and **c** the experimental load–displacement curve obtained during loading for an HTPB-AP-Tepanol sample. Reprinted with permission from Prakash et al. (2018a). Copyright (2018) Elsevier

recorded and observed till it reached the interface. As shown in Fig. 13.7b, and c, the scan area decreases when the crack advanced toward the interface.

As can be seen from Fig. 13.7, the stress distribution around the interface depends on the interface composition. The interface strength was calculated from the stress distribution obtained using NMRS based on the assumption that the maximum stress occurring near the interface, when the delamination starts, is equal to the stress required to delaminate the interface. Cohesive zone parameters for different interfaces are given in Table 2. This shows that by adding the binding agent in the mixture, the interface strength increases.

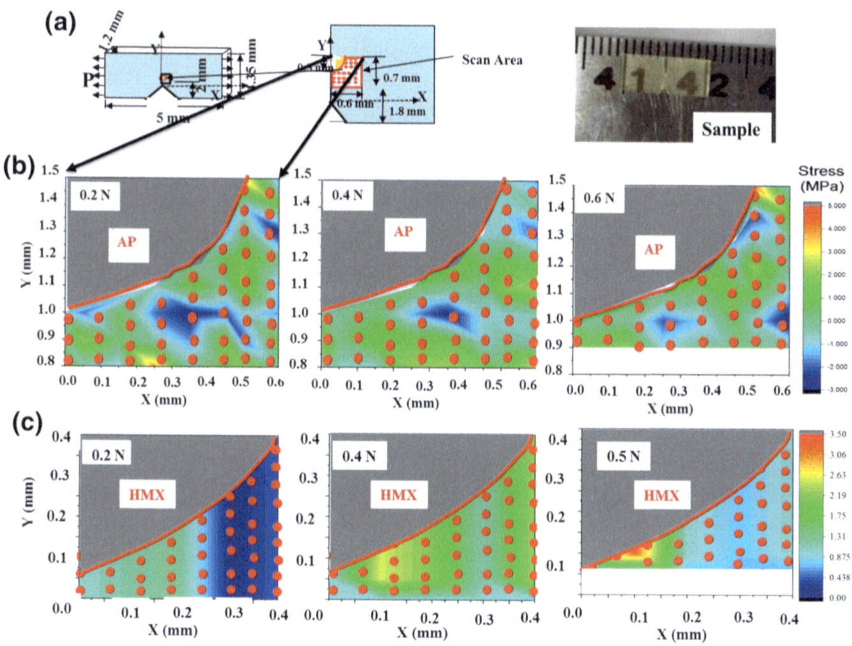

Fig. 13.7 **a** A representative sample (HTPB-AP), sample dimension, and the experimental loading conditions and scan area. Distribution of stress near the interface at different loads for **b** HTPB-AP-Tepanol sample. Reprinted with permission from Prakash et al. (2018a). Copyright (2018) Elsevier and **c** HTPB-HMX-Dantocol sample

Table 13.2 Cohesive zone parameters for particle–matrix interface with and without binding agent

Interface type	Binding agent	Cohesive strength (MPa)	Critical displacement (mm)	Cohesive energy (N/mm)
HTPB-AP	None	1.1	0.12	0.065
	Tepanol	2.91	0.11	0.16
HTPB-HMX	None	1.6	0.16	0.13
	Dantocol	2.2	0.30	0.33

13.4 Conclusions

Mechanical properties of the constituent material and the interface are required in order to simulate the complex mechanisms involved in the impact-induced hot-spot generation in energetic materials. This work presents the technique to obtain constitutive model of bulk phases as well as the interfaces. The effect of binding agent on the constitutive and failure properties of interfaces in energetic materials was analyzed. We obtained a strain-rate-dependent constitutive models of the constituents of the energetic material (HTPB-HMX and HTPB-AP) using a nanoscale dynamic impact experiment which allows us to obtain not only the properties of bulk material but also of the interface which is otherwise difficult to obtain. A bilinear cohesive zone model is identified, and the corresponding model parameters were obtained experimentally. An in situ mechanical Raman spectroscopy is shown to be effective in obtaining the interface cohesive properties from the local failure consideration. The change in the interface chemistry by adding a binding agent is shown to affect the constitutive model behavior by increasing the stiffness of the interface as well as increases the interface failure strength of energetic materials. This increase in interface mechanical behavior affects the impact induces fracture behavior of the particle–matrix interface. This effect of interface chemistry on failure behavior will eventually affect the initiation of hot spots due to frictional contact occurring under impact/shock-induced interface delamination in such material.

Acknowledgements This research was supported by US-AFoSR Grant FA9550-15-1-0202 (Program Manager Dr. Martin Schmidt).

References

Anastassakis E et al (1970) Effect of static uniaxial stress on the Raman spectrum of silicon. Solid State Commun 8(2):133–138

Barua A, Zhou M (2011) A Lagrangian framework for analyzing microstructural level response of polymer-bonded explosives. Modell Simul Mater Sci Eng 19(5):055001

Barua A, Horie Y, Zhou M (2012) Microstructural level response of HMX-Estane polymer-bonded explosive under effects of transient stress waves. Proc R Soc A Math Phys Eng Sci 468(2147):3725–3744

Benson DJ, Conley P (1999) Eulerian finite-element simulations of experimentally acquired HMX microstructures. Model Simul Mater Sci Eng 7:333–354

Brill T, Goetz F (1976) Laser Raman studies of solid oxidizer behavior. In: AIAA 14th aerospace sciences meeting. American Institute of Aeronauticcs and Astronautics, Washington, D.C.

Buback M, Schulz KR (1976) Raman scattering of pure ammonia to high pressures and temperatures. J Phys Chem 80(22):2478–2482

Cady CM et al (2006) Mechanical properties of plastic-bonded explosive binder materials as a function of strain-rate and temperature. Polym Eng Sci 46(6):812–819

Chakraborty T, Khatri SS, Verma AL (1986) Temperature-dependent Raman study of ammonium perchlorate single crystals: The orientational dynamics of the NH_4^+ ions and phase transitions. J Chem Phys 84(12):7018

Chen P, Zhou Z, Huang F (2011) Macro-micro mechanical behavior of a highly-particle-filled composite using digital image correlation method. In: Tesinova P (ed) Advances in composite materials—analysis of natural and man-made materials. InTech

De Wolf I (1996) Micro-Raman spectroscopy to study local mechanical stress in silicon integrated circuits. Semicond Sci Technol 11:139–154

Drodge DR et al (2009) The effects of particle size and separation on Pbx deformation, pp 1381–1384

Fell NF et al (1995) Fourier transform Raman (FTR) spectroscopy of some energetic materials and propellant formulations II. Army Research Laboratory, Aberdeen Proving Ground, MD, p 34

Gan M, Tomar V (2014) An in situ platform for the investigation of Raman shift in micro-scale silicon structures as a function of mechanical stress and temperature increase. Rev Sci Instrum 85(1)

Khasainov BA et al (1997) On the effect of grain size on shock sensitivity of heterogeneous high explosives. Shock Waves 7:89–105

Kimura E, Oyumi Y (1998) Shock instability test for azide polymer propellants. J Energ Mater 16 (2–3):173–185

Lin Y et al (2008) Raman spectroscopy study of ammonia borane at high pressure. J Chem Phys 129(23):234509

Nallasamy P, Anbarasan PM, Mohan S (2002) Vibrational spectra and assignments of cis- and trans-1,4- polybutadiene. Turk J Chem 26:105–111

Olokun AM et al (2018) Interface chemistry dependent mechanical properties in energetic materials using nano-scale impact experiment. In: SEM annual conference. Springer, Greenville

Palmer SJP, Field JE, Huntley JM (1993) Deformation, strengths and strains to failure of polymer bonded explosives. Proc R Soc A Math Phys Eng Sci 440:399–419

Peiris SM, Pangilinan GI, Russell TP (2000) Structural properties of ammonium perchlorate compressed to 5.6 GPa. J Phys Chem A 104:11188–11193

Prakash C et al. (2017) Strain rate dependent failure of interfaces examined via nanoimpact experiments. In: Challenges in mechanics of time dependent materials, vol 2: Conference Proceedings of the Society for Experimental Mechanics Series

Prakash C et al (2018a) Effect of interface chemistry and strain rate on particle-matrix delamination in an energetic material. Eng Fract Mech 191:46–64

Prakash C et al (2018b) Effect of interface chemistry and strain rate on particle-matrix delamination in an energetic material. Eng Fract Mech 191C:46–64

Rae PJ et al (2002a) Quasi-static studies of the deformation and failure of β-HMX based polymer bonded explosives. Proc R Soc A Math Phys Eng Sci 458:743–762

Rae PJ et al (2002b) Quasi-static studies of the deformation and failure of PBX 9501. Proc R Soc A Math Phys Eng Sci 458:2227–2242

Stacer RG, Husband M (1990) Small deformation viscoelastic response of gum and highly filled elastomers. Rheol Acta 29:152–162

Stacer RG, Hubner C, Husband M (1990) Binder/filler interaction and the nonlinear behavior of highly-filled elastomers. Rubber Chem Technol 63(4):488–502

Tan H (2012) The cohesive law of particle/binder interfaces in solid propellants. Prog Propuls Phys 2:59–66

Tan H et al (2005a) The cohesive law for the particle/matrix interfaces in high explosives. J Mech Phys Solids 53(8):1892–1917

Tan H et al (2005b) The Mori-Tanaka method for composite materials with nonlinear interface debonding. Int J Plast 21(10):1890–1918

Tan H, Huang Y, Liu C (2008) The viscoelastic composite with interface debonding. Compos Sci Technol 68(15–16):3145–3149

Verma D, Exner M, Tomar V (2016) An investigation into strain rate dependent constitutive properties of a sandwiched epoxy interface. Mater Des 112:345–356

Verma D, Prakash C, Tomar V (2017) Properties of material interfaces: dynamic local versus nonlocal. In: Voyiadjis G (ed) Handbook of nonlocal continuum mechanics for materials and structures. Springer, Cham, pp 1–16

Verma D, Prakash C, Tomar V (2017b) Interface mechanics and its correlation with plasticity in polycrystalline metals, polymer composites, and natural materials. Procedia Eng 173: 1266–1274

Wiegand JH (1961) Study of mechanical properties of solid propellant. Armed Forces Technical Information Agency, Arlington

Wu X et al (2007) Micro-Raman spectroscopy measurement of stress in silicon. Microelectron J 38(1):87–90

Yeager JD (2011) Microstructural characterization of simulated plastic bonded explosives, in mechanical and materials engineering. Washington State University, Washington

The manufacturer's authorised representative in the EU is Springer Nature Customer Service Centre GmbH, Europaplatz 3, 69115 Heidelberg, Germany. If you have any concerns regarding our products, please contact ProductSafety@springernature.com

Printed and bound by CPI Group (UK) Ltd, Croydon, CR0 4YY
25/09/2026
02222899-0003